P9-CRO-590

Springer Finance

Springer
Berlin
Heidelberg
New York
Barcelona
Hong Kong
London
Milan
Paris
Singapore
Tokyo

Springer Finance

Springer Finance is a new programme of books aimed at students, academics and practitioners working on increasingly technical approaches to the analysis of financial markets. It aims to cover a variety of topics, not only mathematical finance but foreign exchanges, term structure, risk management, portfolio theory, equity derivatives, and financial economics.

Credit Risk: Modelling, Valuation and Hedging
T.R. Bielecki and M. Rutkowski
ISBN 3-540-67593-0 (2001)

Risk-Neutral Valuation: Pricing and Hedging of Finance Derivatives
N.H. Bingham and R. Kiesel
ISBN 1-85233-001-5 (1998)

Visual Explorations in Finance with Self-Organizing Maps
G. Deboeck and T. Kohonen (Editors)
ISBN 3-540-76266-3 (1998)

Mathematics of Financial Markets
R.J. Elliott and P.E. Kopp
ISBN 0-387-98533-0 (1999)

Mathematical Finance – Bachelier Congress 2000 – Selected Papers from the First World Congress of the Bachelier Finance Society, held in Paris, June 29–July 1, 2000
H. Geman, D. Madan, S.R. Pliska and T. Vorst (Editors)
ISBN 3-540-67781-X (2001)

Mathematical Models of Financial Derivatives
Y.-K. Kwok
ISBN 981-3083-25-5 (1998)

Efficient Methods for Valuing Interest Rate Derivatives
A. Pelsser
ISBN 1-85233-304-9 (2000)

Exponential Functionals of Brownian Motion and Related Processes
M. Yor
ISBN 3-540-65943-9 (2001)

Manuel Ammann

Credit Risk Valuation

Methods, Models, and Applications

Second Edition

With 17 Figures
and 23 Tables

Dr. Manuel Ammann
University of St. Gallen
Swiss Institute of Banking and Finance
Rosenbergstrasse 52
9000 St. Gallen
Switzerland

Originally published as volume 470 in the series"Lecture Notes in Economics and Mathematical Systems" with the title "Pricing Derivative Credit Risk".

Mathematics Subject Classification (2001): 60 Gxx, 60 Hxx, 62 PO5, 91 B28

2nd ed. 2001, corr. 2nd printing
ISBN 3-540-67805-0 Springer-Verlag Berlin Heidelberg New York

Library of Congress Cataloging-in-Publication Data applied for
Die Deutsche Bibliothek – CIP-Einheitsaufnahme
Ammann, Manuel: Credit Risk Valuation: Methods, Models, and Applications; with 23 Tables / Manuel Ammann.– 2nd ed. – Berlin; Heidelberg; New York; Barcelona; Hong Kong; London; Milan; Paris; Tokyo: Springer, 2001
(Springer Finance)
Früher u.d.T.: Ammann, Manuel: Pricing Derivative Credit Risk
ISBN 3-540-67805-0

Springer-Verlag Berlin Heidelberg New York
a member of BertelsmannSpringer Science + Business Media GmbH

http://www.springer.de

Printed in Germany

Hardcover-Design: design & production, Heidelberg

SPIN 10956755 42/3111-5 4 3 2 – Printed on acid-free paper

Preface

Credit risk is an important consideration in most financial transactions. As for any other risk, the risk taker requires compensation for the undiversifiable part of the risk taken. In bond markets, for example, riskier issues have to promise a higher yield to attract investors. But how much higher a yield? Using methods from contingent claims analysis, credit risk valuation models attempt to put a price on credit risk.

This monograph gives an overview of the current methods for the valuation of credit risk and considers several applications of credit risk models in the context of derivative pricing. In particular, credit risk models are incorporated into the pricing of derivative contracts that are subject to credit risk. Credit risk can affect prices of derivatives in a variety of ways. First, financial derivatives can be subject to counterparty default risk. Second, a derivative can be written on a security which is subject to credit risk, such as a corporate bond. Third, the credit risk itself can be the underlying variable of a derivative instrument. In this case, the instrument is called a credit derivative. Fourth, credit derivatives may themselves be exposed to counterparty risk. This text addresses all of those valuation problems but focuses on counterparty risk.

The book is divided into six chapters and an appendix. Chapter 1 gives a brief introduction into credit risk and motivates the use of credit risk models in contingent claims pricing. Chapter 2 introduces general contingent claims valuation theory and summarizes some important applications such as the Black-Scholes formulae for standard options and the Heath-Jarrow-Morton methodology for interest-rate modeling. Chapter 3 reviews previous work in the area of credit risk pricing. Chapter 4 proposes a firm-value valuation model for options and forward contracts subject to counterparty risk, under various assumptions such as Gaussian interest rates and stochastic counterparty liabilities. Chapter 5 presents a hybrid credit risk model combining features of intensity models, as they have recently appeared in the literature, and of the firm-value model. Chapter 6 analyzes the valuation of credit derivatives in the context of a compound valuation approach, presents a reduced-form method for valuing spread derivatives directly, and models credit derivatives subject to default risk by the derivative counterpary as a vulnerable exchange option. Chapter 7 concludes and discusses practical implications of this work. The appendix contains an overview of mathematical tools applied throughout the text.

This book is a revised and extended version of the monograph titled *Pricing Derivative Credit Risk*, which was published as vol. 470 of the Lecture Notes of Economics and Mathematical Systems by Springer-Verlag. In June 1998, a different version of that monograph was accepted by the University of St.Gallen as a doctoral dissertation. Consequently, this book still has the "look-and-feel" of a research monograph for academics and practitioners interested in modeling credit risk and, particularly, derivative credit risk. Nevertheless, a chapter on general derivatives pricing and a review chapter introducing the most popular credit risk models, as well as fairly detailed proofs of propositions, are intended to make it suitable as a supplementary text for an advanced course in credit risk and financial derivatives.

St. Gallen, March 2001 *Manuel Ammann*

Contents

1. Introduction

Credit risk can be defined as the possibility that a contractual counterparty does not meet its obligations stated in the contract, thereby causing the creditor a financial loss. In this broad definition, it is irrelevant whether the counterparty is unable to meet its contractual obligations due to financial distress or is unwilling to honor an unenforceable contract.

Credit risk has long been recognized as a crucial determinant of prices and promised returns of debt. A debt contract involving a high amount of credit risk must promise a higher return to the investor than a contract considered less credit-risky by market participants. The higher promised return manifests itself in lower prices for otherwise identical indenture provisions. Table 1.1 illustrates this effect, depicting average credit spreads over the time period from January 1985 until March 1995 for debt of different credit ratings. The credit rating serves as a proxy for the credit risk contained in a security.

1.1 Motivation

Although the effect of credit risk on bond prices has long been known to market participants, only recently were analytical models developed to quantify this effect. Black and Scholes (1973) took the first significant step towards credit risk models in their seminal paper on option pricing. Merton (1974) further developed the intuition of Black and Scholes and put it into an analytical framework. A large amount of research followed the work of Black, Merton, and Scholes.

In the meantime, various other methods for the valuation of credit risk have been proposed, such as reduced-form approaches. Many of the current models, however, rely on the fundamental ideas of the early approaches or are extensions thereof. We give an overview over many of the credit risk models currently in use and discuss their respective advantages and shortcomings. However, we would like to focus our attention to applying credit risk models to derivative securities. The following sections outline the motivation of applying credit risk valuation models to derivative pricing.

Table 1.1. U.S. corporate bond yield spreads 1985-1995

Maturity class	Rating class	Average spread	Standard deviation	Average maturity
Short	Aaa	0.67	0.083	3.8
	Aa	0.69	0.083	4.0
	A	0.93	0.107	4.2
	Baa	1.42	0.184	4.4
Medium	Aaa	0.77	0.102	10.1
	Aa	0.71	0.084	9.2
	A	1.01	0.106	9.4
	Baa	1.47	0.153	9.1
Long	Aaa	0.79	0.088	23.9
	Aa	0.91	0.087	21.3
	A	1.18	0.125	21.7
	Baa	1.84	0.177	21.2

Averages of yield spreads of non-callable and non-puttable corporate bonds to U.S. Treasury debt, standard deviation of absolute spread changes from month to month, and average maturities. Source: Duffee (1998)

1.1.1 Counterparty Default Risk

Most of the work on credit risk appearing to date has been concerned with the valuation of debt instruments such as corporate bonds, loans, or mortgages. The credit risk of financial derivatives, however, has generally been neglected; even today the great majority of market participants uses pricing models which do not account for credit risk. Several reasons can be given for the neglect of credit risk in derivatives valuation:

- Derivatives traded at major futures and options exchanges contain little credit risk. The institutional organization of derivatives trading at exchanges reduces credit risk substantially. Customarily, the exchange is the legal counterparty to all option positions. There is therefore no credit exposure to an individual market participant. Depending on the credit standing of the exchange itself, this may already reduce credit risk significantly. Furthermore, the exchange imposes margin requirements to minimize its risk of substituting for defaulted counterparties.
- For a long time, the volume of outstanding over-the-counter (OTC) derivative positions has been relatively small. Furthermore, most open positions were held in interest rate swaps. Interest rate swaps tend to contain relatively little credit risk[1] because contracts are designed such that only interest payments, or even only differences between interest payments, are exchanged. Principals are not exchanged in an interest rate swap and are therefore not subject to credit risk.

[1] Nonetheless, empirical work, e.g., by Sun, Suresh, and Ching (1993) and Cossin and Pirotte (1997), indicates that swap rates are also affected by credit risk.

- Pricing models which take counterparty risk into account have simply not been available. Credit risk models for derivative instruments are more complex than for standard debt instruments because the credit risk exposure is not known in advance.

Of course, even an exchange may default in unusual market situations[2] and OTC derivative volume has been considerable for a while, so these reasons only partially explain the lack of concern over credit risk in derivative markets. In any case, this lack of concern has given way to acute awareness of the problem, resulting in a slow-down of market activity.[3]

Fig. 1.1. Outstanding OTC interest rate options

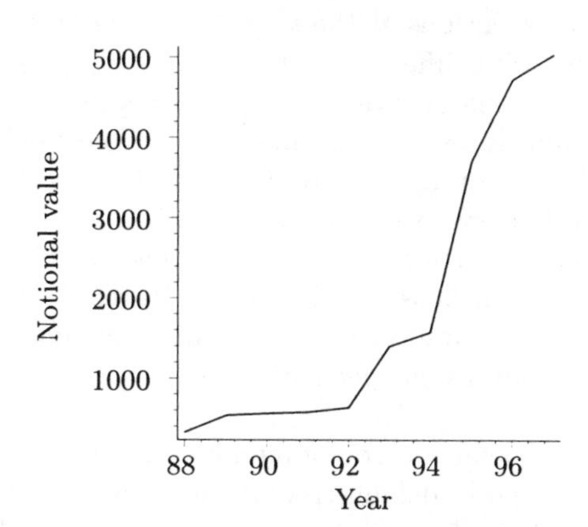

Notional value in billions of U.S. dollars. Data are from the second half of the year except in 1997, where they are from the first half. Data source: International Swaps and Derivatives Association (1988-1997).

An important reason for this change of attitude is certainly the growth of the OTC derivatives market. As Figure 1.1 shows, off-exchange derivatives have experienced tremendous growth over the last decade and now account for a large part of the total derivatives contracts outstanding. Note that Figure 1.1 only shows outstanding interest rate option derivatives and does not include swap or forward contracts.

OTC-issued instruments are usually not guaranteed by an exchange or sovereign institution and are, in most cases, unsecured claims with no collateral posted. Although some attempts have been made to set up OTC clearing

---

[2] In fact, the futures and option exchange in Singapore (Simex) would have been in a precarious position if Barings had defaulted on its margin calls. Cf. Falloon (1995).

[3] Cf. Chew (1992)

houses and to use collateralization to reduce credit risk, such institutional improvements have so far remained the exception. In a reaction recognizing the awareness of the threat of counterparty default in the marketplace, some financial institutions have found it necessary to establish highly rated derivatives subsidiaries to stay competitive or improve their position in the market.[4] It would, however, be overly optimistic to conclude that the credit quality of derivative counterparties has generally improved. In fact, Bhasin (1996) reports a general deterioration of credit quality among derivative counterparties since 1991.

Historical default rates can be found in Figures 1.2 and 1.3. The figures show average cumulated default rates in percent within a given rating class for a given age interval. The averages are based on default data from 1970-1997. Figure 1.2 shows default rates for bonds rated Aaa, Aa, A, Baa. It can be seen that, with a few exceptions at the short end, default rate curves do not intersect, but default rate differentials between rating classes may not change monotonically. A similar picture emerges in Figure 1.3, albeit with tremendously higher default rates. The curve with the highest default rates is an average of defaults for the group of Caa-, Ca-, and C-rated bonds. While the slope of the default rate curves tends to increase with the age of the bonds for investment-grade bonds, it tends to decrease for speculative-grade bonds. This observation indicates that default risk tends to increase with the age of the bond for bonds originally rated investment-grade, but tends to decrease over time for bonds originally rated speculative-grade, given that the bonds survive.

Given the possibility of default on outstanding derivative contracts, pricing models evidently need to take default risk into account. Even OTC derivatives, however, have traditionally been, and still are, priced without regard to credit risk. The main reason for this neglect is today not so much the unquestioned credit quality of counterparties as the lack of suitable valuation models for credit risk. Valuation of credit risk in a derivative context is analytically more involved than in a simple bond context. The reason is the stochastic credit risk exposure.[5] While in the case of a corporate bond the exposure is known to be the principal and in case of a coupon bond also the coupon payments, the exposure of a derivative contract to counterparty risk is not known in advance. In the case of an option, there might be little exposure if the option is likely to expire worthless. Likewise, in the case of swaps or forward contracts, there might be little exposure for a party because the contract can have a negative value and become a liability.

Table 1.1 depicts yield spreads for corporate bonds of investment grade credit quality. Because the yield spread values are not based on the same data set as the default rates, the figures are not directly comparable, but they can still give an idea of the premium demanded for credit risk. Although a yield

[4] Cf. Figlewski (1994).

[5] Cf. Hull and White (1992).

Fig. 1.2. Average cumulated default rates for U.S. investment-grade bonds

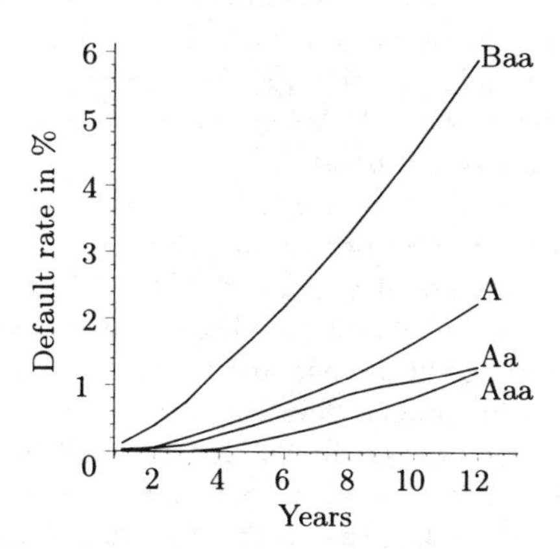

Average cumulated default rates during 1970-1997 depending on the age (in years) of the issue for investment-grade rating classes. Data source: Moody's Investors Services.

Fig. 1.3. Average cumulated default rates for U.S. speculative-grade bonds

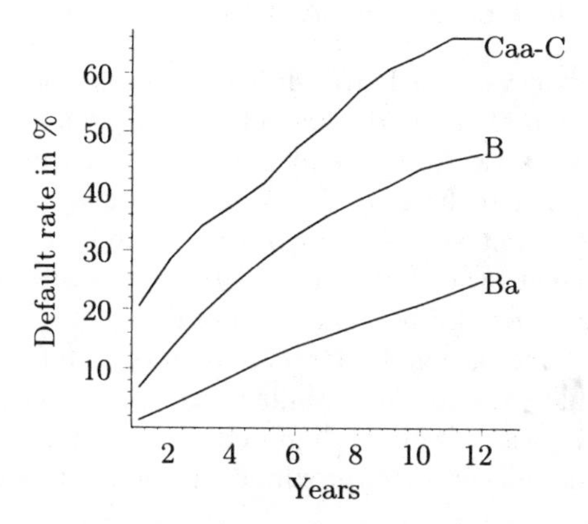

Average cumulated default rates during 1970-1997 depending on the age (in years) of the issue for speculative-grade rating classes. Rating class Caa-C denotes the average of classes Caa, Ca, C. Data source: Moody's Investors Services.

spread of, for instance, 118 basis points over Treasury for A-rated long term bonds seems small at first sight, it has to be noted that, in terms of bond price spreads, this spread is equivalent to a discount to the long-term Treasury of approximately 21% for a 20-year zero-coupon bond. Although not all of this discount may be attributable to credit risk,[6] credit risk can be seen to have a large impact on the bond price. Although much lower, there is a significant credit spread even for Aaa-rated bonds.[7]

Moreover, many counterparties are rated below Aaa. In a study of financial reports filed with the Securities and Exchange Commission (SEC), Bhasin (1996) examines the credit quality of OTC derivative users. His findings contradict the popular belief that only highly rated firms serve as derivative counterparties. Although firms engaging in OTC derivatives transactions tend to be of better credit quality than the average firm, the market is by no means closed to firms of low credit quality. In fact, less than 50% of the firms that reported OTC derivatives use in 1993 and 1994 had a rating of A or above and a significant part of the others were speculative-grade firms.[8]

If credit risk is such a crucial factor when pricing corporate bonds and if it cannot be assumed that only top-rated counterparties exist, it is difficult to justify ignoring credit risk when pricing derivative securities which may be subject to counterparty default. Hence, derivative valuation models which include credit risk effects are clearly needed.

1.1.2 Derivatives on Defaultable Assets

The valuation of derivatives which are subject to counterparty default risk is not the only application of credit risk models. A second application concerns default-free derivatives written on credit-risky bond issues. In this case, the counterparty is assumed to be free of any default risk, but the underlying asset of the derivative contract, e.g., a corporate bond, is subject to default risk. Default risk changes the shape of the price distribution of a bond. By pricing options on credit-risky bonds as if the underlying bond were free of any risk of default, distributional characteristics of a defaultable bond are neglected. In particular, the low-probability, but high-loss areas of the price distribution of a credit-risky bond are ignored. Depending on the riskiness of the bond, the bias introduced by approximating the actual distribution with

---

[6] It is often argued that Treasury securities have a convenience yield because of higher liquidity and institutional reasons such as collateral and margin regulations and similar rules that make holding Treasuries more attractive. The real default-free yield may therefore be slightly higher than the Treasury yield. On the other hand, even Treasuries may not be entirely free of credit risk.

[7] Hsueh and Chandy (1989) reported a significant yield spread between insured and uninsured Aaa-rated securities.

[8] Although derivatives can be a wide range of instruments with different risk characteristics, according to Bhasin (1996), the majority of instruments were interest-rate and currency swaps, for investment-grade as well as for speculative-grade users.

the default-free distribution can be significant. A credit risk model can help correct such a bias.

1.1.3 Credit Derivatives

Very recently, derivatives were introduced the payoff of which depended on the credit risk of a particular firm or group of firms. These new instruments are generally called credit derivatives. Although credit derivatives have long been in existence in simpler forms such as loan and debt insurance, the rapid rise of interest and trading in credit derivatives has given credit risk models an important new area of application.

Table 1.2. Credit derivatives use of U.S. commercial banks

Notional value	1997 1Q	1997 2Q	1997 3Q	1997 4Q	1998 1Q	1998 2Q	1998 3Q	1998 4Q
Billion USD	19	26	39	55	91	129	162	144
%	0.09	0.11	0.16	0.22	0.35	0.46	0.50	0.44
Notional value	**1999 1Q**	**1999 2Q**	**1999 3Q**	**1999 4Q**	**2000 1Q**	**2000 2Q**	**2000 3Q**	**2000 4Q**
Billion USD	191	210	234	287	302	362	379	426
%	0.58	0.64	0.66	0.82	0.80	0.92	0.99	1.05

Absolute outstanding notional amounts in billion USD and percentage values relative to the total notional amount of U.S. banks' total outstanding derivatives positions. Figures are based on reports filed by all U.S. commercial banks having derivatives positions in their books. Data source: Office of the Comptroller of the Currency (1997-2000).

Table 1.2 illustrates the size and growth rate of the market of credit derivatives in the United States. The aggregate notional amount of credit derivatives held by U.S. commercial banks has grown from less than $20 billion in the first quarter of 1997 to as much as $426 billion in the fourth quarter of 2000. This impressive growth rate indicates the increasing popularity of these new derivative instruments. In relative terms, credit derivatives' share in derivatives use has been increasing steadily since the first quarter of 1997, when credit derivatives positions were first reported to the Office of the Comptroller of the Currency (OCC). Nevertheless, it should not be overlooked that credit derivatives still account for only a very small part of the derivatives market. Only in the fourth quarter of 2000 has the share of credit derivatives surpassed 1% of the total notional value of derivatives held by commercial banks. Moreover, only the largest banks tend to engage in credit derivative transactions.

Because the data collected by the OCC includes only credit derivative positions of U.S. commercial banks, the figures in Table 1.2 do not reflect actual market size. A survey of the London credit derivatives market undertaken by the British Bankers' Association (1996) estimates the client market share of commercial banks to be around 60%, the remainder taken up by securities firms, funds, corporates, insurance companies, and others. The survey also gives an estimate of the size of the London credit derivatives market. Based on a dealer poll, the total notional amount outstanding was estimated to be approximately $20 billion at the end of the third quarter of 1996. The same poll also showed that dealers were expecting continuing high growth rates. It can be expected that, since 1996, total market size has increased at a pace similar to the use of credit derivatives by commercial banks shown in Table 1.2.

Clearly, with credit derivatives markets becoming increasingly important both in absolute and relative terms, the need for valuation models also increases. However, another aspect of credit derivatives should not be overlooked. Credit derivatives are OTC-issued financial contracts that are subject to counterparty risk. With credit derivatives playing an increasingly important role for the risk management of financial institutions as shown in Table 1.2, quantifying and managing the counterparty risk of credit derivatives, just as any other derivatives positions, is critical.

1.2 Objectives

This monograph addresses four valuation problems that arise in the context of credit risk and derivative contracts. Namely,

- The valuation of *derivative securities which are subject to counterparty default risk.* The possibility that the counterparty to a derivative contract may not be able or willing to honor the contract tends to reduce the price of the derivative instrument. The price reduction relative to an identical derivative without counterparty default risk needs to be quantified. Generally, the simple method of applying the credit spread derived from the term structure of credit spreads of the counterparty to the derivative does not give the correct price.
- The valuation of default-free *options on risky bonds.* Bonds subject to credit risk have a different price distribution than debt free of credit risk. Specifically, there is a probability that a high loss will occur because the issuer defaults on the obligation. The risk of a loss exhibits itself in lower prices for risky debt. Using bond option pricing models which consider the lower forward price, but not the different distribution of a risky bond, may result in biased option prices.
- The valuation of *credit derivatives.* Credit derivatives are derivatives written on credit risk. In other words, credit risk itself is the underlying variable of the derivative instrument. Pricing such derivatives requires a model of credit risk behavior over time, as pricing stock options requires a model of stock price behavior.

- The valuation of *credit derivatives* that are themselves subject to counterparty *default risk*. Credit derivatives, just as any other OTC-issued derivative intruments, can be subject to counterparty default. If counterparty risk affects the value of standard OTC derivatives, it is probable that it also affects the value of credit derivatives and should therefore be incorporated in valuation models for credit derivatives.

This book emphasizes the first of the above four issues. It turns out that if the first objective is achieved, the latter problems can be solved in a fairly straightforward fashion.

The main objective of this work is to propose, or improve and extend where they already exist, valuation models for derivative instruments where the credit risk involved in the instruments is adequately considered and priced. This valuation problem will be examined in the setting of the firm value framework proposed by Black and Scholes (1973) and Merton (1974). It will be shown that the framework can be extended to more closely reflect reality. In particular, we will derive closed-form solutions for prices of options subject to counterparty risk under various assumptions. In particular, stochastic interest rates and stochastic liabilities of the counterparty will be considered.

Furthermore, we will propose a credit risk framework that overcomes some of the inherent limitations of the firm value approach while retaining its advantages. While we still assume that the rate of recovery in case of default is determined by the firm value, we model the event of default and bankruptcy by a Poisson-like bankruptcy process, which itself can depend on the firm value. Credit risk is therefore represented by two processes which need not be independent. We implement this model using lattice structures.

Large financial institutions serving as derivative counterparties often also have straight bonds outstanding. The credit spread between those bonds and comparable treasuries gives an indication of the counterparty credit risk. The goal must be to price OTC derivatives such that their prices are consistent with the prices, if available, observed on bond markets.

Secondary objectives are to investigate the valuation of credit derivative instruments and default-free options on credit-risky bonds. Ideally, a credit risk model suitable for pricing derivatives with credit risk can be extended to credit derivatives and options on risky bonds. We analyze credit derivatives and options on risky bonds within a compound option framework that can accommodate many underlying credit risk models.

In this monograph we restrict ourselves to pricing credit risk and instruments subject to credit risk and having credit risk as the underlying instrument. Hedging issues are not discussed, nor are institutional details treated in any more detail than immediately necessary for the pricing models. Methods for parameter estimation are not covered either. Other issues such as optimal behavior in the presence of default risk, optimal negotiation of contracts, financial restructuring, collateral issues, macroeconomic influence on credit risk, rating interpretation issues, risk management of credit portfolios, and similar problems, are also beyond the scope of this work.

1.3 Structure

Chapter 2 presents the standard and generally accepted contingent claims valuation methodology initiated by the work of Black and Scholes (1973) and Merton (1973). The goal of this chapter is to provide the fundamental valuation methodologies which later chapters rely upon. The selection of the material has to be viewed in light of this goal. In this chapter we present the fundamental asset pricing theorem, contingent claims pricing results of Black and Scholes (1973) and Merton (1974), as well as extensions such as the exchange option result by Margrabe (1978) and discrete time approaches as suggested by Cox, Ross, and Rubinstein (1979). Moreover, we present some of the basics of term structure modeling, such as the framework by Heath, Jarrow, and Morton (1992) in its continuous and discrete time versions. We also treat the forward measure approach to contingent claims pricing, as it is crucial to later chapters.

Chapter 3 reviews the existing models and approaches of pricing credit risk. Credit risk models can be divided into three different groups: firm value models, first passage time models, and intensity models. We present all three methodologies and select some proponents of each methodology for a detailed analysis while others are treated in less detail. In addition, we also review the far less numerous models that have attempted to price the counterparty credit risk involved in derivative contracts. Moreover, we survey the methods available for pricing derivatives on credit risk.

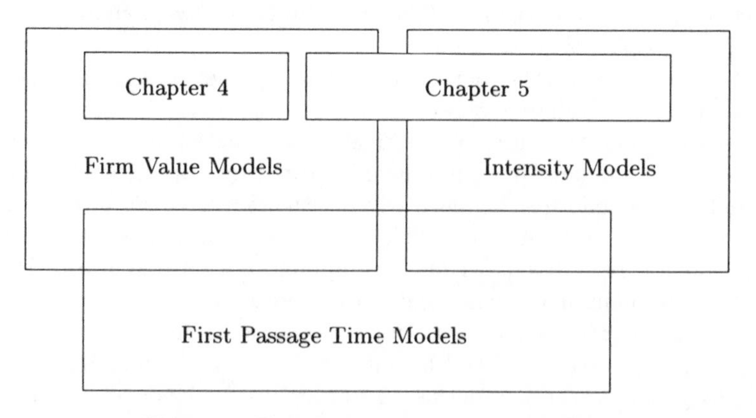

Fig. 1.4. Classification of credit risk models

In Chapter 4, we propose a pricing model for options which are subject to counterparty credit risk. In its simplest form, the model is an extension of

Merton (1974). It is then extended to allow for stochastic counterparty liabilities. We derive explicit pricing formulae for vulnerable options and forward contracts. In a further extension, we derive analytical solutions for the model with stochastic interest rates in a Gaussian framework and also give a proof for this more general model.

In Chapter 5, we set out to alleviate some of the limitations of the approach from Chapter 4. In particular, we add a default process to better capture the timing of default. The model proposed in Chapter 5 attempts to combine the advantages of the traditional firm value-based models with the more recent default intensity models based on Poisson processes and applies them to derivative instruments. As Figure 1.4 illustrates, it is a hybrid model being related to both firm value-based and intensity-based models. It turns out that the model presented in this chapter is not only suitable for pricing derivatives with counterparty default risk, but also default-free derivatives on credit-risky bonds. The latter application reveals largely different option prices in some circumstances than if computed with traditional models.

In Chapter 6, we propose a valuation method for a very general class of credit derivatives. The model proposed in Chapter 5 lends itself also to the pricing of credit derivatives. Because the model from Chapter 5 takes into account credit risk in a very general form, credit derivatives, which are nothing else than derivative contracts on credit risk, can be priced as compound derivatives. Additionally, we present a reduced-form approach for for valuing spread derivatives modeling the credit spread directly. Furthermore, we show that credit derivatives can be viewed as exchange options and, consequently, credit derivatives that are subject to counterparty default risk can be modeled as vulnerable exchange options.

In Chapter 7, we summarize the results from previous chapters and state conclusions. We also discuss some practical implications of our work.

The appendix contains a brief overview on some of the stochastic techniques used in the main body of the text. Many theorems crucial to derivatives pricing are outlined in this appendix.

A brief note with respect to some of the terminology used is called for at this point. In standard usage, *riskless* often refers to the zero-variance money market account. In this work, *riskless* is often used to mean free of credit risk and does not refer to the money market account. *Default-free* is used synonymously with riskless or risk-free. Similarly, within a credit risk context, *risky* often refers to credit risk, not to market risk. *Default* and *bankruptcy* are used as synonyms throughout since we do not differentiate between the event of default and subsequent bankruptcy or restructuring of the firm. This is a frequent simplification in credit risk pricing and is justified by our focus on the risk of loss and its magnitude in case of a default event rather than on the procedure of financial distress.

2. Contingent Claim Valuation

This chapter develops general contingent claim pricing concepts fundamental to the subjects treated in subsequent chapters.

We start with finite markets. A market is called finite if the sample space (state space) and time are discrete and finite. Finite markets have the advantage of avoiding technical problems that occur in markets with infinite components.

The second section extends the concept from the finite markets to continuous-time, continuous-state markets. We omit the re-derivation of all the finite results in the continuous world because the intuition is unchanged but the technicality of the proofs greatly increases.[1] However, we do establish two results upon which much of the material in the remaining chapters relies. First, the existence of a unique equivalent martingale measure in a market implies absence of arbitrage. Second, given such a probability measure, a claim can be uniquely replicated by a self-financing trading strategy such that the investment needed to implement the strategy corresponds to the conditional expectation of the deflated future value of the claim under the martingale measure. Therefore, the price of a claim has a simple representation in terms of an expectation and a deflating numéraire asset.

In an arbitrage-free market, it can be shown that completeness is equivalent to the existence of a unique martingale measure.[2] We always work within the complete market setting. If the market is incomplete, the martingale measure is no longer unique, implying that arbitrage cannot price the claims using a replicating, self-financing trading strategy. For an introduction to incomplete markets in a general equilibrium setting, see Geanakoplos (1990). A number of authors have investigated the pricing and hedging of contingent claims in incomplete markets. A detailed introduction can be found in Karatzas and Shreve (1998).

We also review some applications of martingale pricing theory, such as the frameworks by Black and Scholes (1973) and Heath, Jarrow, and Morton (1992).

---

[1] Cf. Musiela and Rutkowski (1997) for an overview with proofs.

[2] See, for example, Harrison and Pliska (1981), Harrison and Pliska (1983), or Jarrow and Madan (1991).

2.1 Valuation in Discrete Time

In this section we model financial markets in discrete time and state space. Harrison and Kreps (1979) introduce the martingale approach to valuation in discrete time. Most of the material covered in this section is based on and presented in the spirit of work by Harrison and Kreps (1979) and Harrison and Pliska (1981). Taqqu and Willinger (1987) give a more rigorous approach to the material. A general overview on the martingale approach to pricing in discrete time can be found in Pliska (1997).

2.1.1 Definitions

The time interval under consideration is denoted by $\mathcal{T}$ and consists of m trading periods such that t_0 denotes the beginning of the first period and t_m the end of the last period. Therefore, $\mathcal{T} = \{t_0, \ldots, t_m\}$. For simplicity we often write $\mathcal{T} = \{0, \ldots, T\}$

The market is modeled by a family of probability spaces $(\Omega, \mathcal{F}, \mathbf{P})$. $\Omega = (\omega_1, \ldots, \omega_d)$ is the set of outcomes called the sample space. $\mathcal{F}$ is the σ-algebra of all subsets of Ω. $\mathbf{P}$ is a probability measure defined on $(\Omega, \mathcal{F})$, i.e., a set function mapping $\mathcal{F} \to [0,1]$ with the standard augmented filtration $\mathbf{F} = \{\mathcal{F}_t : t \in \mathcal{T}\}$. In this notation, $\mathcal{F}$ is equal to $\mathcal{F}_T$. In short, we have a filtered probability space $(\Omega, \mathcal{F}, (\mathcal{F}_t)_{t\in\mathcal{T}}, \mathbf{P})$ or abbreviated $(\Omega, (\mathcal{F}_t)_{t\in\mathcal{T}}, \mathbf{P})$.

We assume that the market consists of n primary securities such that the $\mathcal{F}_t$-adapted stochastic vector process in $\mathbb{R}^n_+$ $S_t = (S_t^1, \ldots, S_t^n)$ models the prices of the securities. $\mathbb{R}^n$ denotes the n-dimensional space of real numbers and $+$ implies non-negativity. The security S^n is defined to be the *money market account*. Its price is given by $B_t = S_t^n = \prod_{k=0}^{t-1}(1 + r_k)$, $\forall t \in \mathcal{T}$. r_k is an adapted process and can be interpreted as the interest rate for a credit risk-free investment over one observation period. B_t is a predictable process, i.e., it is $\mathcal{F}_{t-1}$-measurable. Therefore, B_t is sometimes called the (locally) *riskless asset*. Security prices in terms of the numéraire security are called *relative* or deflated prices and are defined as $S'_t = S_t B_t^{-1}$.

We generally assume that the market is without frictions, meaning that all securities are perfectly divisible and that no short-sale restrictions, transaction costs, or taxes are present.

A *trading strategy* is a predictable process with initial investment $V_0(\theta) = \theta_0 \cdot S_0$ and *wealth process* $V_t(\theta) = \theta_t \cdot S_t$. Every trading strategy has an associated *gains process* defined by $G_t(\theta) = \sum_k^{t-1} \theta_k \cdot (S_{k+1} - S_k)$. We define the relative wealth and gains processes such that $V'_t = V_t B_t^{-1}$ and $G'_t(\theta) = \sum_k^{t-1} \theta_k \cdot (S'_{k+1} - S'_k)$. The symbol "$\cdot$" denotes the inner product of two vectors. No specific symbol is used for matrix products.

A trading strategy θ is called *self-financing* if the change in wealth is determined solely by capital gains and losses, i.e., if and only if $V_t(\theta) = V_0(\theta) + G_t(\theta)$. The class of self-financing trading strategies is denoted by Θ.

A trading strategy θ is called an *arbitrage opportunity* (or simply an arbitrage) if $V_0(\theta) = 0$ almost surely (a.s.), $V_T(\theta) \geq 0$ a.s., and $\mathbf{P}(V_T(\theta) > 0) > 0$. In other words, there is arbitrage if, with strictly positive probability, the trading strategy generates wealth without initial investment and without risk of negative wealth. This is sometimes referred to as an arbitrage of the first type. Note that $V_T(\theta) \geq 0$ a.s., and $\mathbf{P}(V_T(\theta) > 0) > 0$ implies that $\mathbf{E_0}[V_T] > 0$. Further, a trading strategy θ with $V_t(\theta) < 0$ and $V_T(\theta) = 0$ is sometimes called an arbitrage of the second type. A trading strategy is also an arbitrage of the first type if the initial proceeds can be invested such that $V_t = 0$ and $V_T \geq 0$ and $\mathbf{P}(V_T > 0) > 0$.

A (European) *contingent claim* maturing at time T is a $\mathfrak{F}_F$-measurable random variable X. The class of all claims in the market is in $\mathbb{R}^d$ (since Ω is also in $\mathbb{R}^d$) and is written $\mathcal{X}$.

A claim is called *attainable* if there exists at least one trading strategy $\theta \in \Theta$ such that $V_T(\theta) = X$. Such a trading strategy is called a *replicating strategy*. A claim is *uniquely* replicated in the market if, for any arbitrary two replicating strategies $\{\theta, \phi\}$, we have $V_t(\theta) = V_t(\phi)$ almost everywere (a.e.). This means that the initial investment required to replicate the claim is the same for all replicating strategies with probability 1.

A *market* is defined as a collection of securities (assets) and self-financing trading strategies and written $\mathcal{M}(S, \Theta)$. $\mathcal{M}(S, \Theta)$ is called *complete* if there exists a replicating strategy for every claim $X \in \mathcal{X}$.

We say that $\mathcal{M}(S, \Theta)$ admits an equivalent martingale measure (or simply a martingale measure) if, for any trading strategy $\theta \in \Theta$, the associated wealth process V_t measured in terms of the numéraire is a martingale under the equivalent measure.

A market $\mathcal{M}(S, \Theta)$ is called *arbitrage-free* if none of the elements of Θ is an arbitrage opportunity.

A *price system* is a linear map $\pi : \mathcal{X} \to \mathbb{R}^+$. For any $X \in \mathcal{X}$, $\pi(X) = 0$ if and only if $X = 0$.

2.1.2 The Finite Setting

A market in a discrete-time, discrete-state-space setting is called finite if the time horizon is finite. A finite time horizon implies that the state space, the number of securities and the number of trading periods are finite, i.e., $d < \infty$, $n < \infty$ and $m < \infty$. Ω and $\mathcal{T} = \{0, \ldots, T\}$ are finite sets.

Lemma 2.1.1. *If the market admits an equivalent martingale measure, then there is no arbitrage.*

Proof. The deflated gains process is given by $G'(\phi) = \sum_k^{t-1} \phi_k \cdot (S'_{k+1} - S'_k)$. Since S'_t is a martingale under the martingale measure, by the discrete version of the martingale representation theorem, $G'(\phi)$ is a martingale for a predictable process ϕ. Thus, if $\mathcal{M}(S, \Theta)$ admits a martingale measure $\mathbf{Q}$, it follows that for any trading strategy $\theta \in \Theta$, $\mathbf{E_Q}[V'_T|\mathfrak{F}_t] = V'_t$. This means that $\mathbf{E_Q}[G'_T|\mathfrak{F}_t] = 0$. An arbitrage opportunity requires that $G'_T \geq 0$, $\mathbf{P}-a.s.$ Since $\mathbf{P}$ and $\mathbf{Q}$ are equivalent, we have $G'_T \geq 0$, $\mathbf{Q} - a.s.$ Together with the condition that $G'_T > 0$ with positive probability, we obtain $\mathbf{E_Q}[G'_T|\mathfrak{F}_t] > 0$. Therefore arbitrage opportunities are inconsistent with the existence of a martingale measure.

Lemma 2.1.2. *If there is no arbitrage, then the market admits a price system* π.

Proof. Define the subspaces of $\mathcal{X}$

$$\mathcal{X}^+ = \{X \in \mathcal{X} | X \geq 0 \text{ and } \mathbf{E_0}[V_T] > 0\}$$
$$\mathcal{X}^0 = \{X \in \mathcal{X} | X = V_t(\phi) \text{ and } V_0(\phi) = 0\}.$$

There are no arbitrage opportunities if and only if $\mathcal{X}^+ \cap \mathcal{X}^0 = \emptyset$. Since $\mathcal{X}^0$ and $\mathcal{X}^+$ are linear and closed convex subspaces, respectively, the theorem of separating hyperplanes can be applied. Thus, there exists a mapping $f : X \to \mathbb{R}$ such that

$$f(X) : \begin{cases} f(X) = 0 & \text{if } X \in \mathcal{X}^0 \\ f(X) > 0 & \text{if } X \in \mathcal{X}^+. \end{cases}$$

It can be seen that $\pi = \frac{f(X)}{f(B_T)}$ is linear and non-negative and therefore a price system. To show that π is consistent, define two trading strategies such that

$$\theta_t = \begin{cases} \phi_t^k & \text{if } k = 1, \ldots, m \\ \phi_t^n - V_0(\phi) & \text{if } k = n. \end{cases}$$

This means that strategy θ has zero investment, i.e., $V_0(\theta) = 0$. It follows that $V_T(\theta) = V_T(\phi) - V_0(\phi)B_T$. If there is no arbitrage, then $\pi(\theta) = 0$ since $\theta \in \mathcal{X}^0$. By the linear property of π, $0 = \pi(\theta) = \pi(V_T(\phi) - V_0(\phi))\pi(B_T)$. Clearly, $\pi(B_T) = 1$, and thus $\pi(V_T(\phi)) = V_0(\phi)$ holds.

Remark 2.1.1. This proof is originally from Harrison and Pliska (1981). See also Duffie (1996) for a version of this proof. In the following, we sketch a different proof by Taqqu and Willinger (1987). Yet another proof comes from the duality theorem found in linear programming. Cf. Ingersoll (1987).

For a given $m \times n$ matrix M it can be shown that either

$$\exists \pi \in \mathbb{R}^n \text{ s.t. } M\pi = 0, \ \pi > 0, \text{ or}$$
$$\exists \theta \in \mathbb{R}^m \text{ s.t. } \theta M \geq 0, \ \theta M \neq 0,$$

but not both. This is a theorem of alternatives for linear systems and can be proved by Farka's lemma.

M can be interpreted as the payoff matrix, θ is a trading strategy, and π is a price vector. The conditions of the second alternative clearly coincide with an arbitrage opportunity. Therefore, a strictly positive price vector exists if and only if there is no arbitrage.

We show that arbitrage opportunities and price systems are incompatible. Any equilibrium, however, requires a price system. Consequently, a market which allows arbitrage cannot be in equilibrium. This insight is intuitively clear as the existence of arbitrage opportunities would entice arbitragers to take positions of infinite size.

We do not address the issue of viability in this context. A market is usually called viable if there is an optimal portfolio for an agent. This notion of viability is introduced by Harrison and Kreps (1979), and is closely related to arbitrage and equilibrium. Obviously, a market cannot be viable if arbitrage opportunities exist. It can be shown that viability and absence of arbitrage are equivalent in finite markets. Karatzas and Shreve (1998) therefore define viability directly as the absence of arbitrage opportunities.

Theorem 2.1.1. *For a finite market $\mathcal{M}$, the following statements are equivalent:*

i) $\mathcal{M}$ admits an equivalent martingale measure.
ii) $\mathcal{M}$ is arbitrage-free.
iii) $\mathcal{M}$ admits a price system.

Proof. By Lemma 2.1.1, (i) $\Rightarrow$ (ii), and by Lemma 2.1.2, (ii) $\Rightarrow$ (iii). Therefore, to prove the theorem, we need only show that (iii) $\Rightarrow$ (i).

Assume that π is a consistent price system and define

$$\mathbf{Q}(A) = \pi(B_T \mathbf{1}_{\{A\}})$$

for any $A \in \mathcal{F}$. Now consider a trading strategy θ that holds all funds in the money market account. The consistency property implies $V_t(\theta) = \pi(V_T(\theta))$. In case of the money market account, this is $1 = \pi(B_T \mathbf{1}_{\{\Omega\}})$ since the money market payoff is independent of the realization of an event. We thus have $\mathbf{Q}(\Omega) = 1$. $\mathbf{Q}$ is therefore a probability measure equivalent to $\mathbf{P}$ and thus $\pi(X) = \mathbf{E_Q}[B_T^{-1} X]$ for any claim $X \in \mathcal{X}$. Next, consider a strategy which holds its funds in an arbitrary stock k until stopping time τ, when it switches into the money market account, i.e., $\theta_t^k = \mathbf{1}_{\{t \le \tau\}}$ and $\theta_t^n = \frac{S_\tau^k}{B_\tau} \mathbf{1}_{\{T > \tau\}}$. Then we have $V_t(\theta) = S_0^k$ and $V_T(\theta) = \frac{S_\tau^k}{B_\tau} B_T$ such that

$$S_0^k = \pi(S_\tau^k \frac{B_T}{B_\tau}) = \mathbf{E_Q}[B_T^{-1} S_\tau^k \frac{B_T}{B_\tau}] = \mathbf{E_Q}[B_\tau^{-1} S_\tau^k].$$

This proof is from Harrison and Pliska (1981).

Remark 2.1.2. The equivalence of absence of arbitrage and existence of a martingale measure is called the *fundamental theorem of asset pricing*.

Corollary 2.1.1. *If a market $\mathcal{M}(S, \Theta)$ does not admit arbitrage, then any attainable claim $X \in \mathcal{X}$ can be uniquely replicated in $\mathcal{M}$.*

Proof. We give an informal proof by counterexample. Assume that there are two trading strategies $\phi \in \Theta$ and $\theta \in \Theta$ such that we have for the associated wealth processes $V(\theta)$ and $V(\phi)$, $V_t(\phi) < V_t(\theta)$ and $V_T(\phi) = V_T(\theta)$ for any $T > t$. The trading strategy $\delta = \phi - \theta$, however, results in $V_t(\delta) < 0$ and $V_T(\delta) = 0$. This is, by its definition, an arbitrage opportunity.

Remark 2.1.3. There can be a market where all claims are uniquely replicated, but arbitrage is not excluded. The reverse of the proposition is therefore not generally true.

Theorem 2.1.2. *If a market $\mathcal{M}(S, \Theta)$ does not admit arbitrage, then the price of any attainable claim $X \in \mathcal{X}$ is given by*

$$X_t = B_t \, \mathbf{E_Q}[X_T B_T^{-1} | \mathcal{F}_t] \qquad \forall t \in \mathcal{T}.$$

Proof. From Theorem 2.1.1, the absence of arbitrage is equivalent to the existence of a martingale measure. The existence of a martingale measure implies that all deflated price processes are martingales. The wealth process of a self-financing strategy ϕ is given by $V_{t+1}(\phi) = V_t(\phi) + \phi \cdot (S'_{t+1} - S'_t)$, where S'_t denotes the deflated asset price process $S'_t = S_t B_t^{-1}$. We therefore have $V_T(\phi) - V_t(\phi) = G'(\phi)$ with a gains process $G'(\phi) = \sum_k^{t-1} \phi_k \cdot (S'_{k+1} - S'_k)$. Since S'_t is a martingale under the martingale measure, the discrete version of the martingale representation theorem applies, stating that $G'(\phi)$ is a martingale for a predictable process ϕ. By the martingale property, we therefore have $V_t(\phi) = B_t \mathbf{E_Q}[X B_T^{-1} | \mathcal{F}_t]$. We have already established $\pi(V_T) = V_t$ in the proof of Lemma 2.1.2 and therefore the theorem is proved.

2.1.3 Extensions

Various authors have extended the fundamental theorem of asset pricing presented above to different degrees of generality within the finite market model. Kabanov and Kramkov (1994) give a formal version of the original proof by Harrison and Pliska (1981). Kreps (1981) introduces economies with infinitely many securities. Back and Pliska (1991) and Dalang, Morton, and Willinger (1990) present extensions to infinite state spaces. See also Rogers (1994) for work on infinite state spaces in discrete time. For infinite time horizons, however, absence of arbitrage no longer implies the existence of a martingale measure. See Pliska (1997), Section 7.2., for an accessible exposition of the problem. Schachermayer (1994) proves the asset pricing theorem by replacing the no-arbitrage condition with the somewhat stronger assumption "no-free-lunch-with-bounded-risk".

2.2 Valuation in Continuous Time

In this section, we give an overview of the main results of general arbitrage pricing theory in a continuous-time economy. In essence, we review the implication of no-arbitrage if a market admits an equivalent probability measure such that relative security prices are martingales. Consequently, the price of an arbitrary European claim is given by the expected deflated value of the claim under the martingale measure. We also show that the numéraire need not be the money market account but can be an arbitrary security. An extensive treatment of contingent claim valuation in continuous time can be found in Zimmermann (1998).

2.2.1 Definitions

Many definitions are equal or very similar to their discrete time counterparts since the setup of the continuous market model is similar to that of the finite market.

We specify a filtered probability space $(\Omega, \mathcal{F}, (\mathcal{F}_t)_{0\leq t\leq T}, \mathbf{P})$. The trading interval is continuous on $t \in [0, T]$.

We assume again that the market consists of n primary securities such that the $\mathcal{F}_t$-adapted stochastic vector process in $\mathbb{R}^n_+$ $S_t = (S_t^1, \ldots, S_t^n)$ models the prices of the securities. We further assume that we have a perfect market where trading takes place in continuous time and there are no frictions.

Let the non-negative process $r_t \in \mathcal{L}^1$ denote the riskless *short rate*. A process defined by

$$B_t = \exp\left(\int_0^t r_s ds\right) \qquad t \in [0, T]$$

satisfies $dB_t = r_t B_t dt$ with $B_0 = 1$ and is called *money market account*. As r is the riskless return obtained over an infinitesimally short period of time, B is often considered a riskless investment. It is riskless in the sense that it is a predictable process. This means that the return over the next instant is known.

Let W_t be a Brownian motion in $\mathbb{R}^d$ defined on the filtered probability space $(\Omega, \mathcal{F}, (\mathcal{F}_t)_{0\leq t\leq T}, \mathbf{P})$. Suppose there are n securities defined as $\mathcal{F}_t$-adapted processes X_t.

A *trading strategy* θ is defined as a $\mathbb{R}^n$-valued $\mathcal{F}_t$-predictable vector process $\theta_t = (\theta_t^1, \ldots, \theta_t^n)$, such that $V_t = \sum_{i=1}^n \theta_t^i X_t^i = \theta_t \cdot X_t$, where V_t is the total wealth at time t. The collection of trading strategies is written as Θ. A *market* $\mathcal{M}(X, \Theta)$ is a collection of assets and a collection of trading strategies.

We define the $\mathbb{R}$-valued $\mathcal{F}_t$-adapted process G_t to be $G_t = \int_0^t \theta_s \cdot dX_s$. A trading strategy θ is called *self-financing* if is satisfies $V_t = V_0 + G_t(\theta)$. A (European) *contingent claim* maturing at time T is a $\mathcal{F}_t$-measurable random variable g. An American contingent claim is an $\mathcal{F}_t$-adapted process for $0 \leq t \leq T$. A trading strategy θ is called *replicating* if it is self-financing and $V_T(\theta) = g$.

A trading strategy θ is called an *arbitrage opportunity* (or simply an *arbitrage*) if either

$$V_0 < 0 \quad \text{and} \quad \mathbf{P}(V_T(\theta) \geq 0) = 1$$

$$\text{or}$$

$$V_0 = 0 \quad \text{and} \quad \mathbf{P}(V_T(\theta) > 0) > 0.$$

2.2.2 Arbitrage Pricing

Theorems 2.2.1 to 2.2.3 below represent the core insights of arbitrage pricing. These theorems are of a very general nature and remain the foundations of most results in derivative pricing. Most of the pricing results in the remainder of this monograph use these theorems.

Theorem 2.2.1. *If there exists an equivalent martingale measure for the market $\mathcal{M}$, then there are no arbitrage opportunities.*

Proof. As in the discrete case, we invoke the martingale representation theorem. Since under the martingale measure, deflated asset prices are martingales, we have

$$V_T(\phi) = V_t(\phi) + \int_t^T \phi_s \cdot dS'_s.$$

However, $V_T(\phi)$ is only a martingale if ϕ is suitably bounded, i.e., if $\phi \in (\mathcal{H}^2)^n$. Assuming this is the case, if $V_t(\phi) = 0$ and $V_T(\phi) \geq 0$ a.s., then $V_T(\phi) = 0$ a.s. and therefore arbitrage is excluded from $\mathcal{M}$.

Remark 2.2.1. If ϕ is unrestricted, doubling strategies are possible, as pointed out by Harrison and Kreps (1979). There are several ways to enforce a boundedness condition on the self-financing strategy ϕ. Limiting the number of securities that can be bought or sold infringes on the assumption of frictionless markets. The suggestion by Harrison and Kreps (1979) to allow trading only at finite points in time is undesirable for the same reason. Alternatively, the amount of credit granted can be restricted, as Dybvig and Huang (1988) suggest. The self-financing strategy would thus be bounded from below. This is called a *tame* strategy. Cox and Huang (1989) and Heath and Jarrow (1987) show that margin requirements have a similar effect. From all those restrictions, it follows that the trading strategy ϕ with respect to deflated prices is in $\mathcal{H}^2$ as postulated in Theorem A.3.2.

The following theorem points out the relationship between the equivalent martingale measure and contingent claims prices. It turns out that the price of a contingent claim can be interpreted as a deflated, sometimes called discounted, expectation of the final value of the claim if the expectation is taken with respect to the equivalent martingale measure.

Theorem 2.2.2. *If the market admits unique martingale measures $\mathbf{Q^i}$ and $\mathbf{Q}$ for numéraires $S_i(t)$ and B_t, respectively, then the price of an attainable* *contingent claim* $X_u \in \mathcal{X}$ *(in* $\mathcal{L}^2$*) coincides for both numéraires and is given by*

$$X_t = S_i(t)\mathbf{E}_{\mathbf{Q}^i}[S_i(u)^{-1}X_u|\mathcal{F}_t] = B_t\mathbf{E}_{\mathbf{Q}}[B_u^{-1}X_u|\mathcal{F}_t]$$

for any $t, u \in [0, T]$ *such that* $t \leq u$.

Proof. Since the claim is attainable, there is a self-financing strategy θ such that $V_u = V_t + \int_t^u \theta_s \cdot dS_s$. Deflating, we have $V_u' = V_t' + \int_t^u \theta_s' \cdot dS_s'$. According to the martingale representation theorem, the adapted process V_u' is a $\mathbf{Q}$-martingale since S' is a $\mathbf{Q}$-martingale. Thus, from the martingale property $V_t' = \mathbf{E}_{\mathbf{Q}}[V_u'|\mathcal{F}_t]$. From the replicating property of θ, we have $V_s' = X_s'$, $\forall s \in [t, u]$. Since $V_t' = V_t B_t^{-1}$ and $V_u' = V_u B_u^{-1}$, we obtain $X_t = B_t\mathbf{E}_{\mathbf{Q}^i}[B_u^{-1}X_u|\mathcal{F}_t]$.

It is left to show that other numéraires with corresponding martingale measures give the same contingent claims prices. We define a new deflator such that $V_t^i = V_t S_i^{-1}(t)$. From the integration by parts formula, we know that $dV_t^i = S_i^{-1}(t)\, dV + V_t\, d(S_i^{-1}) + d\langle V, S_i^{-1}\rangle$. Substituting for V gives $dV^i = S_i^{-1}(t)(\theta_t \cdot dS) + (\theta_t S_t)\, d(S_i^{-1}) + d\langle \theta \cdot S, S_i^{-1}\rangle$. By the linearity of variation, we have

$$dV^i = \theta_t \cdot \left(S_i^{-1}(t)\, dS + S_t\, d(S_i^1) + d\langle S, S_i^{-1}\rangle\right) = \theta_t \cdot d(SS_i^{-1}).$$

This shows that the replicating process, and thus the price, is not affected by the choice of numéraire.

Theorem 2.2.1 states that arbitrage is excluded from the market if an equivalent martingale measure exists and Theorem 2.2.2 gives the corresponding arbitrage-free prices under this martingale measure. Next, Theorem 2.2.3 gives conditions on the existence of such an equivalent martingale measure. It also defines the new measure for a given numéraire process. While no assumptions were made in Theorems 2.2.1 and 2.2.2 regarding the definition of the process S other than it being a semimartingale, Theorem 2.2.3 applies only to Itô processes. More general conditions can be found in Christopeit and Musiela (1994).

The dynamic behavior of $S(t)$ is defined by the following integral equation:

$$S(t) = S(0) + \int_0^t \alpha(s, S)\, ds + \int_0^t \sigma(s, S)\, dW(s), \tag{2.1}$$

where $W(t)$ is a vector Brownian motion in $\mathbb{R}^d$ defined on the probability space $(\Omega, \mathcal{F}, \mathbf{P})$. S itself takes values in $\mathbb{R}^n$. $\alpha(t)$ is a predictable vector process valued in $\mathbb{R}^n$, and $\sigma(t)$ a predictable matrix process valued in $\mathbb{R}^{n\times d}$. $\sigma(t)$ and $\alpha(t)$ satisfy the integrability properties $\alpha(t) \in (\mathcal{L}^1)^n$ and $\sigma(t) \in (\mathcal{L}^2)^{n\times d}$. $\alpha(t, S)$ and $\sigma(t, S)$ satisfy global Lipschitz and growth conditions, i.e., $\|\alpha(t, x) - \alpha(t, y)\| + \|\sigma(t, x) - \sigma(t, y)\| \leq K\|x - y\|$ and $\|\alpha(t, x)\|^2 + \|\sigma(t, x)\|^2 \leq K(1 + \|x\|^2)$, for an arbitrary positive constant

K. As a consequence, there is an adapted process $S(t)$ which is a strong solution to integral equation (2.1).

We often make the assumption that asset prices are governed by the following equations. For any asset $i \in \{1..n\}$,

$$\begin{aligned} S_i(t) &= S_i(0) + \int_0^t S_i(s)\alpha_i(s)\,ds + \int_0^t S_i(t)\sigma_i(t) \cdot dW(s) \\ &= S_i(0)\exp\Big(\int_0^t (\alpha_i(s) - \frac{1}{2}\|\sigma_i(s)\|^2)\,ds + \int_0^t \sigma_i(s) \cdot dW(s)\Big) \\ dS_i(t) &= S_i(t)\big(\alpha_i(t)\,dt - \sigma_i(t) \cdot dW(t)\big). \end{aligned} \tag{2.2}$$

$S_i(t)$ and $\alpha_i(t)$ are valued in $\mathbb{R}$, $\sigma_i(t)$ in $\mathbb{R}^d$. For notational simplicity, we often omit subscripts i. The three expressions are essentially equivalent ways of representing the process $S(t)$. The third expression is generally referred to as a stochastic differential equation (SDE) representation. Of course, we may also use vector notation as in (2.1). The SDE, for example, is then written $dS(t) = S(t)(\alpha(t)dt - \sigma(t)dW(t))$ with $\alpha(t)$ a $\mathbb{R}^n$ vector and $\sigma(t)$ a $\mathbb{R}^{n\times d}$ matrix.

The specification of $S(t)$ in (2.1) is more general than in (2.2). As can easily be seen, (2.2) corresponds to (2.1) with $\alpha(t,S) = S(t)\alpha(t)$ and $\sigma(t,S) = S(t)\sigma(t)$.

Consider the security market $\mathcal{M}_i$ with asset process $S(t)$ defined by (2.1) and numéraire asset $S_i(t)$. Let $Z(t)$ be an $\mathbb{R}^n$-process defined by $Z_k(t) = \frac{S_k(t)}{S_i(t)}$, $\forall k \in \{1..n\}$. $Z(t)$ allows the representation

$$Z(t) = Z(0) + \int_0^t \beta(s)\,ds + \int_0^t \nu(s)\,dW(s). \tag{2.3}$$

$\beta(t)$ and $\nu(t)$ are predictable processes in $\mathbb{R}^n$ and $\mathbb{R}^{n\times d}$, respectively, satisfying the same boundedness conditions as $\alpha(t)$ and $\sigma(t)$ in expression (2.1). $Z(t)$ is called a *relative* price process or a *deflated* price process.

The following theorem states the conditions under which a martingale measure exists in a security market. Note that the martingale measure is not invariant with respect to the chosen numéraire.

Theorem 2.2.3. *Let the relative $\mathbb{R}^n$-valued price process of expression* (2.3) *be given. If*

$$-\nu(t)\gamma(t) = \beta(t)$$

has a unique non-trivial solution for the $\mathbb{R}^d$ process $\gamma(t) \in (\mathcal{L}^2)^d$, then the market $\mathcal{M}_i$ admits a unique martingale measure.

Remark 2.2.2. The process $\gamma(t)$ is called the *relative market price of risk* process. It is relative because risk is determined with respect to a numéraire asset for comparison. It is a market price of risk because the excess drift of the asset price process over the benchmark asset (numéraire) is proportional to the excess risk of the asset relative to the benchmark. $\gamma(t)$ is the factor of proportion and can therefore be interpreted as a price per unit risk.

Proof. Assume the matrix $\nu(t)$ is regular, i.e., $\text{rank}(\nu) = d$, then the linear equation system has a unique non-trivial solution $\gamma(t)$. By substituting $-\nu(t)\gamma(t)$ for $\beta(t)$, we obtain $Z(t) = Z(0) + \int_0^t -\nu(s)\gamma(s)\,ds + \int_0^t \nu(s)\,dW(s)$. We define an equivalent martingale measure such that

$$\frac{d\mathbf{Q}}{d\mathbf{P}} = \exp\left(\int_0^T \gamma(s) \cdot dW(s) - \frac{1}{2}\int_0^T \|\gamma(s)\|^2\,ds\right).$$

By Girsanov's Theorem, $\tilde{W}(t) = W(t) - \int_0^T \gamma(s)ds$ is a standard Brownian motion under $\mathbf{Q}$. $Z(t) = Z(0) + \int_0^t -\nu(s)\gamma(s)\,ds + \int_0^t \nu(s)\,(d\tilde{W}(s) + \gamma(s)\,ds)$ simplifies to $Z(t) = Z(0) + \int_0^t \nu(s)\,d\tilde{W}(s)$. This is a martingale for suitably bounded $\nu(t)$, e.g., if Novikov's condition is satisfied.

Now consider a security market $\mathcal{M}$ with asset dynamics defined by (2.2) and numéraire asset S_i. Integrating by parts (cf. the examples accompanying Theorem A.5.2) we obtain for the deflated asset prices $Z_j(t)$, $\forall j \in \{1..n\}$,

$$Z_j(t) = \frac{S_j(0)}{S_i(0)} \exp\left(\int_0^t \beta_j(s) - \frac{1}{2}\|\nu_j(s)\|^2\,ds + \int_0^t \nu_j(s)\,dW(s)\right), \tag{2.4}$$

where $\nu_j(t) = \sigma_j(t) - \sigma_i(t)$ and $\beta_j(t) = \alpha_j(t) - \alpha_i(t) + \|\sigma_i(t)\|^2 - \sigma_i(t) \cdot \sigma_j(t)$ and with corresponding SDE

$$\frac{dZ_j}{Z_j} = \beta_j(t)\,dt + \nu_j(t) \cdot dW(t).$$

Corollary 2.2.1. *Consider an $\mathbb{R}^n$-valued price process as given by (2.3). If this process allows for a solution γ_t in accordance with Theorem 2.2.3, the martingale measure $\mathbf{Q^i}$ with respect to numéraire S_i is defined by*

$$\frac{d\mathbf{Q^i}}{d\mathbf{P}} = \exp\left(\int_0^T \gamma(s) \cdot dW(s) - \frac{1}{2}\int_0^T \|\gamma(s)\|^2\,ds\right),$$

where

$$\gamma_t = \Omega^\top(\Omega\,\Omega^\top)^{-1}(\alpha_i(t)\mathbf{1}_n - \alpha(t))$$

and

$$\Omega = \sigma(t) - \mathbf{1}_n\sigma_i(t),$$

where $\mathbf{1}_x$ is the x-dimensional identity vector.

Proof. We integrate by parts and apply Girsanov's Theorem. Under the new measure $\mathbf{Q^i}$, we have for the process Z_j, $\forall j \in \{1..n\}$,

$$dZ_j(t) = \frac{S_j}{S_i}\Big(\alpha_j(t) - \alpha_i(t) + \|\sigma_i(t)\|^2 - \sigma_i(t)\cdot\sigma_j(t)\,dt \\ + (\sigma_j(t) - \sigma_i(t))\cdot(d\tilde{W}(t) + \gamma(t)dt)\Big).$$

This SDE simplifies to

$$dZ_j(t) = \frac{S_j(t)}{S_i(t)}(\sigma_j(t) - \sigma_i(t))\cdot d\tilde{W}(t).$$

If the process $\sigma_j(t) - \sigma_i(t)$ is suitably bounded, $Z_j(t)$ is a martingale under measure $\mathbf{Q^i}$.

Example 2.2.1. A frequently used numéraire is the money market account B_t. In this case expression (2.4) simplifies to

$$Z_j(t) = \frac{S_j(0)}{B(0)}\exp\left(\int_0^t \alpha_j(s) - r(s) - \frac{1}{2}\|\sigma_j(s)\|^2\,ds + \int_0^t \sigma_j(s)\cdot dW(s)\right)$$

We define an equivalent measure

$$\frac{d\mathbf{Q}}{d\mathbf{P}} = \exp\left(\int_0^T \gamma(s)\cdot dW(s) - \frac{1}{2}\int_0^T \|\gamma(s)\|^2\,ds\right),$$

with

$$\gamma_t = \sigma(t)^\top(\sigma(t)\,\sigma(t)^\top)^{-1}(r(t)\mathbf{1}_n - \alpha(t)).$$

Applying Girsanov's theorem gives, $\forall j \in \{1..n\}$,

$$\begin{aligned} Z_j(t) &= \frac{S_j(0)}{B(0)}\exp\Big(\int_0^t \alpha_j(s) - r(s) - \frac{1}{2}\|\sigma_j(s)\|^2\,ds \\ &\quad + \int_0^t \sigma_j(s)\cdot\left(d\tilde{W}(s) + \sigma_j(t)^\top(\sigma_j(t)\,\sigma_j(t)^\top)^{-1}(r(s) - \alpha_j(s))\,ds\right)\Big) \\ &= \frac{S_j(0)}{B(0)}\exp\left(\int_0^t -\frac{1}{2}\|\sigma_j(s)\|^2\,ds + \int_0^t \sigma_j(s)\cdot d\tilde{W}(s)\right). \end{aligned}$$

Itô's formula gives the corresponding SDE,

$$dZ_j(t) = \frac{S_j(t)}{B(t)}\sigma_j(t)\cdot d\tilde{W}(t).$$

The asset process under $\mathbf{Q}$ is therefore

$$S_j(t) = S_j(t) \exp\left(\int_0^t (r(s) - \frac{1}{2}\|\sigma_j(s)\|^2)\, ds + \int_0^t \sigma_j(s) \cdot d\tilde{W}(s) \right).$$

Measure $\mathbf{Q}$ is generally called the *risk-neutral* measure since, under this measure, any asset's drift is equal to the appreciation rate of the money market account regardless of risk.

Changing numéraire from $B(t)$ back to $S_i(t)$ is straightforward. We simply define a new measure change by

$$\frac{d\mathbf{Q}^{\mathbf{i}}}{d\mathbf{Q}} = \exp\left(\int_0^T \gamma(s) \cdot dW(s) - \frac{1}{2} \int_0^T \|\gamma(s)\|^2\, ds \right),$$

with

$$\gamma(t) = -\sigma_i(t).$$

As can easily be seen by Girsanov's theorem, this change of measure results in the S_i-deflated asset price process being a martingale under measure $\mathbf{Q}^{\mathbf{i}}$, completing the example.

2.2.3 Fundamental Asset Pricing Theorem

Although the existence of an equivalent martingale measure still implies absence of arbitrage in the continuous-time market model, the opposite is no longer true. A technical problem very similar to that of the infinite investment horizon setting in the discrete market occurs. Numerous authors have presented versions of the fundamental theorem of asset pricing under various assumptions and degrees of generality, replacing no-arbitrage with slightly different notions to re-establish the theorem. For example, the condition of no-arbitrage is replaced by "no approximate arbitrage", "no-free-lunch-with-bounded-risk" by Delbaen (1992) and further generalized to "no-free-lunch-with-vanishing-risk" by Delbaen and Schachermayer (1994a) and Delbaen and Schachermayer (1994b) and "no asymptotic arbitrage" by Kabanov and Kramkov (1998).

2.3 Applications in Continuous Time

In this section, we give an overview of some important results of contingent claims valuation. We derive the Black-Scholes option pricing formulae and their extension to exchange options by Margrabe (1978). We also present the interest-rate modeling framework by Heath, Jarrow, and Morton (1992) and introduce the concept of the forward measure.

2.3.1 Black-Scholes Model

We begin by applying martingale pricing theory to the valuation of Black-Scholes equity options.

Let the market $\mathcal{M}(S, \Theta)$ be defined by a set of primary securities and a set of self-financing trading strategies. In this case, we assume that the market consists of two securities, a stock S and a money market account B. Our trading interval is $\mathcal{T} = [0, T]$.

The dynamics of the money market account are assumed to be given by the ordinary differential equation $dB_t = rB_t dt$ with $B_0 = 1$. r is the instantaneous riskless interest rate. It is assumed constant over $[0, \mathcal{T}]$. The value of the money market account is therefore given by $B_t = e^{rt}$, $\forall t \in \mathcal{T}$.

The dynamics of the stock are assumed to be given by the stochastic differential equation

$$\begin{aligned} dS_t &= \alpha S_t dt + \sigma S_t dW_t\,, \\ S_0 &= s. \end{aligned} \tag{2.5}$$

The process coefficients α and σ are assumed constant on $\mathcal{T}$. s denotes an arbitrary positive starting value. W_t is a standard Brownian motion (also called Wiener process) defined on the filtered probability space $(\Omega, \mathcal{F}, (\mathcal{F}_t)_{0 \leq t \leq T}, \mathbf{P})$.

Proposition 2.3.1. *The solution to the SDE in expression* (2.5) *is given by*

$$S_t = S_0 e^{(\alpha - \frac{1}{2}\sigma^2)t + \sigma W_t}. \tag{2.6}$$

Proof. This is the Doléans-Dade solution applied to (2.5). Since the coefficients are constants, the integrability conditions are satisfied. An application of Itô's formula confirms this conjecture. We consider the process $X_t = \alpha t + \sigma W_t$. Clearly, this is a solution to $dX_t = \alpha dt + \sigma dW_t$. After making the transformation $Y = e^X$, an application of Itô's formula gives the SDE for Y, $dY_t = Y_t(\alpha + \frac{1}{2}\sigma^2)dt + Y_t \sigma dW_t$. Now we consider the process $X_t = (\alpha - \frac{1}{2}\sigma^2)t + \sigma W_t$ and make the same transformation. Itô's formula confirms that $dY_t = Y_t \alpha dt + Y_t \sigma dW_t$.

Corollary 2.3.1. *The process* $S_t = e^{(\alpha - \frac{1}{2}\sigma^2)t + \sigma W_t}$ *has expectation*

$$\mathbf{E}_{\mathbf{P}}[S_t | \mathcal{F}_s] = S_s e^{\alpha(t-s)},$$

for any $s < t$.

Proof. By Itô's formula.

We now show that the market $\mathcal{M}$ admits an equivalent martingale measure.

Proposition 2.3.2. *The market $\mathcal{M}(S,\Theta)$ admits a martingale measure* $\mathbf{Q}$, *which is called the* risk-neutral *measure. Under* $\mathbf{Q}$, *the price process of* (2.6) *becomes*

$$S_t = S_0 e^{(r-\frac{1}{2}\sigma^2)t+\sigma W_t}. \tag{2.7}$$

Proof. We need to find a process γ,

$$\frac{d\mathbf{Q}}{d\mathbf{P}} = \exp\left(\int_0^T \gamma_s\, dW_s - \frac{1}{2}\int_0^T \gamma_s^2\, ds\right)$$

such that the price processes are martingales if measured in terms of the numéraire security B_t.

An application of Girsanov's Theorem changes the stochastic differential equation from (2.5) to

$$dS_t = \alpha S_t\, dt + \sigma S_t (d\tilde{W}_t + \gamma_t\, dt). \tag{2.8}$$

$\tilde{W}_t$ is defined on $(\Omega, \mathcal{F}, (\mathcal{F}_t)_{0\le t\le T}, \mathbf{Q})$.

For the SDE of S_t to become $dS_t = rS_t dt + \sigma S_t d\tilde{W}_t$, we need to set $\gamma_t = \frac{r-\alpha}{\sigma}$. Substituting γ_t into (2.8) gives

$$dS_t = \alpha S_t dt + \sigma S_t \left(d\tilde{W}_t + \frac{r-\alpha}{\sigma} dt\right).$$

This simplifies to

$$dS_t = rS_t dt + \sigma S_t d\tilde{W}_t.$$

By Proposition 2.3.1, this SDE yields the process given in the proposition. We now need to show that in its deflated form, this process is a martingale. Define Z to be the stock price process in terms of the numéraire B_t,

$$Z_t = Z_0 B_t^{-1} e^{(r-\frac{1}{2}\sigma^2)t+\sigma\tilde{W}_t}.$$

Since r is assumed constant, we have

$$Z_t = Z_0 e^{-\frac{1}{2}\sigma^2 t+\sigma\tilde{W}_t}. \tag{2.9}$$

We recognize (2.9) as the Doléans-Dade exponential to the driftless SDE $dZ_t = Z_t \sigma d\tilde{W}_t$. If Novikov's condition holds, Z is a martingale. Since σ is a constant,

$$\mathbf{E_Q}[\exp\left(\frac{1}{2}\int_0^T \sigma_t^2 dt\right)] = \mathbf{E_Q}[\exp\left(\frac{1}{2}\sigma^2 T\right)] < \infty.$$

It follows that Z is a martingale. Trivially, the discounted process of the second security in the market, B_t, is also a martingale. $\mathbf{Q}$ is therefore a martingale measure equivalent to $\mathbf{P}$.

Proposition 2.3.3. *The price of a contingent claim $X \in \mathcal{X}$ (suitably bounded) to pay out $X(S)$ at time T is given by $X_t = \mathbf{E_Q}[e^{-r(T-t)}X_T|\mathcal{F}_t]$.*

Proof. By Proposition 2.3.2, there is a martingale measure. By Theorem 2.2.2, the risk-neutral valuation formula applies. Since $B_t = e^{rt}$, this proposition is a special case of Theorem 2.2.2, completing the proof.

Proposition 2.3.4 (Black-Scholes). *A claim with payoff $X_T = (S_T - K)^+$ is called a* call *option. Its price is given by*

$$X_t = S_t N(d) - Ke^{T-t}N(d - \sigma\sqrt{T-t}),$$

where

$$d = \frac{\ln \frac{S_t}{K} + (r + \frac{\sigma^2}{2})(T-t)}{\sigma\sqrt{T-t}}.$$

The price of a put option with payoff $Y_T = (x - S_T)^+$ is

$$Y_t = -S_t N(-d) + Ke^{T-t}N(-d + \sigma\sqrt{T-t}).$$

Remark 2.3.1. N is the cumulated standard normal distribution function such that $N(x) = \int_{-\infty}^{x} \exp(-\frac{1}{2}z^2)dz$. $(\cdot)^+$ is sometimes written $\max(\cdot, 0)$ or $\cdot \vee 0$.

Proof. The claim is bounded and attainable (see above). The price is given by $X_t = \mathbf{E_Q}[e^{-r(T-t)}(S_T - K)^+|\mathcal{F}_t]$. Substituting for S_T gives

$$X_t = \mathbf{E_Q}[(S_t e^{(-\sigma^2/2)(T-t)+\sigma(\tilde{W}_T - \tilde{W}_t)} - e^{-r(T-t)}K)^+)|\mathcal{F}_t].$$

$\tilde{W}_T - \tilde{W}_t$ is independent of $\mathcal{F}_t$. By the definition of conditional expectation, $X_t = E_1 - E_2$ with

$$E_1 = \mathbf{E_Q}[S_t e^{(-\sigma^2/2)(T-t)+\sigma(\tilde{W}_T - \tilde{W}_t)}\mathbf{1}_{\{S_T > K\}}]$$
$$E_2 = \mathbf{E_Q}[Ke^{-r(T-t)}\mathbf{1}_{\{S_T > K\}}].$$

$\tilde{W}_T - \tilde{W}_t$ is a normal random variable with law $N(0, T-t)$. By normalizing we obtain

$$\tilde{z} = \frac{\tilde{W}_T - \tilde{W}_t}{\sqrt{T-t}}, \tag{2.10}$$

which has law $N(0,1)$. Thus,

$$\begin{aligned} E_1 &= \int_{-\infty}^{d} S_t e^{-\frac{1}{2}\sigma^2(T-t)+\sigma\tilde{z}\sqrt{T-t}} \frac{1}{\sqrt{2\pi}} e^{-\frac{1}{2}\tilde{z}^2} d\tilde{z} \\ &= \int_{-\infty}^{d} S_t \frac{1}{\sqrt{2\pi}} e^{-\frac{1}{2}(\tilde{z}-\sigma\sqrt{T-t})^2} d\tilde{z}. \end{aligned} \tag{2.11}$$

d depends on the evaluation of the indicator function and is unspecified at this time. We have a normal distribution with law $N(\sigma\sqrt{T-t}, 1)$.

We define an equivalent probability measure

$$\frac{d\dot{\mathbf{Q}}}{d\mathbf{Q}} = \exp\left(-\sigma\tilde{W}_T - \frac{1}{2}\sigma^2 T\right).$$

By Girsanov's Theorem, $\dot{W}_t = \tilde{W}_t - \sigma t$ is a standard Brownian motion. Clearly, $\tilde{W}_T - \tilde{W}_t = \dot{W}_T - \dot{W}_t + \sigma(T-t)$. Combining this result with (2.10) we obtain

$$\tilde{z} = \frac{\tilde{W}_T - \tilde{W}_t}{\sqrt{T-t}} = \frac{\dot{W}_T - \dot{W}_t + \sigma(T-t)}{\sqrt{T-t}} = \dot{z} + \sigma\sqrt{T-t}.$$

Seeing that (2.11) has law $N(0,1)$ under the equivalent measure $\dot{\mathbf{Q}}$, $E_1 = S_t N(d)$.

To obtain d, we now evaluate the indicator function under the measure $\dot{\mathbf{Q}}$.

$$\begin{aligned}
\mathbf{E}_{\dot{\mathbf{Q}}}[\mathbf{1}_{\{S_T>K\}}] &= \dot{\mathbf{Q}}(S_T > K) \\
&= \dot{\mathbf{Q}}\left(S_t e^{(r-\frac{1}{2}\sigma^2)(T-t)+\sigma(\dot{W}_T-\dot{W}_t+\sigma(T-t))} > K\right) \\
&= \dot{\mathbf{Q}}\left(\sigma(\dot{W}_T - \dot{W}_t) > \ln K - \ln S_t - (r + \frac{1}{2}\sigma^2)(T-t)\right) \\
&= \dot{\mathbf{Q}}\left(\dot{z} < \frac{\ln S_t - \ln K + (r+\frac{1}{2}\sigma^2)(T-t)}{\sigma\sqrt{T-t}}\right).
\end{aligned}$$

Therefore,

$$d = \frac{\ln\frac{S_t}{K} + (r+\frac{1}{2}\sigma^2)(T-t)}{\sigma\sqrt{T-t}}.$$

E_2 is evaluated similarly, but under the measure $\mathbf{Q}$.

$$\begin{aligned}
E_2 &= Ke^{-r(T-t)}\mathbf{Q}(S_T > K) \\
&= Ke^{-r(T-t)}\mathbf{Q}\left(S_t e^{(r-\frac{1}{2}\sigma^2)(T-t)+\sigma(\tilde{W}_T-\tilde{W}_t)}) > K\right) \\
&= Ke^{-r(T-t)}\mathbf{Q}\left(\tilde{z} < \frac{\ln S_t - \ln K + (r-\frac{1}{2}\sigma^2)(T-t)}{\sigma\sqrt{T-t}}\right) \\
&= Ke^{-r(T-t)}N(d - \sigma\sqrt{T-t}).
\end{aligned}$$

The original derivation of Black and Scholes (1973) and Merton (1973) took a partial differential equation (PDE) approach to solving the valuation problem. A derivative instrument on the underlying S is a function of S and time, i.e., $F = F(S,t)$. Itô's formula gives

$$dF = F^S dS + F^t dt + \frac{1}{2} F^{SS} (dS)^2.$$

Superscripts of F denote partial derivatives. From the SDE for S in (2.5), we substitute for $(dS)^2$. By the multiplication rules derived in Section A.4, we have $(dt)^2 = 0$, $dW\,dt = 0$, $(dW)^2 = dt$. Thus,

$$dF = F^S dS + F^t dt + \frac{1}{2} \sigma^2 S^2 F^{SS} dt.$$

Now a dynamically adjusted hedge portfolio is constructed, consisting of $\Delta = -F^S$ units of S. The SDE of the hedged total position $H = F + \Delta S$ is therefore $dH = dF + \Delta dS = (F^t + \frac{1}{2}\sigma^2 S^2 F^{SS})dt$. Because H is a riskless position, i.e., independent of the process of S, the riskless rate r is earned on the portfolio. Therefore, $dH = rHdt$, resulting in the PDE

$$0 = F^t + \frac{1}{2}\sigma^2 S^2 F^{SS} + rSF^S - rF. \tag{2.12}$$

This is the Black-Scholes PDE. Its boundary conditions are specific to the derivative instrument under consideration. Given suitable boundary conditions, it can be solved using Fourier transforms or a similarity transformation. For a call option with condition $(S(T) - K)^+$, Proposition 2.3.4 is the result.

The link between the PDE approach and the direct evaluation used in Proposition 2.3.4 is provided by a special case of the Feynman-Kac representation of parabolic differential equations. The Feynman-Kac solution of the PDE in (2.12) with $dS = S(rdt + \sigma dW(t))$ and boundary condition $g(S(T))$ is a process $F(S,t) = E[e^{-r(T-t)} g(S(T))]$. More generally,[3] for SDE $dS = r(S,t)dt + \sigma(S,t)dW(t)$ and PDE $0 = F^t + \frac{1}{2}\sigma^2(S,t)F^{SS} + r(S,t)SF^S - r(S,t)F$ there is a process

$$F = F(S,t) = E\left[\exp\left(-\int_t^T r(S,s)\,ds\right) g(S(T))\right],$$

where $r(S,t)$ and $\sigma(S,t)$ must satisfy Lipschitz and growth conditions.

2.3.2 Margrabe's Model

A generalization of the Black-Scholes formula is a solution derived by Margrabe (1978) for an option that gives the right to exchange one asset S_2 into another asset S_1. In other words, the option gives the right to buy asset S_1 for the price of asset S_2.

[3] The Feynman-Kac formula applies to an even more general PDE called the Cauchy problem. For this generalization as well as for the technical conditions, refer to Duffie (1996), Appendix E, or Karatzas and Shreve (1991), Section 5.7.

Proposition 2.3.5 (Margrabe). *The price of a claim* $X_T = (S_1(T) - S_2(T))^+$ *is given by*

$$X_t = S_1(t)N(d) - S_2(t)N(d - \sigma\sqrt{T-t}),$$

where

$$d = \frac{\ln \frac{S_1(t)}{S_2(t)} + \frac{\sigma^2}{2}(T-t)}{\sigma\sqrt{T-t}}$$

and

$$\sigma = \sqrt{\sigma_1^2 + \sigma_2^2 - 2\rho\sigma_1\sigma_2}.$$

Proof. By Theorem 2.2.2, the arbitrage-free price for a bounded and attainable claim with payoff X can be written $X_t = B_t \mathbf{E}_{\mathbf{Q}}[B_T^{-1} X_T | \mathfrak{F}_t]$. However, the numéraire need not be the money market account but can be any asset price process. It turns out that, for this case, it is convenient to use asset S_2 as numéraire. By Theorem 2.2.2, the price is therefore $X_t = S_2(t)\mathbf{E}_{\mathbf{Q}^2}[S_2^{-1}(T) X_T | \mathfrak{F}_t]$ where $\mathbf{Q}^2$ is the equivalent martingale measure for numéraire S_2. Since the exchange option has payoff $(S_1(T) - S_2(T))^+$, the price is given by

$$X_t = S_2(t)\mathbf{E}_{\mathbf{Q}^2}[(\delta_T - 1)^+ | \mathfrak{F}_t], \tag{2.13}$$

where $\delta_t = \frac{S_1(t)}{S_2(t)}$. To evaluate the expectation expression, we need to determine the process of δ and the martingale measure $\mathbf{Q}^2$.

Since processes $S_1(t)$ and $S_2(t)$ are defined by (2.5), the SDE for δ_t is

$$\frac{d\delta}{\delta} = (\alpha_1 - \alpha_2 + \sigma_2^2 - \rho\sigma_1\sigma_2)\, dt + \sigma_1\, dW_1 - \sigma_2\, dW_2.$$

This SDE can be derived using integration by parts as shown in one of the examples accompanying Theorem A.5.2. In conformity with Corollary 2.2.1, we introduce a new measure defined by

$$\frac{d\mathbf{Q}^2}{d\mathbf{P}} = \exp\left(\gamma\, W_T - \frac{1}{2}\gamma^2 T\right),$$

with $\gamma_i = \frac{r - \alpha_i}{\sigma_i} + \rho_{i2}\sigma_2$ for $i \in 1, 2$. ρ_{i2} is the correlation between asset i and the numéraire, asset S_2. By Girsanov's theorem, $d\tilde{W} = dW - \gamma\, dt$. Therefore,

$$\begin{aligned}\frac{d\delta}{\delta} &= (\alpha_1 - \alpha_2 + \sigma_2^2 - \rho\sigma_1\sigma_2)\, dt \\ &\quad + \sigma_1\left(d\tilde{W}_1 + (\frac{r - \alpha_1}{\sigma_1} + \rho\sigma_2)\, dt\right) \\ &\quad - \sigma_2\left(d\tilde{W}_2 + (\frac{r - \alpha_2}{\sigma_2} + \sigma_2)\, dt\right),\end{aligned}$$

which is seen to be equal to $\frac{d\delta}{\delta} = \sigma_1 d\tilde{W}_1 - \sigma_2 d\tilde{W}_2$. Given a martingale boundedness condition, δ is a martingale under the measure $\mathbf{Q^2}$. For convenience, we set $\sigma\tilde{W}_t = \sigma_1\tilde{W}_1 - \sigma_2\tilde{W}_2$. By the property $\langle ax + by\rangle = a^2\langle x\rangle + b^2\langle y\rangle + 2ab\langle x, y\rangle$ for processes x and y, we have $\sigma = \sqrt{\sigma_1^2 + \sigma_2^2 - 2\rho\sigma_1\sigma_2}$. Since we have determined the measure, we can now evaluate (2.13). The following analysis is very similar to the proof to the Black-Scholes formula.

By the linearity property of the expectation operator, the definition of conditional expectation, and substitution of δ_T, we can can write $X_t = S_2(t)\,(E_1 - E_2)$ with

$$
\begin{aligned}
E_1 &= \mathbf{E}_{\mathbf{Q^2}}\left[\delta_t e^{-\frac{1}{2}\sigma^2(T-t)+\sigma(\tilde{W}_T-\tilde{W}_t)}\mathbf{1}_{\{\delta_T>1\}}\right] \\
E_2 &= \mathbf{E}_{\mathbf{Q^2}}\left[\mathbf{1}_{\{\delta_T>1\}}\right].
\end{aligned}
$$

$\tilde{W}_T - \tilde{W}_t$ is a random variable with law $N(0, T-t)$. Normalizing, we obtain $\tilde{z} = \frac{\tilde{W}_T - \tilde{W}_t}{\sqrt{T-t}}$ which has law $N(0,1)$. Thus,

$$
\begin{aligned}
E_1 &= \int_{-\infty}^{d} \delta_t e^{-\frac{1}{2}\sigma^2(T-t)+\sigma\tilde{z}\sqrt{T-t}}\frac{1}{\sqrt{2\pi}}e^{-\frac{1}{2}\tilde{z}^2}\,d\tilde{z} \\
&= \int_{-\infty}^{d} \delta_t \frac{1}{\sqrt{2\pi}}e^{-\frac{1}{2}(\tilde{z}-\sigma\sqrt{T-t})^2}\,d\tilde{z}.
\end{aligned} \tag{2.14}
$$

d depends on the evaluation of the indicator function and is unspecified at this time. We have a normal distribution with law $N(\sigma\sqrt{T-t}, 1)$. We proceed to define an equivalent probability measure

$$
\frac{d\dot{\mathbf{Q}}^{\mathbf{2}}}{d\mathbf{Q^2}} = \exp\left(\sigma\,\tilde{W}_T - \frac{1}{2}\sigma^2 T\right).
$$

By Girsanov's Theorem, $\dot{W}_t = \tilde{W}_t - \sigma t$ is a standard Brownian motion and therefore (2.14) has law $N(0,1)$ under the equivalent measure $\dot{\mathbf{Q}}^{\mathbf{2}}$. Thus, $E_1 = \delta_t N(d)$ where N is the cumulated normal distribution function.

To obtain d, we now evaluate the indicator function under the measure $\dot{\mathbf{Q}}$.

$$
\begin{aligned}
\mathbf{E}_{\dot{\mathbf{Q}}^{\mathbf{2}}}\left[\mathbf{1}_{\{\delta_T>1\}}\right] &= \dot{\mathbf{Q}}^{\mathbf{2}}(\delta_T > 1) \\
&= \dot{\mathbf{Q}}^{\mathbf{2}}\left(\delta_t e^{(-\frac{1}{2}\sigma^2)(T-t)+\sigma(\dot{W}_T-\dot{W}_t+\sigma(T-t))} > 1\right) \\
&= \dot{\mathbf{Q}}^{\mathbf{2}}\left(\sigma(\dot{W}_T - \dot{W}_t) > -\ln\delta_t - \frac{1}{2}\sigma^2(T-t)\right) \\
&= \dot{\mathbf{Q}}^{\mathbf{2}}\left(\dot{z} < \frac{\ln\delta_t + \frac{1}{2}\sigma^2(T-t)}{\sigma\sqrt{T-t}}\right).
\end{aligned}
$$

Therefore,

$$d = \frac{\ln \delta_t + \frac{1}{2}\sigma^2(T-t)}{\sigma\sqrt{T-t}}.$$

Obviously, E_2 can be written $N(d_2)$ with d_2 to be determined by the indicator function. Since no additional measure change is necessary, $E_2 = N(d - \sigma\sqrt{T-t})$. Alternatively, E_2 can be calculated analogously to E_1, but under measure $\mathbf{Q^2}$ instead of $\dot{\mathbf{Q}}^\mathbf{2}$. Multiplying $E_1 + E_2$ by $S_2(t)$ yields the proposition.

2.3.3 Heath-Jarrow-Morton Framework

This section reviews the interest-rate framework developed by Heath, Jarrow, and Morton (1992).

Definitions. Consider the time interval $\mathcal{T} = [0, T']$. The price at time t of a zero-coupon bond maturing at time T is written $P(t,T)$ for any $t, T \in \mathcal{T}$ such that $t \leq T$.

The instantaneous *forward rate* $f(t,T)$ is given by

$$f(t,T) = -\frac{\partial}{\partial T} \ln P(t,T), \quad \forall t, T \in \mathcal{T} s.t.\ t \leq T.$$

$f(t,T)$ is the rate of return available at time t for an instantaneous investment at time T.

Alternatively, the time t price of a zero-coupon bond maturing at time T can be expressed in terms of forward rates

$$P(t,T) = \exp\left(-\int_t^T f(t,s)ds\right). \tag{2.15}$$

$f(t,s)$ is in $\mathcal{L}^1$ to ensure integrability.

The *short rate* is a special forward rate, namely

$$r_t = f(t,t). \tag{2.16}$$

The money market account, which serves as the numéraire, is defined as in the economy with deterministic interest rates, i.e., $dB_t = r_t B_t dt$ with $B_0 = 1$. This gives

$$B_t = \exp\left(\int_0^t r_s\, ds\right). \tag{2.17}$$

The forward rate structure is assumed to follow a family of Itô processes such that for any fixed $T \in [0, T']$ such that $\alpha(t) \in \mathcal{L}^1$ and $\sigma(t) \in \mathcal{L}^2$,

$$f(t,T) = f(0,T) + \int_0^t \alpha(s,T)\, ds + \int_0^t \sigma(s,T) \cdot dW_s, \tag{2.18}$$

$\forall T$ s.t. $0 \leq t \leq T$, where W_t takes values in $\mathbb{R}^d$ and is defined on a filtered probability space $(\Omega, \mathcal{F}, (\mathcal{F}_t)_{0 \leq t \leq T}, \mathbf{P})$. The processes $\alpha(t,T)$ and $\sigma(t,T)$ take values in $\mathbb{R}$ and $\mathbb{R}^d$, respectively. We require that $\alpha(t,T) \in \mathcal{L}^{1,1}$ and $\sigma(t,T) \in (\mathcal{L}^{2,1})^d$.

From (2.16) and (2.18), the short rate obeys

$$r_t = f(0,t) + \int_0^t \alpha(s,t)\,ds + \int_0^t \sigma(s,t) \cdot dW_s, \qquad \forall t \in [0,T']. \tag{2.19}$$

For the money market account process to be well defined, we require $f(0,t) \in \mathcal{L}^1$, in addition to the regularity conditions on α and σ specified above. Then, from expressions (2.17) and (2.19),

$$B_t = \exp\left(\int_0^t f(0,u)\,du + \int_0^t \int_0^u \alpha(s,u)\,ds\,du + \int_0^t \int_0^u \sigma(s,u) \cdot dW_s\,du\right),$$

An application of Fubini's theorem in its standard and stochastic versions, respectively, gives

$$B_t = \exp\left(\int_0^t f(0,u)\,du + \int_0^t \int_s^t \alpha(s,u)\,du\,ds + \int_0^t \int_s^t \sigma(s,u)\,du \cdot dW_s\right). \tag{2.20}$$

Similarly, the bond price can be written in terms of forward rates. By (2.15) and (2.18) we have after an application of Fubini's theorems,

$$P(t,T) = \exp\left(-\int_t^T f(0,s)\,ds - \int_0^t \int_t^T \alpha(s,u)\,du\,ds - \int_0^t \int_t^T \sigma(s,u)\,du \cdot dW_s\right). \tag{2.21}$$

Proposition 2.3.6. *The bond price process in terms of the numéraire money market account, $Z(t,T) = B(t)^{-1}P(t,T)$ is given by*

$$\ln Z(t,T) = -\int_0^T f(0,s)\,ds - \int_0^t \int_s^T \alpha(s,u)\,du\,ds - \int_0^t \int_s^T \sigma(s,u)\,du \cdot dW_s\,,$$

with corresponding SDE

$$\frac{dZ(t,T)}{Z(t,T)} =$$

$$-\left(\int_t^T \alpha(t,u)\,du\,dt - \frac{1}{2}\int_t^T \|\sigma(s,u)\|^2\,du\,dt + \int_t^T \sigma(t,u)\,du \cdot dW_t\right).$$

Proof. The logarithm of the deflated bond price can be written $\ln Z(t,T) = \ln P(t,T) - \ln B(t)$. By expressions (2.20) and (2.21),

$$\ln Z(t,T) =$$

$$-\int_t^T f(0,s)\,ds - \int_0^t\int_t^T \alpha(s,u)\,du\,ds - \int_0^t\int_t^T \sigma(s,u)\,du \cdot dW_s$$

$$-\int_0^t f(0,u)\,du - \int_0^t\int_s^t \alpha(s,u)\,du\,ds - \int_0^t\int_s^t \sigma(s,u)\,du \cdot dW_s\,.$$

Clearly, by combining like integrals we obtain the proposition. The SDE is derived by setting $Z = \exp(\ln Z)$ and applying Itô's formula.

Definition 2.3.1. *For notational simplicity, we define*

$$\alpha'(t,T) = \int_t^T \alpha(t,s)\,ds \quad \text{and} \quad \sigma'(t,T) = \int_t^T \sigma(t,s)\,ds.$$

Proposition 2.3.7. *If the linear equation system*

$$-\alpha'(t,T) + \frac{1}{2}\|\sigma'(t,T)\|^2 = \sigma'(t,T)\cdot\gamma_t \tag{2.22}$$

admits a unique solution for the $\mathbb{R}^d$-valued process $\gamma_t \in \mathcal{L}^2$ such that Novikov's condition is satisfied, then there exists an equivalent martingale measure for Z.

Proof. Consider the SDE for $Z(t,T)$ as derived in Proposition 2.3.6. By the shorthand notation introduced in Definition 2.3.1, we can write the SDE for $Z(t,T)$ as

$$dZ(t,T) = -Z(t,T)\left(\alpha'(t,T)\,dt + \sigma'(t,T)\cdot dW_t - \frac{1}{2}\|\sigma'(t,T)\|^2\,dt\right).$$

If γ_t is a unique solution to expression (2.22), then

$$dZ(t,T) = -Z(t,T)(-\sigma'(t,T)\cdot\gamma_t + \sigma'(t,T)\cdot dW_t).$$

Define an equivalent probability measure such that

$$\frac{d\mathbf{Q}}{d\mathbf{P}} = \exp\left(\int_0^T \gamma_s\cdot dW_s - \frac{1}{2}\int_0^T \|\gamma_s\|^2 ds\right).$$

By Girsanov's Theorem, $\tilde{W}_t = W_t - \int_0^t \gamma_s \, ds$ is a standard Brownian motion under $\mathbf{Q}$. Thus, $dZ(t,T) = -Z(t,T)(-\sigma'(t,T) \cdot \gamma_t dt + \sigma'(t,T) \cdot (d\tilde{W}_t + \gamma_t dt))$. Clearly, this is $dZ(t,T) = -Z(t,T)\sigma'(t,T) \cdot d\tilde{W}_t$ which is a $\mathbf{Q}$-martingale if Novikov's condition is satisfied.

Remark 2.3.2. Proposition 2.3.7 is a special case of Theorem 2.2.3.

Example 2.3.1. Consider the special case of a one-factor model where W_t is a $\mathbb{R}$-valued Brownian motion. Define an equivalent probability measure

$$\frac{d\mathbf{Q}}{d\mathbf{P}} = \exp\left(\int_0^T \gamma_s dW_s - \frac{1}{2}\int_0^T \gamma_s^2 ds\right)$$

such that

$$\gamma_t = -\sigma'(t,T)^{-1}\,\alpha'(t,T) + \frac{1}{2}\sigma'(t,T).$$

By Girsanov's theorem,

$$dZ(t,T) = \\ -Z(t,T)\left(\alpha'(t,T)\,dt + \sigma'(t,T)(d\tilde{W}_t + \gamma_t\,dt) - \frac{1}{2}\sigma'(t,T)^2\,dt\right).$$

Substituting for γ gives $dZ(t,T) = -Z(t,T)\sigma'(t,T)\,d\tilde{W}_t$ under the equivalent probability measure $\mathbf{Q}$.

Proposition 2.3.8. *If, for arbitrary $T \in [0,T']$,*

$$\alpha(t,T) = \sigma(t,T) \cdot \sigma'(t,T), \qquad \forall t \in [0,T'] \text{ s.t. } t \leq T$$

under the equivalent probability measure $\mathbf{Q}$ and s.t. $\alpha(t,T) \in \mathcal{L}^{1,1}$, then the bond price dynamics are arbitrage-free.

Proof. By Proposition 2.3.7, there is a unique equivalent martingale measure if $-\alpha(t,T)' + \frac{1}{2}\|\sigma'(t,T)\|^2 = \sigma'(t,T) \cdot \gamma_t$ has a unique solution γ_t in $\mathbb{R}^d$. Therefore, from the proof of Proposition 2.3.7, the drift of the process Z_t is $-\alpha'(t,T)\,dt + \frac{1}{2}\|\sigma'(t,T)\|^2\,dt = \sigma'(t,T) \cdot \gamma_t\,dt$. By Girsanov, changing the probability measure from $\mathbf{P}$ to $\mathbf{Q}$ decreases the drift by $\gamma_t dt$. Thus, the γ_t terms "cancel out" such that the drift coefficients $-\alpha'(t,T)\,dt + \frac{1}{2}\|\sigma'(t,T)\|^2 dt = 0$. Differentiation gives $\alpha(t,T) = \sigma(t,T) \cdot \sigma'(t,T)$.

We can now determine the processes and SDEs for the the interest rate and bond processes. The process of the forward rates under $\mathbf{Q}$ immediately follows from Proposition 2.3.8.

$$f(t,T) = f(0,T) + \int_0^t \sigma(s,T) \cdot \sigma'(s,T)\,ds + \int_0^t \sigma(s,T) \cdot d\tilde{W}_s, \tag{2.23}$$

$\forall t$ s.t. $0 \le t \le T$. In differential notation,

$$df(t,T) = \sigma(t,T) \cdot \sigma'(t,T)\,dt + \sigma(t,T) \cdot d\tilde{W}_t.$$

To determine the bond price process under **Q**, we need to make some transformations. The definition of the bond in terms of forward rates is given by expression (2.21). Consider the following integral decomposition of the logarithm of the bond price.

$$\begin{aligned}
\ln P(t,T) &= \\
&- \int_0^T f(0,u)\,du - \int_0^t \int_s^T \alpha(s,u)\,du\,ds - \int_0^t \int_s^T \sigma(s,u)\,du \cdot dW_s \\
&+ \int_0^t f(0,u)\,du + \int_0^t \int_s^t \alpha(s,u)\,du\,ds + \int_0^t \int_s^t \sigma(s,u)\,du \cdot dW_s,
\end{aligned}$$

or in shorthand notation

$$\begin{aligned}
\ln P(t,T) = &- \int_0^T f(0,u)\,du - \int_0^t \alpha'(s,T)\,ds - \int_0^t \sigma'(s,T) \cdot dW_s \\
&+ \int_0^t f(0,u)\,du + \int_0^t \alpha'(s,t)\,ds + \int_0^t \sigma'(s,t) \cdot dW_s.
\end{aligned}$$

By expression (2.17) and (2.20), the second half of the above expressions is equal to $\int_0^t r_s ds$. Additionally, we have $-\int_0^T f(0,u)\,du = \ln P(0,T)$.

We now impose the arbitrage condition from Proposition 2.3.7 and change probability measures from **P** to **Q** using the substitution $dW_t = d\tilde{W} + \gamma_t dt$. We obtain

$$\begin{aligned}
\ln P(t,T) = \ln P(0,T) &- \frac{1}{2}\int_0^t \|\sigma'(s,T)\|^2\,ds - \int_0^t \sigma'(s,T) \cdot d\tilde{W}_s \\
&+ \int_0^t f(0,u)\,du + \frac{1}{2}\int_0^t \|\sigma'(s,t)\|^2\,ds + \int_0^t \sigma'(s,t) \cdot d\tilde{W}_s.
\end{aligned}$$

Since $r_t = f(t,t)$ and $f(t,T)$ under **Q** is given by (2.23), r_t under **Q** can be written

$$r_t = f(0,t) + \int_0^t \sigma(s,t) \cdot \sigma'(s,t)\,ds + \int_0^t \sigma(s,t) \cdot d\tilde{W}_s. \tag{2.24}$$

Integrating (2.24) gives again

$$\int_0^t r_s ds = \int_0^t f(0,u)\,du + \frac{1}{2}\int_0^t \|\sigma'(s,t)\|^2\,ds + \int_0^t \sigma'(s,t) \cdot d\tilde{W}_s.$$

Therefore, the bond price process under measure **Q** is given by

$$P(t,T) = P(0,T)\exp\left(\int_0^t r_s ds - \frac{1}{2}\int_0^t \|\sigma'(s,T)\|^2 ds - \int_0^t \sigma'(s,T)\cdot d\tilde{W}_s\right). \tag{2.25}$$

An application of Itô's formula gives the corresponding SDE

$$dP(t,T) = P(t,T)\left(r_t dt - \sigma'(t,T)\cdot d\tilde{W}_t\right). \tag{2.26}$$

As expected, the drift of the bond price is the short rate under probability measure $\mathbf{Q}$.

2.3.4 Forward Measure

So far, the numéraire has usually been the money market account B_t. The associated equivalent probability measure $\mathbf{Q}$ is called the risk-neutral measure. Under the risk-neutral measure, asset prices are martingales in terms of the numéraire B_t, i.e.,

$$\mathbf{E}_{\mathbf{Q}}[B_T^{-1}S_T|\mathcal{F}_t] = B_t^{-1}S_t,$$

for arbitrary $t \le T$.

As established in Theorem 2.2.2, the numéraire need not be the locally riskless asset, but can be chosen from a larger class of price processes.[4] Since the equivalent martingale measure is not independent of the numéraire, it is necessary to show that there exists a martingale measure for the price processes in terms of the new numéraire.[5]

In this section, the bond price $P(t,T)$ is used as a numéraire. We show that there is an equivalent probability measure such that asset prices are martingales in terms of the new numéraire. If the new numéraire is a zero-coupon bond, this probability measure $\mathbf{F}$ is sometimes called *forward neutral* since asset prices measured in terms of a bond price have the financial interpretation of forward prices.

If there exists an equivalent probability measure such that prices measured in terms of bond prices are martingales, then

$$\mathbf{E}_{\mathbf{F}}[P(T,T)^{-1}S_T|\mathcal{F}_t] = \mathbf{E}_{\mathbf{F}}[S_T|\mathcal{F}_t] = S_t,$$

for arbitrary $t \le T$. In other words, forward prices are unbiased estimates of future prices under the forward neutral measure.

According to Theorem 2.2.2, the price of a claim that pays X_T depending on asset S_T at time T can be written

$$X_t = P(t,T)\mathbf{E}_{\mathbf{F}}[P(T,T)^{-1}X_T(S(T,T))|\mathcal{F}_t].$$

[4] Cf. Geman, Karoui, and Rochet (1995) for a general exposition of the change of numéraire concept.

[5] Delbaen and Schachermayer (1995) give necessary and sufficient conditions under which a numéraire price process allows for a martingale measure.

Since $P(T,T) = 1$ we have

$$X_t = P(t,T)\mathbf{E_F}[X_T(S(T,T))|\mathcal{F}_t].$$

Definition 2.3.2. *The price of a bond $P(t,\tau)$ in terms of the numéraire bond is given by*

$$F(t,T,\tau) = \frac{P(t,\tau)}{P(t,T)}.$$

$F(t,T,\tau)$ is called the forward price *of $P(t,\tau)$.*

Definition 2.3.3. *Define a probability measure $\mathbf{F}$ equivalent to $\mathbf{Q}$ such that*

$$\frac{d\mathbf{F}}{d\mathbf{Q}} = \frac{P(T,T)P(0,T)^{-1}}{B_T B_0^{-1}} = \frac{1}{P(0,T)B_T}.$$

Measure $\mathbf{F}$ is called forward measure. *Define ζ_t to be the Radon-Nikodým process such that*

$$\zeta_t = \frac{P(t,T)}{P(0,T)B_t}.$$

Consider the bond price process under measure $\mathbf{Q}$ from (2.25) and the money market process from (2.17) and (2.24). Given these processes, the Radon-Nikodým process to change measure from $\mathbf{Q}$ to $\mathbf{F}$ can be determined by straightforward substitution of $P(t,T)$, $P(0,T)$, and B_t. Thus,

$$\zeta_t = \frac{P(t,T)}{P(0,T)B_t} = \exp\left(-\frac{1}{2}\int_0^t \|\sigma(s,T)\|^2\, ds - \int_0^t \sigma'(s,T)\cdot d\tilde{W}s\right). \quad (2.27)$$

By Itô's formula, we obtain the corresponding SDE,

$$d\zeta_t = -\zeta_t\sigma'(t,T)\cdot d\tilde{W}_t.$$

Proposition 2.3.9. *The price of a bond in terms of the numéraire bond $F(t,T,\tau)$ is a martingale under the forward measure from Definition 2.3.3 if Novikov's condition is satisfied.*

Proof. The forward price is $F(t,T,\tau) = \frac{P(t,\tau)}{P(t,T)}$ as given by Definition 2.3.2. The process of a quotient of semimartingales is given in the example accompanying Theorem A.5.2. Differentiating, we obtain

$$dF_t = d\left(\frac{X_t}{Z_t}\right) = \frac{1}{Z}\,dX - \frac{X}{Z^2}\,dZ + \frac{X}{Z^3}\,d\langle Z\rangle - \frac{1}{Z^2}\,d\langle X,Z\rangle.$$

Setting F, X, Z to $F(t,T,\tau)$, $P(t,\tau)$ and $P(t,T)$, respectively, and dividing the expression by $F(t,T,\tau) = \frac{X}{Z}$ gives

$$\frac{dF(t,T,\tau)}{F(t,T,\tau)} = \frac{dP(t,\tau)}{P(t,\tau)} - \frac{dP(t,T)}{P(t,T)} + \left(\frac{dP(t,T)}{P(t,T)}\right)^2 - \frac{dP(t,\tau)}{P(t,\tau)}\frac{dP(t,T)}{P(t,T)}.$$

Inserting for $\frac{dP(t,T)}{P(t,T)}$ and $\frac{dP(t,\tau)}{P(t,\tau)}$ from expression (2.26) gives

$$\frac{dF(t,T,\tau)}{F(t,T,\tau)} = -\sigma'(t,\tau)\cdot d\tilde{W}_t + \sigma'(t,T)\cdot d\tilde{W}_t + \|\sigma'(t,T)\|^2\,dt - \sigma'(t,T)\cdot\sigma'(t,\tau)\,dt.$$

In full notation,

$$\frac{dF(t,T,\tau)}{F(t,T,\tau)} = -\int_t^\tau \sigma(t,u)\,du\cdot d\tilde{W}_t + \int_t^T \sigma(t,u)\,du\cdot d\tilde{W}_t + \int_t^T \|\sigma(t,u)\|^2\,du\,dt - \int_t^T \sigma(t,u)\,du\cdot\int_t^\tau \sigma(t,u)\,du\,dt.$$

Since $t \le T \le \tau$, decomposition of the last term of this integral and simplification gives

$$\frac{dF(t,T,\tau)}{F(t,T,\tau)} = -\int_T^\tau \sigma(t,u)\,du\cdot d\tilde{W}_t - \int_t^T \sigma(t,u)\,du\cdot\int_T^\tau \sigma(t,u)\,du\,dt.$$

Changing probability measure from **Q** to **F**, we have, by Girsanov's theorem and expression 2.27,

$$d\hat{W}_t = d\tilde{W}_t + \int_t^T \sigma(t,u)\,du\,dt.$$

Substituting for $\tilde{W}$ and simplifying after applying Fubini's theorem give

$$\begin{aligned}\frac{dF(t,T,\tau)}{F(t,T,\tau)} &= -\int_T^\tau \sigma(t,u)\,du\cdot\left(d\hat{W}_t - \int_t^T \sigma(t,u)\,du\,dt\right)\\ &\quad - \int_t^T \sigma(t,u)\,du\cdot\int_T^\tau \sigma(t,u)\,du\,dt\\ &= -\int_T^\tau \sigma(t,u)\,du\cdot d\hat{W}_t.\end{aligned} \tag{2.28}$$

Therefore, $F(t,T,\tau)$ is a martingale if Novikov's condition holds.

It is now straightforward to show that forward rates themselves also follow a driftless process under the forward measure.

Proposition 2.3.10. *Under measure* **F**, *forward rates satisfy the SDE*

$$df(t,T) = \sigma(t,T)\cdot d\hat{W}_t.$$

Proof. From expression (2.23), the SDE for the forward rates under $\mathbf{Q}$ is

$$df(t,T) = \sigma(t,T) \cdot d\tilde{W}_t + \sigma(t,T) \cdot \int_t^T \sigma(t,u)\, du\, dt.$$

Invoking Girsanov's theorem with $d\tilde{W}_t = d\hat{W}_t - \int_t^T \sigma(t,u)\, du\, dt$, we obtain the proposition.

2.4 Applications in Discrete Time

This section gives two examples of discrete-time processes used to approximate their continuous-time equivalents. We give a discrete-time specification of geometric Brownian motion as used in the Black and Scholes (1973) model and discrete-time specification of the forward rate process by Heath, Jarrow, and Morton (1992).

2.4.1 Geometric Brownian Motion

We begin by presenting the discrete-time equivalent of geometric Brownian motion. The Black-Scholes model for the discrete case is originally derived in Cox, Ross, and Rubinstein (1979).

In the discrete case, the Brownian motion consisting of a standard normal random variable and a time scaling factor $\sqrt{h}$ is replaced by $\epsilon\sqrt{h}$, where h is the discrete time step from t_n to t_{n+1} and ϵ represents a random variable taking values $\epsilon \in \{1, -1\}$. The normal distribution is approximated by the two values $+1, -1$ in each step.

The goal is to specify the discrete binomial processes such that they converge to their continuous equivalents in distribution. In the case of geometric Brownian motion, such weak convergence can be obtained by specifying the mean and the variance of the discrete process such that they match the mean and variance of the continuous process asymptotically. Of course, the lattice needs to be specified such that under the equivalent martingale measure, the expected price of a security in terms of the money market numéraire is a martingale.

The risk-neutral process in expression (2.7) on page 27 follows geometric Brownian motion. It is therefore lognormally distributed with first two moments[6]

$$\begin{aligned} \mathbf{E_Q}[\frac{S_{t+h}}{S_t}|\mathcal{F}_t] &= \exp(rh) \\ \mathbf{E_Q}[(\frac{S_{t+h}}{S_t})^2|\mathcal{F}_t] - \mathbf{E_Q}[\frac{S_{t+h}}{S_t}|\mathcal{F}_t]^2 &= \exp(\sigma^2 + 2r). \end{aligned} \tag{2.29}$$

[6] Cf. Abramowitz and Stegun (1972).

Let u, d denote the multiplicative move sizes in the tree. For an up-move, $S_{n+1} = S_n u$, for a down move $S_{n+1} = S_n d$. The probability of an up-move, i.e., $\epsilon = +1$, is denoted by p and the probability of a down-move is therefore $1 - p$. p is the risk-neutral probability because it refers to the risk-neutral distribution characterized by (2.29).

To approximate the moments of the lognormal distribution of (2.29), we obtain the following conditions

$$\begin{aligned} pu + (1-p)d &= \exp(rh) \\ pu^2 + (1-p)d^2 &= \exp((\sigma^2 + 2r)h). \end{aligned} \tag{2.30}$$

Because we have two conditions with three unknown variables, we need to specify a third condition. This third condition is somewhat arbitrary. Cox, Ross, and Rubinstein (1979) choose a restriction on the move sizes, $u = d^{-1}$. Another convenient restriction is $p = \frac{1}{2}$.

Given the additional constraint, the equation system can be solved for p, u, and d. Although solutions for p, u, d can be obtained that solve (2.29) exactly,[7] asymptotic solutions are often chosen. With such asymptotic solutions, the moments of the distribution implied by the tree converge to the actual values of (2.29).

For example, if we use a simple discretization of the continuous process given in Proposition 2.3.1, i.e.,

$$S_{t+h} = S_t e^{(\alpha - \frac{1}{2}\sigma^2)h + \sigma\epsilon\sqrt{h}}, \tag{2.31}$$

we obtain, because $\epsilon = (1, -1)$, for the multiplicative move sizes u and d

$$\begin{aligned} u &= e^{(r - \frac{1}{2}\sigma^2)h + \sigma\sqrt{h}} \\ d &= e^{(r - \frac{1}{2}\sigma^2)h - \sigma\sqrt{h}}. \end{aligned} \tag{2.32}$$

From (2.30),

$$p = \frac{\exp(rh) - u}{u - d}.$$

Substituting (2.32) into this equation gives

$$p = \frac{\exp(\frac{1}{2}\sigma^2 h) - \exp(-\sigma\sqrt{h})}{\sinh(\sigma\sqrt{h})}.$$

It can be shown that this expression is $\frac{1}{2}$ in the limit as h tends to zero. It can also be shown that it approximates the conditions in (2.30). We therefore have a simple discrete approximation for geometric Brownian motion.

[7] Cf. Wilmott, Howison, and Dewynne (1995)

2.4.2 Heath-Jarrow-Morton Forward Rates

In this section a discrete-time approximation of the Heath-Jarrow-Morton evolution of forward rates is presented as an alternative to Heath, Jarrow, and Morton (1990). Let $P_t(T)$ denote the time t price of a zero-coupon bond maturing at time T. One discrete time step is of length h. The bond price evolution from t to $t+h$ is characterized by $u_t(T)$, i.e.,

$$P_{t+h}(T) = u_t(T)P_t(T).$$

The continuously compounded forward rate for an investment from time T to $T+h$ is defined as

$$f_t(T) = \ln\left(\frac{P_t(T)}{P_t(T+h)}\right)/h \qquad \forall T > t.$$

In terms of bond prices at t,

$$e^{f_{t+h}(T)h} = \frac{P_{t+h}(T)}{P_{t+h}(T+h)} = \frac{P_t(T)u_t(T)}{P_t(T+h)u_t(T+h)}.$$

The bond price change can therefore be expressed as

$$u_t(T+h) = e^{(f_t(T)-f_{t+h}(T))h}u_t(T). \tag{2.33}$$

Substituting for $u_t(T)$,

$$u_t(T+h) = e^{(f_t(T)-f_{t+h}(T))h}e^{(f_t(T-h)-f_{t+h}(T-h))h}u_t(T-h).$$

This substitution is repeated for all u_t. The last substitution is $u_t(t+h) = e^{r_t}h$, where r is the short rate. Let a be the forward rate change, i.e.,

$$f_{t+h}(T) = f_t(T) + a_t(T).$$

Then we obtain for the change of the bond price

$$u_t(T) = e^{-\sum_{j=t+h}^{T-h} a_t(j)}e^{r_t h}. \tag{2.34}$$

Summing is done in increments of h. We parameterize the process a according to

$$a_t(T) = \mu_t(T)h + \epsilon\,\sigma_t(T)\sqrt{h}, \tag{2.35}$$

where ϵ is the source of uncertainty and $\epsilon \in [-1,1]$. Substituting (2.35) into (2.34) gives

$$\begin{aligned} u_t(T) &= \exp\left(-\sum_{j=t+h}^{T-h}(\mu_t(j)h + \epsilon\,\sigma_t(j)\sqrt{h}) + r_t h\right) \\ &= \exp\left(-\Sigma^{T-h}\mu h - \epsilon\,\Sigma^{T-h}\sigma\sqrt{h} + r_t h\right), \end{aligned} \tag{2.36}$$

where

$$\Sigma^{T-h}\mu h = \sum_{j=t+h}^{T-h} \mu_t(j)h \qquad \Sigma^{T-h}\sigma\sqrt{h} = \sum_{j=t+h}^{T-h} \sigma_t(j)\sqrt{h}.$$

The no-arbitrage condition is

$$\begin{aligned} E_Q(u_t(T)) &= pe^{-\Sigma^{T-h}\mu h - \Sigma^{T-h}\sigma\sqrt{h}}e^{r_t h} + (1-p)e^{-\Sigma^{T-h}\mu h + \Sigma^{T-h}\sigma\sqrt{h}}e^{r_t h} \\ &= e^{r_t h} \end{aligned} \tag{2.37}$$

under a measure **Q**. If this condition holds, the discounted bond price process is a martingale under **Q**.

If $p = \frac{1}{2}$, expression (2.37) simplifies to

$$E_Q(u_t(T)) = \frac{1}{2}e^{-\Sigma^{T-h}\mu h}(e^{\Sigma^{T-h}\sigma\sqrt{h}} + e^{-\Sigma^{T-h}\sigma\sqrt{h}}) = 1.$$

Therefore, the arbitrage-free drift of the bond price process for bond $P_t(T)$ is

$$e^{\Sigma^{T-h}\mu h} = \cosh \Sigma^{T-h}\sigma\sqrt{h}.$$

Thus, expression (2.36) can be written as

$$u_t(T) = \frac{e^{-\epsilon\,\Sigma^{T-h}\sigma\sqrt{h}}\, e^{r_t h}}{\cosh(\Sigma^{T-h}\sigma_t\sqrt{h})}. \tag{2.38}$$

From expression (2.33) we know that

$$(f_t(T) - f_{t+h}(T))h = \ln \frac{u_t(T)}{u_t(T+h)}. \tag{2.39}$$

$u_t(T+h)$ is calculated analogously to $u_t(T)$, yielding

$$u_t(T+h) = \frac{e^{-\epsilon\,\Sigma^{T}\sigma\sqrt{h}}\, e^{r_t h}}{\cosh(\Sigma^{T}\sigma_t(T)\sqrt{h})}. \tag{2.40}$$

Substituting for $u_t(T)$ and $u_t(T+h)$ in expression (2.39) gives

$$f_t(T)h - f_{t+h}(T)h = \ln \frac{\cosh(\Sigma^{T}\sigma\sqrt{h})}{\cosh(\Sigma^{T-h}\sigma\sqrt{h})} - \epsilon\,\Sigma^{T-h}\sigma\sqrt{h} + \epsilon\,\Sigma^{T}\sigma\sqrt{h}.$$

By simplifying and rearranging, we obtain for the arbitrage-free evolution of the forward rates,

$$f_{t+h}(T) = f_t(T) + \ln\left(\frac{\cosh\sum_{j=t+h}^{T}\sigma_t(j)\sqrt{h}}{\cosh\sum_{j=t+h}^{T-h}\sigma_t(j)\sqrt{h}}\right)/h + \epsilon\,\sigma_t(T)\sqrt{h}/h,$$

where $\epsilon = +1$ and $\epsilon = -1$ occur each with a probability of $\frac{1}{2}$.

Given this forward rate process, the discounted bond price process is a martingale under $\mathbf{Q}$, as has been shown. Arbitrage is therefore excluded. The evolution of the short rate can be derived from the forward rate process because the short rate is a special forward rate.

2.5 Summary

This chapter gives an overview of some important results in contingent claim valuation theory. Some of the main results used throughout this book can be summarized in the following three statements:

- The existence of a martingale measure for the security market implies the absence of arbitrage opportunities. Furthermore, in the finite setting, absence of arbitrage also implies the existence of a martingale measure.
- Given a martingale measure, any European contingent claim can be valued as the expectation under the martingale measure of its deflated payoff at maturity. This method of valuing derivative securities was called risk-neutral valuation because the payoffs of derivative instruments can be dynamically replicated in a unique fashion independent of individual risk preferences. In other words, the martingale measure is unique for the entire market and all its participants.
- Often, a zero-volatility money market account is used as the deflating asset, called numéraire. This is, however, only one possible choice. It was demonstrated that other assets can also be used as numéraire.

We also derived the Black-Scholes option pricing formula and the Margrabe formula for exchange options. Both are foundations to our derivation of pricing formulae for vulnerable derivatives in subsequent chapters. Additionally, we introduced interest rate modeling with the Heath-Jarrow-Morton methodology. In subsequent chapters we will rely on this methodology when modeling stochastic interest rates.

3. Credit Risk Models

This chapter reviews the most common credit risk models. A number of approaches have been developed to price credit-risky bonds. Credit risk pricing approaches usually fall into two categories. One group is based on the evolution of the firm value to determine default and recovery rate, called firm value models. The more recently introduced intensity models, on the other hand, specify an exogenous default process which governs default. The default process is usually a Poisson-like process and the recovery rate is often exogenous to the model. Of course, this distinction is not clear-cut and some models use elements of both approaches.

Most work in credit risk pricing focuses on pricing credit-risky bonds or similar instruments, such as mortgages. We therefore begin by reviewing the most common models used to price credit-risky bonds. Within this section, we focus on firm value models, first passage time models – a variant of firm value models – and intensity models. Second, we discuss models that can be applied to derivative instruments. Third, we review the few existing models that were explicitly developed with credit derivatives in mind. Other surveys of credit risk pricing models include Cooper and Martin (1996), Cossin (1997), and Lando (1997).

3.1 Pricing Credit-Risky Bonds

In this section we give an overview of the large body of literature concerned with valuing bonds that are subject to default.

Credit risk models can be categorized in many ways. We choose to distinguish between *traditional methods*, *firm value models*, *first passage time models*, and *intensity models*.[1] Traditional methods center on the concept of valuing credit risk by gathering historical default data and inferring the credit spread from this data. These approaches are usually not model-based. Firm value models use the approach developed by Merton (1974), in which credit risk is considered to be a put option on the value of the firm's assets.

[1] Firm value models and first passage time models are sometimes also called *structural* models, while intensity models are called *reduced-form* models. Cf. Duffie and Singleton (1995).

First passage time models also use the concept of firm value, but the firm value is used to determine the time of default. The recovery rate is often not based on the firm value. Intensity models use an arbitrage-free bankruptcy process that triggers default. The intensity process is calibrated to fit market data.

3.1.1 Traditional Methods

The traditional approaches to pricing credit risk focus on deriving default information from historical data. Risky bonds are then priced such that the investor is compensated for the expected loss on the bond based on estimates from historical data. For example, Fons (1994) determines prices of credit-risky bonds such that they are consistent with past losses from defaults of comparable securities. Aside from the problem of projecting historical default data into the future, this approach assumes true risk-neutrality of investors, unless, of course, credit risk is not systematic. If it is not, credit-risky bonds do not command an expected return in excess of the riskless rate. In other words, their yield spread over riskless bonds reflects only the expected loss from default under the empirical probability measure, but no risk premium.

Empirical findings, however, suggest that credit risk is systematic. For example, Altman (1989) finds that low-rated bonds have higher long-run average returns than high-rated bonds, although he hints that this might be an anomaly and not compensation for systematic risk taken.

Litterman and Iben (1991) and Hurley and Johnson (1996) take an approach in the opposite direction from Fons (1994). They derive implied default probabilities from observed credit spreads. Litterman and Iben (1991), however, point out that the implied probabilities are not true empirical probabilities but must be interpreted as risk-neutral probabilities.

3.1.2 Firm Value Models

Firm value models derive the price of default risk by modeling the the value of a firm's assets relative to its liabilities. Because they model the evolution of the firm's capital structure, they are sometimes also called structural models.

3.1.2.1 Merton's Model. One of the first models for pricing credit-risky bonds or similar instruments[2] is developed by Merton (1974). While already Black and Scholes (1973) give the intuition of interpreting capital structure in terms of option contracts, Merton provides the analytical framework.

It is assumed that the ability of a firm to redeem its debt is determined by the total value of its assets V. Consider the firm to have a single liability with promised terminal payoff K. This claim can be interpreted as a zero-coupon bond. According to Black and Scholes (1973), by issuing debt, equity

[2] For a survey of Merton's model applied to mortgages, see Kau and Keenan (1995).

holders sell the firm's assets to the bond holders while keeping a call option to buy back the assets. This is equivalent to saying that equity holders own the firm's assets and buy a put option from the bond holders. If the assets of the firm are worth less than the amount owed to the debt holders, the equity holders can balance the debt due to be redeemed with their payoff from the put option. Thus, a corporate bond can be viewed as a default-free bond minus a put option with strike price K written on the assets of the firm. In this case the payoff ϕ of the bond with promised payoff of K is

$$\phi = K - \max(K - V, 0) = \min(V, K). \tag{3.1}$$

The dynamics of the firm value under the risk-neutral probability measure $\mathbf{Q}$ are specified as a standard geometric Brownian motion with

$$\frac{dV}{V} = r\,dt + \sigma_V d\tilde{W}(t),$$

where r is the constant riskless interest rate and σ_V the instantaneous standard deviation of the firm value. Although V itself is not a traded asset, a derivative of V is — the stock of the firm. Merton (1974) shows for this case that valuation of derivatives of V is independent of investors' risk preferences. We can therefore assume risk-neutrality without loss of generality. This implies that there exists a unique equivalent probability measure $\mathbf{Q}$ under which V is a martingale if discounted at the continuously compounded riskless rate. Thus, the price of an option with payoff $\max(K - V, 0)$ is given by the Black-Scholes option pricing formula.

Given these dynamics, the price of a defaultable bond is obtained by deducting a standard Black-Scholes option from the risk-free value of the bond. The time t price of the risky bond can be expressed as

$$P^d(t, T) = P(t, T) - p(t). \tag{3.2}$$

Since the value of the put option is given by Proposition 2.3.4, we can write

$$P^d(t, T) = P(t, T) - P(T, T)e^{-r(T-t)}N(-d + \sigma_V\sqrt{T-t}) + VN(-d),$$

with

$$d = \frac{\ln \frac{V}{P(T,T)} + r + \frac{1}{2}\sigma_V(T - t)}{\sigma_V\sqrt{T-t}}.$$

Since we have $P(t, T) = P(T, T)e^{-r(T-t)}$ and $1 - N(d) = N(-d)$, we can write

$$P^d(t, T) = P(t, T)N(d - \sigma_V\sqrt{T-t}) + VN(-d),$$

with

$$d = \frac{\ln \frac{V}{P(t,T)} + \frac{1}{2}\sigma_V^2(T-t)}{\sigma_V\sqrt{T-t}}.$$

Setting $\Gamma = P(t,T)^{-1}V$, we have

$$P^d(t,T) = P(t,T)\Big(N(d - \sigma_V\sqrt{T-t}) + \Gamma N(-d)\Big), \tag{3.3}$$

with

$$d = \frac{\ln \Gamma + \frac{1}{2}\sigma_V^2(T-t)}{\sigma_V\sqrt{T-t}}.$$

Γ is a pseudo assets-to-debt ratio. It is not the real assets-to-debt ratio because the value of the debt is computed as if it were riskless. Nonetheless, it is evident from the formula that the magnitude of credit risk is immediately derived from the ratio of assets to debt. In fact, this ratio can be interpreted as the underlying of the option. Sometimes, the term *distance to default* is used in this context. A large distance to default means a high assets-to-debt ratio, i.e., the option is far out-of-the-money. A firm on the brink of bankruptcy has a short distance to default; in other words, the option is at-the-money.

Expression (3.3) also gives the bond price spread,

$$\frac{P^d(t,T)}{P(t,T)} = N(d - \sigma_V\sqrt{T-t}) + \Gamma N(-d).$$

Furthermore, since $P(t,T) = e^{-r(T-t)}$ and $P^d(t,T) = e^{-r^d(T-t)}$,

$$r^d - r = -\frac{\ln\big(N(d - \sigma_V\sqrt{T-t}) + \Gamma N(-d)\big)}{T-t} \tag{3.4}$$

is the credit spread of the risky bond.

This simple credit risk model makes a number of rather stringent assumptions. For example, it assumes that default cannot occur before maturity of the debt. The debt of the firm can only consist of a single class; different maturity dates or seniority levels are excluded. Because there is only one class of debt, deviations from absolute priority are not an issue. Moreover, coupons cannot be handled. Bankruptcy is also assumed to be free of any associated cost. Along similar lines, it is assumed that bankruptcy can occur only if the firm value is below the face value of the debt. Bankruptcy induced by a liquidity crunch is therefore excluded. Of course, all the assumptions of the Black-Scholes option pricing model also apply, such as constant interest rates.

As in the Black-Scholes analysis for standard options, Merton's original derivation included a PDE for defaultable bonds. Starting with firm value dynamics

$$\frac{dV}{V} = (r - C)\,dt + \sigma_V d\tilde{W}(t),$$

he obtained, by Itô's formula, for a derivative $F(V,t)$ the PDE

$$0 = \frac{1}{2}\sigma_V V^2 F^{VV} + (rV - c_E)F^V - rF + F^t + c_B,$$

where c_f denote payouts of the firm and c_F payouts to the security F. By imposing suitable initial and boundary conditions, he solved the PDE and obtained the formulae above.

3.1.2.2 Extensions and Applications of Merton's Model. Merton's credit risk model for zero-coupon bonds has been extended in several ways. Extensions include adaptations to different types of securities such as coupon bonds, callable bonds, mortgages, convertible bonds, variable rate bonds. Other extensions treat the valuation of claims with different maturity dates, classes of seniority, or special indenture provisions.

Geske (1977) and Geske and Johnson (1984), for example, derive closed-form solutions for risky coupon bonds. Geske assumes that equity holders make the coupon payment and therefore effectively own a compound option. By paying the coupon, they exercise the compound option and receive the option to the value of the firm for the price of the coupon payment. The price of this compound option can be expressed in terms of multivariate normal distributions with the dimensionality depending on the number of coupon payments left until maturity. Geske (1977) also provides a formula for subordinate debt within this compound option framework.

Ho and Singer (1982) analyze the effect on different indenture provisions such as time to maturity, financing restriction on the firm, priority rules, and payment schedules on the credit risk of bonds within the Merton framework. Furthermore, Ho and Singer (1984) analyze the effect of sinking fund provisions on the price of risky debt.

Cox, Ingersoll, and Ross (1980) apply the Merton approach to the valuation of credit-risky variable-rate debt to identify variable coupon payout structures that eliminate or reduce interest rate risk. Claessens and Pennacchi (1996) derive an implicit default probability from the prices of Brady bonds using the Merton model.

Chance (1990) examines the duration of defaultable zero bonds within the framework of Merton's model. He shows that default-prone bonds have lower duration than their riskless counterparts with equal time to maturity. Therefore, credit-risky bonds are less sensitive to interest rate changes. This analytical result is consistent with empirical observations[3] that interest rate changes explain bond returns to a lesser degree the riskier the bond is.

Shimko, Tejima, and Deventer (1993) derive closed-form solutions for risky zero-coupon bonds when interest rates evolve according to Vasicek

[3] Cf. Cornell and Green (1991).

(1977). Since, in the Merton model, the risky bond contains an option, the formula by Shimko, Tejima, and Deventer (1993) is similar to the formulae derived by Jamshidian (1989) for options under Vasicek interest rates.

3.1.2.3 Bankruptcy Costs and Endogenous Default. Financial distress is accompanied by costs. Such costs often acerbate the situation of the firm and may even lead to its bankruptcy. Such costs are called bankruptcy costs. See, for example, Franks and Torous (1994) for a study of the effect of bankruptcy costs on asset prices.

The issue of moral hazard is another interesting problem when pricing risky debt. Unlike normal options, where the holder of the option has no control over the process of the underlying security, the option held by the owners of the firm is not usually free from manipulation of its underlying. Since the owners of the firm hold a call option on asset value, they will be tempted to change investment decisions to maximize the value of their option.

For example, they might increase the value of their holding by increasing the volatility of the underlying asset price process, as first noted by Jensen and Meckling (1976). Alternatively, the owners of the firm might forego positive net present value projects if they benefit bondholders, as is the case when new equity capital is injected into the firm. This is the under-investment problem, originally identified by Myers (1977).

Recently, some models have been introduced that do not assume bankruptcy to be an exogenous event, but model bankruptcy as an endogenous event driven by various factors such as agency and bankruptcy cost. As already noted by Franks and Torous (1989), Franks and Torous (1994), Eberhart, Moore, and Roenfeldt (1990), Weiss (1990), among others, strict priority is rarely upheld in bankruptcy proceedings. In situations of financial distress, complicated renegotiation of debt takes place and outcomes quite different from strict priority are the rule. Not only does junior debt often fare better than predicted by strict priority, but equity holders usually manage to obtain large concessions from bondholders by threatening bankruptcy with its associated costs.

Anderson and Sundaresan (1996) and Mella-Barral and Perraudin (1997) investigate this form of strategic debt service and conclude that it can explain a substantial part of credit risk premiums on risky debt.

In a series of articles, Leland (1994a), Leland (1994b), and Leland and Toft (1996) investigate corporate debt prices in the context of optimal capital structure. Leland and Toft (1996) also examine bankruptcy as an optimal decision by equity holders and derive endogenous conditions which must be satisfied for equity holders to declare bankruptcy. By taking into account agency costs, taxes and bankruptcy costs, they are able to make statements on the optimality of capital structure, with respect to amount and maturity of debt, while explaining debt risk premia. A common conclusion of these studies is that, in the presence of bankruptcy costs and strategic debt service, credit spreads are significantly higher than if computed by the simple Merton model.

3.1.3 First Passage Time Models

First passage time models attempt to solve the problem of premature bankruptcy. The model of Merton (1974) did not allow bankruptcy before maturity of the bond.[4] First passage time models assume that bankruptcy occurs if the firm value crosses a specified and often time-dependent boundary. Unlike the pure Merton model, which does not allow bankruptcy before maturity of the debt, first passage time models allow premature bankruptcy. On the other hand, modeling the recovery rate becomes more difficult. If the firm is assumed to go bankrupt when the boundary is first crossed by the firm value, and the boundary is a deterministic function of time, then the recovery rate itself is also a deterministic function of time. As mentioned below, some authors circumvent this problem by assuming an exogenously given recovery rate which is independent of the firm value. Table 3.1 gives an overview of proposed first passage time models and their specifications as to default triggering and recovery rate.

First passage time models were introduced by Black and Cox (1976), modifying Merton's firm value model to facilitate the modeling of so-called safety covenants in the indenture provisions. A safety covenant allows the bondholders to force bankruptcy if certain conditions are met. Frequent safety covenants include the right of bondholders to declare the entire amount of the debt issue due to be redeemed if the debtor is unable to meet interest obligations. In this case, the entire amount of debt becomes due immediately and forces restructuring or the firm's bankruptcy. The aim of such provisions is to protect debt holders from further devaluations of the firm value.

Black and Cox (1976) modeled such a safety covenant as an exogenous, time-dependent boundary. They set the boundary to be a an exponential function in the form of

$$V^d(t) = ke^{-\gamma(T-t)},$$

with k and γ exogenous constants. As soon as the firm value equals $V^d(t)$, the firm is forced into restructuring or bankruptcy in which case the bond holders take over the firm's assets $V = V^d$. A consequence of such a provision is a restriction upon stockholders' ability to transfer wealth from bondholders to themselves by increasing the volatility of the firm's assets. Since Black and Cox did not include bankruptcy cost in their analysis, such safety covenants

[4] It has been argued (e.g., Longstaff and Schwartz (1995a)) that bankruptcy can occur before maturity if all assets have been exhausted. Although this is true, for firm value processes following continuous geometric Brownian motion, such an event is "almost impossible", i.e., it occurs with probability zero, and is therefore not a relevant scenario.

were shown to reduce credit risk by a potentially substantial amount, depending on the specification of the boundary. The existence of an exogenous boundary below which the firm value cannot fall without triggering bankruptcy changes the original Merton model into a first-passage-time model. Black and Cox obtained a closed-form solution for bond prices with continuous dividends proportional to the firm value.

Brennan and Schwartz (1980) also take an approach to default which is similar to the Black and Cox (1976) model. They, however, use a constant default boundary in their valuation model for convertible bonds and solve the resulting PDE numerically.

Black and Cox provide an argument that allows capital structures with senior and junior debt. The price of a junior bond can be derived from the prices of senior bonds by the following argument: since the junior bond holders are only served if the senior bond holders have been paid off, the value of a junior bond can be expressed as the difference between two senior bonds. Let $D = P + J$ denote the amount due at maturity to bondholders of senior and junior issues combined. P and J are the amounts promised to senior and junior bond holders, respectively. Furthermore, let $B(P, t)$ denote the price of the senior bond at time t. All debt issues are zero-coupon bonds. Because senior and junior bond holders have to be paid off before anything can be paid out to shareholders, the price of total debt $B(D, t)$ can be determined as before. However, because junior debt cannot be redeemed until senior debt is redeemed, the price of senior debt can be determined as if no junior debt existed. The price of the junior debt issue is the difference between total debt and senior debt, i.e.,

$$B(J, t, T) = B(D, t, T) - B(P, t, T).$$

Of course, if strict absolute priority is not upheld in bankruptcy proceedings, this argument does not apply. Empirical evidence, as provided by Franks and Torous (1989), Franks and Torous (1994), Eberhart, Moore, and Roenfeldt (1990), Weiss (1990), and others, shows that strict absolute priority is a very strong assumption that does not reflect reality.

The model of Longstaff and Schwartz (1995a) is an adaption of the Black-Cox model to a more realistic setting. First, they allow interest rates to be stochastic, with dynamics as proposed by Vasicek (1977). Second, they do not require the recovery rate to be equal to the boundary value upon first passage, but assume an exogenously given rate w. Similarly, they also implement different classes of seniority by the exogenously given recovery rate. Senior debt will have a higher recovery rate than junior debt. Both are exogenous to the model and have to be estimated. The default boundary is also an exogenous constant denoted by $V^d(t) = k$.

The price of a credit-risky zero bond can be written

$$P^d(t, T) = P(t, T)(1 - wQ(X, r, t, T)),$$

Table 3.1. First passage time models

Model	Default Boundary	Recovery Rate	Interest Rates
BC76	$V^d = ke^{-\gamma(T-t)}$	V	non-stochastic
LS95	$V^d = k$	w	Vasicek (1977)
KRS93	$V^d = k$ [a]	$\min(V, F)$ [c] $\min(w(t)P(t,T), V)$ [d]	Cox, Ingersoll, Ross (1985)
NSS93	$V^d = k(t)$ [b]	$wP(t,T)$	Vasicek (1977)
BV97	$V^d = kFP(t,T)$	w_1V, if $V(T) < F$ [c] $w_2V^d(t)$ [d]	extended Vasicek (1977)

[a] Set to $k = c\gamma^{-1}$, where γ denote the rate of cash-flows out of the firm. c is the bond coupon rate.
[b] Increases at the rate of the return on the risky bond.
[c] Applies if default does not occur before maturity of the bond.
[d] Applies if default has occurred before maturity.

where $P(t,T)$ is the price of a default-free bond given by a closed-form expression derived by Vasicek (1977) for the Vasicek short rate model.[5] w is the write-down in percent of the face value in case of default. $X = Vk^{-1}$ is the ratio between firm value and default threshold. Q denotes the risk-neutral first passage time distribution, i.e., the risk-neutral probability that default occurs. In other words, the price for the risky bond is the price for a risk-free bond multiplied with the probability-weighted recovery rate under the risk-neutral measure.

Longstaff and Schwartz (1995a) give an approximation to the solution[6] of the first passage time integral Q using a discretizing procedure. It is a simple recursive algorithm such that

$$Q(X, r, t, T) = \sum_{i=1}^{n} \left(N(a_i) - \sum_{j=1}^{i-1} q_j N(b_{ij}) \right),$$

where N denotes the cumulated distribution function as in the Black-Scholes case. Explicit expressions for a_i and b_{ij} are derived by Longstaff and Schwartz, but are omitted here.

The Longstaff-Schwartz model, however, does not generally exclude arbitrage opportunities. The use of the Vasicek interest rate model precludes the

[5] Vasicek (1977) models the short rate according to the SDE $dr = (\alpha - \beta r)dt + \sigma d\tilde{W}$, where α, β, and σ are constants. The process of r is mean-reverting with Gaussian increments and a stationary distribution with mean $\alpha\beta^{-1}$ and variance $\sigma^2(2b)^{-1}$.

[6] They use a result on predictable stopping times derived by Buoncuore, Nobile, and Ricciardi (1987). The stopping time is fully predictable in the Longstaff-Schwartz case because the firm value process is a continuous process without discrete random jumps. In technical terms, there is an increasing sequence of stopping times that converges to the time of default. The default time is therefore predictable.

Longstaff-Schwartz model from being matched to arbitrary observed market term structures. There can therefore be situations where arbitrage will not be excluded. Moreover, the first passage time distribution requires several parameters to be estimated. Estimating those parameters will not likely result in a perfect match of the observed credit spreads. The Longstaff-Schwartz model therefore follows the tradition of models that are not free of arbitrage started by Merton (1974) and Black and Cox (1976).

Kim, Ramaswamy, and Sundaresan (1993) also extend the original first passage time approach by introducing stochastic interest rates. They use the square-root process proposed by Cox, Ingersoll, and Ross (1985). Similar to Longstaff and Schwartz (1995a), but unlike Black and Cox (1976), they assume the default boundary $V^d(t) = k$ to be constant over time. The level of the boundary depends on the coupon payment due to bond holders. If the boundary is not reached during the lifetime of the bond, the final payoff is $\min(V, F)$, where V is the firm value and F the face value of the bond. This means that even though default may not have occurred prematurely, it may occur when the bond is due. If, on the other hand, the boundary is reached, then default is assumed to occur immediately and the recovery rate is $\min(w(t)P(t,T), V)$, where $w(t)$ is the write-down on the default-free bond $P(t,T)$. This is different from Longstaff and Schwartz (1995a), where default cannot occur unless the boundary is reached and, if it is reached, the recovery rate is a constant. Kim, Ramaswamy, and Sundaresan (1993) solve the resulting PDE numerically, and find that the introduction of stochastic interest rates leads to higher credit spreads than predicted by models with deterministic rates.

Nielsen, Saà-Requejo, and Santa-Clara (1993) further generalize the first passage time approach by allowing for a stochastic boundary. Similar to the other models, default occurs if this boundary is reached. The recovery rate in case of default is specified to be an exogenously set fraction of an equivalent default-free bond. In this respect, the model corresponds to Black and Cox (1976). Nielsen, Saà-Requejo, and Santa-Clara (1993), however, assume interest rates to evolve according to Vasicek (1977) or the extended Vasicek model with time-varying parameters as specified by Hull and White (1990). They derive closed-form solutions for the prices of zero-coupon bonds if interest rates follow Vasicek (1977).

Briys and de Varenne (1997) propose yet another default triggering and payout rule. Because the model by Longstaff and Schwartz (1995a) and Nielsen, Saà-Requejo, and Santa-Clara (1993) allow a payout upon default which is greater than the firm value in some circumstances, Briys and de Varenne (1997) suggest that default threshold and recovery rate be specified such that the firm cannot pay more than its assets are worth. The default boundary is given by $v(t) = kFP(t,T)$, where k is an exogenous constant, F the face value of the bond, and $P(t,T)$ the usual default free bond. Briys and de Varenne (1997) distinguish between default at and prior to maturity.

If default occurs at maturity, i.e., if the firm's assets are worth less than its liabilities, then a fraction w_1 of the asset value is paid out. This fraction is exogenously specified. If, on the other hand, the default threshold is reached before maturity, then a fraction w_2 of the threshold value $v(t)$ is paid out. This fraction is also exogenous. They derive a closed-form solution for a discount bond for the extended Vasicek interest rate model proposed by Hull and White (1990).

Mason and Bhattacharya (1981) extend the first passage time approach by Black and Cox (1976) using a jump process to model firm value. While the default time is deterministic in the Black-Cox framework because of their assumption of a continuous process, the jump process results in a random default time. Mason and Bhattacharya showed that this effect can greatly influence risky bond prices. Zhou (1997) generalized the approach of Mason and Bhattacharya (1981) using a jump-diffusion process.

Although first passage time models solve the problem of default prior to maturity of the debt, they are not without shortcomings. First, a realistic default threshold is difficult to estimate for the value of the firm. Some models assume a constant threshold (Kim, Ramaswamy, and Sundaresan (1993), Longstaff and Schwartz (1995a)), others a constant fraction of the value of a default free bond (Briys and de Varenne (1997)).

Furthermore, with the exception of Black and Cox (1976), all models separate firm value and recovery rate. Usually, the firm value determines the time of default while the recovery rate is an exogenously specified function which may depend on firm value. The usually exogenous recovery rate of first passage time models can be seen as a limitation of those models. As Altman and Kishore (1996) have shown, standard deviations of recovery rate are generally high. Also, the recovery rate may correlate with other variables such as interest rates. Moreover, by separating firm value and recovery rate, these models allow the inconsistent situation in which the firm pays out more than its total assets. Avoiding this entails complicated provisions such as in Kim, Ramaswamy, and Sundaresan (1993) and Briys and de Varenne (1997).

Another problem of some first passage time models is that they do not generally exclude the existence of arbitrage opportunities. First passage time models that use the Vasicek (1977) and Cox, Ingersoll, and Ross (1985) models cannot be calibrated to match the observed risk-free term structure exactly. This limitation of those models is generally recognized in term structure modeling. However, another source of possible arbitrage opportunities may exist. With the exception of Black and Cox (1976), all first passage time models discussed above exhibit discontinuities in the sense that the payoff on the bond abruptly changes if the default boundary is reached. None of the authors proves that the relevant market assets are still martingales under the risk-neutral measure. In fact, if the firm value is modeled as

$$dV = V(r(t)dt - \sigma_V d\tilde{W}),$$

where $\tilde{W}$ denotes a Brownian motion under the risk-neutral measure, V the firm value, r the money market rate and σ_V the instantaneous volatility of the firm value, then the implication of a discontinuous change in firm value upon reaching the default boundary results in $VB(t)^{-1}$ not following a martingale under the risk-neutral measure. In such a case, absence of arbitrage cannot be guaranteed. Hence, discontinuities in asset returns can pose problems that need to be addressed.

3.1.4 Intensity Models

Intensity models take an approach completely different from firm value-based models. Default or bankruptcy is modeled by a bankruptcy process. The default process is usually defined as a one-jump process which can jump from no-default to default. The probability of a jump in a given time interval is governed by the default intensity, usually denoted by λ. Because the default process models only default time, not the severity of the loss in case of default, the recovery rate is often assumed to be exogenously given. Recently, several such intensity-based models have been proposed. In this section, we give an overview of intensity approaches and point out some of the possibilities of specifying an intensity-based model.

3.1.4.1 Jarrow-Turnbull Model. We consider a simple discrete-time intensity model-based on Jarrow and Turnbull (1995). Assume the default-free short rate has the Markov property and therefore can be modeled by a recombining lattice. From each node in the lattice, there are two nodes to which the short rate can jump. The sizes of the jumps and their risk-neutral probabilities can be computed by standard term structure modeling techniques. Denote the probability of an up-jump at time t with π_t. Consequently, the risk-neutral probability of a down-jump is $1-\pi_t$. We have a standard binomial lattice, as is frequently used for discrete-time term structure modeling.

We now superimpose a bankruptcy process onto the default-free term structure lattice. In each period, default can occur with risk-neutral probability λ_t. This default process has the effect of a second state variable. At each node, while the default-free bond price can jump up or down, default may also occur in each period. Hence, four rather than two states can be attained from each node in the lattice. This situation is illustrated by Figure 3.1 for a three-period economy. The credit-risky bond can be recursively valued in this lattice in a similar fashion as in standard lattices if the risk-neutral probabilities are known.

In Figure 3.1, a node denoted by ud/n has been reached by one down-jump and one up-jump of interest rates, and has not jumped into default. Alternatively, the node uu/b indicates that the firm has jumped into bankruptcy while interest rates jumped up twice. At maturity, only two values are attainable, default or no-default. If default has occurred, the recovery rate δ is paid out. If the firm has not defaulted, the full promised payout of 1 is made. Because a bond in default stays in default, once default has occurred, only two tree branches emerge from all subsequent nodes.

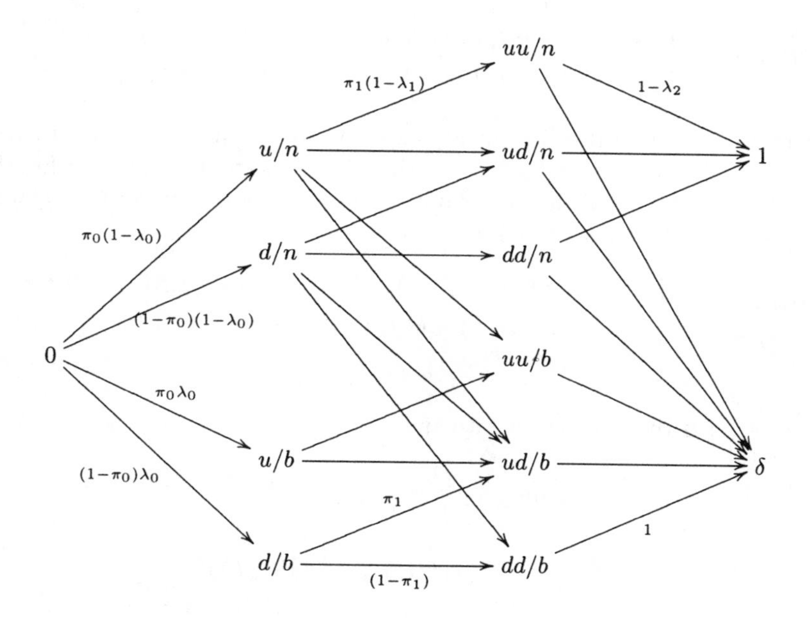

Fig. 3.1. Lattice with bankruptcy process

Jarrow and Turnbull (1995) also provide the continuous-time equivalent of the defaultable term structure model illustrated by Figure 3.1. This represents the first formal continuous-time model of the defaultable term structure based on a bankruptcy process.

They assume that the time of default or bankruptcy, τ is exponentially distributed with parameter λ, called *default intensity* or *hazard rate*. In this model, λ is constant, i.e., the default intensity is independent of any state variables such as the interest rate. The default-free term structure is assumed to be defined as in Section 2.3.3. The forward rate process is given by (2.18) and the corresponding bond price process by (2.21). To simplify notation, we use Definition 2.3.1. For further simplification, we define

$$\alpha^+(t,T) = -\alpha'(t,T) + \frac{1}{2}\|\sigma'(t,T)\|^2$$
$$b(t,T) = -\sigma'(t,T).$$

In this notation the bond price SDE is

$$\frac{dP(t,T)}{P(t,T)} = (r(t) + \alpha^+(t,T))\,dt + b(t,T)dW(t),$$

where $r(t)$ is the short rate and $W(t)$ has the usual definition of a Brownian motion in $\mathbb{R}^d$. All technical conditions are omitted in this section.[7] The forward rate SDE for the credit-risky bond $P^d(t,T)$ is given by

$$df^d(t,T) = \begin{cases} \left(\alpha^d(t,T) - \theta(t,T)\lambda\right) dt + \sigma(t,T)\, dW(t) & \text{if } t < \tau \\ \left(\alpha^d(t,T) - \theta(t,T)\lambda\right) dt + \sigma(t,T)\, dW(t) + \theta(t,T) & \text{if } t = \tau \end{cases}$$

where $\theta(t,T)$ is a drift adjustment to account for the jump at time of bankruptcy, i.e., when $t = \tau$. If $t > \tau$, then the forward rate process for the defaultable bond equals its no-default counterpart. The SDE for the defaultable bond can be shown to be

$$\frac{dP^d(t,T)}{P^d(t,T)} = \begin{cases} \left(r^d(t) + \alpha^{d+}(t,T) - \Theta(t,T)\lambda\right) dt + b(t,T)\, dW(t) & \text{if } t < \tau \\ \left(r^d(t) + \alpha^{+}(t,T) - \Theta(t,T)\lambda\right) dt + b(t,T)\, dW(t) & \\ \qquad +(\delta e^{\Theta(t,T)} - 1) & \text{if } t = \tau \end{cases}$$

where $r^d(t)$ is the risky short rate and

$$\Theta(t,T) = -\int_t^T \theta(t,s)\, ds$$

$$\alpha^{d+}(t,T) = -\alpha'^{d}(t,T) + \frac{1}{2}\|\sigma'(t,T)\|^2.$$

The risky money market account is

$$\frac{dB^d(t)}{B(t)} = \begin{cases} r^d(t)\, dt & \text{if } t < \tau \\ r^d(t)\, dt + (\delta - 1) & \text{if } t = \tau. \end{cases}$$

Thus, at the time of default, two jumps occur. First, there is a jump in the forward rates which changes the bond price from $P^d(t,T)$ to $P^d(t,T)e^{\Theta(t,T)}$. Second, there is a jump caused by the write-off upon bankruptcy, which reduces the payoff ratio from unity to δ. The new value is therefore $P(\tau,T) = P^d(t,T)\delta \exp(\Theta(\tau,T))$.

The SDEs above can be intuitively interpreted within what Jarrow and Turnbull (1995) call a *foreign exchange analogy.* They in fact model the defaultable bond price as $P^d(t,T) = e(t)p(t,T)$, where $e(t)$ is unity before default and changes to δ at default. $e(t)$ can therefore be interpreted as a quasi exchange rate and $p(t,T)$ as a default-free bond in the foreign currency. Since the the currency $e(t)$ is expected to devalue from 1 to $\delta < 1$ at some future point in time (default time), investments in this currency earn a higher rate of return; hence, the adjustment term $\theta(t,T)\lambda$. The adjustment term is not needed after default because then the payoff is known to be δ and therefore the bond contains no more credit risk.

[7] Refer to Jarrow and Turnbull (1995) and especially to Jarrow and Madan (1995) for technical conditions in the context of jump processes.

As was established in Proposition 2.3.7, the default-free economy admits an equivalent martingale measure and is therefore free of arbitrage if the equation system

$$\alpha^+(t,T) = -b(t,T) \cdot \gamma(t) \tag{3.5}$$

has a unique solution $\gamma(t)$. The vector $\gamma(t)$ is interpreted as the price of risk vector. In other words, drift and volatility are proportional with a proportionality factor of $\gamma(t)$.

In the defaultable economy, there is an additional arbitrage condition for $t < \tau$ arising from default risk. As in the non-default case, the market price of risk with respect to jump risk is derived by requiring the drift of the deflated process to be proportional to the volatility induced by a jump. The volatility rate from jumps is simply the jump return multiplied by the instantaneous probability of a jump, λ. The drift of the deflated process is $r^d(t) - r(t) + \alpha^{d+}(t,T) - \alpha^+(t,T) - \Theta(t,T)\lambda$. Therefore, the jump arbitrage condition is

$$\begin{aligned} r^d(t) - r(t) + \alpha^{d+}(t,T) - \alpha^+(t,T) - \Theta(t,T)\lambda \\ = r^d(t) - r(t) + \alpha^{d+}(t,T) + b(t,T)\gamma(t) - \Theta(t,T)\lambda \\ = -(\delta e^{\Theta(t,T)} - 1)\lambda\gamma^d(t), \qquad \text{if } t < \tau, \end{aligned} \tag{3.6}$$

where $\gamma^d(t)$ is the price of jump risk. Similarly, for the short rate, we have

$$r^d(t) - r(t) = (1-\delta)\lambda\gamma^d(t). \tag{3.7}$$

If (3.5), (3.6), (3.7) admit a unique solution for the pair $(\gamma(t), \gamma^d(t))$, then the economy is free of arbitrage opportunities. Drift restrictions implied by the arbitrage condition in (3.6) can be found in in Appendix B of Duffie and Singleton (1995).

Jarrow and Turnbull (1995) assume that $\gamma^d(t) = \gamma^d$ is constant. In this case, since τ is exponentially distributed with parameter λ under the empirical measure, τ is also exponentially distributed under the risk-neutral measure, with parameter $\lambda\gamma^d$.[8] As default intensity under the risk neutral measure is deterministic, a default event is independent of interest-rates. From the properties of the exponential distribution, this implies that probability of survival until time $T > t$ is given by

$$\mathbf{Q}(\tau > T | \tau > t) = e^{-\lambda\gamma^d(T-t)}$$

By the general result from Proposition 2.2.2, the price of a credit-risky zero bond with payoff $e(T)$ at maturity T can be written

[8] General technical conditions regarding the existence of a default intensity under an equivalent probability measure given that an intensity exists under the empirical measure are given by Artzner and Delbaen (1995).

$$P^d(t,T) = B(t)\mathbf{E}_{\mathbf{Q}}[B^{-1}(T)e(T)|\mathcal{F}_t], \qquad t < \tau,$$

where $e(T) = 1$ if $T < \tau$ and $e(T) = \delta$ if $T \geq \tau$. Because the price of a risk-free zero bond is $P(t,T) = B(t)\mathbf{E}_{\mathbf{Q}}[B^{-1}(T)|\mathcal{F}_t]$ and the bankruptcy process is assumed to be independent from the bond price process, the expression simplifies to

$$P^d(t,T) = P(t,T)\mathbf{E}_{\mathbf{Q}}[e(T)|\mathcal{F}_t].$$

Evaluating the expectation gives

$$\begin{aligned} P^d(t,T) &= P(t,T)\mathbf{E}_{\mathbf{Q}}[\mathbf{1}_{\{\tau > T\}} + \delta \mathbf{1}_{\{\tau \leq T\}}|\mathcal{F}_t] \\ &= P(t,T)\left(\mathbf{Q}(\tau > T|\tau > t) + \delta(1 - \mathbf{Q}(\tau > T|\tau > t))\right) \\ &= P(t,T)\left(e^{-\lambda\gamma^d(T-t)} + \delta(1 - e^{-\lambda\gamma^d(T-t)})\right). \end{aligned} \tag{3.8}$$

There are two critical assumptions made in the model of Jarrow and Turnbull (1995). First, the recovery rate δ is an exogenous constant. This assumption implies that the recovery rate is independent of any state variables and can be estimated. Second, the default intensity is an exogenous constant under both empirical and risk-neutral probabilities. A consequence of this assumption is that default is always equally probable during the entire lifetime of a bond. The default intensity also has to be estimated to derive bond prices. These strong assumptions make the model somewhat inflexible for practical use. On the other hand, simple closed-form solutions can be derived not only for zero-coupon bonds, but also for a number of derivative instruments, as will be shown in a subsequent section.

3.1.4.2 Jarrow-Lando-Turnbull Model. To overcome some of the shortcoming of the Jarrow-Turnbull model, Jarrow, Lando, and Turnbull (1997) propose a model that relates default probabilities to credit ratings. Jarrow, Lando, and Turnbull (1997) specify a time-homogeneous finite state space Markov chain with a generator matrix

$$\Lambda = \begin{pmatrix} \lambda_1 & \lambda_{12} & \dots & \lambda_{1,K-1} & \lambda_{1K} \\ \lambda_{21} & \lambda_2 & \dots & \lambda_{2,K-1} & \lambda_{2K} \\ & & & & \\ \lambda_{K-1,1} & \lambda_{K-1,2} & \dots & \lambda_{K-1,K-1} & \lambda_{K-1,K} \\ 0 & 0 & \dots & 0 & 0 \end{pmatrix}.$$

This matrix has the interpretation of a generator transition matrix. Jarrow, Lando, and Turnbull assume that each state λ_i corresponds to a rating class. For example, a bond rated AAA, the highest rating available, would be in position 1. The transition rates are given by the first row of λ, i.e., the transition rate of staying in its class is λ_1, the transition rate of being downrated by one rating class is λ_{12}, by two rating classes λ_{13}, etc. The transition rate of going straight to default is λ_{1K}. Similarly, for a bond currently rated in the second rating class, the transition rates of the second row apply. An issue of a firm that is bankrupt would be placed in the bottom row. This means that the issue's transition rate is 0, i.e., it cannot work its way out of default. State K is therefore an absorbing state. In this case the recovery rate is specified as an exogenous constant.

The time spent in one rating class is assumed to be exponentially distributed with parameter λ. Given a transition, the probability of the chain jumping from i to j is $q_{ij} = \lambda_j \lambda_i^{-1}$. By the properties of the exponential distribution, the transition probability for state i to state j is given by $q(t) = \exp(\lambda_{ij} t)$. The $K \times K$-transition matrix for a given time period i$Q(t) = \exp(t\Lambda)$. Element q_{ij} of the transition matrix can be interpreted as the probability that the chain is in state j after a time interval of t given that it has started out from state i.

The zero-coupon bond price is

$$P_i^d(t,T) = P(t,T)(\delta + (1-\delta)\mathbf{Q}^i(\tau > T|\tau > t)). \tag{3.9}$$

The difference between the recovery rate δ and the no-default payoff ratio 1 is weighted with the probability of survival under the martingale measure.

The survival probability can be easily computed from the risk-neutral transition matrix. It is

$$\mathbf{Q}^i(\tau > T|\tau > t) = \sum_{i \neq K} q_{ij}(t,T) = 1 - q_{iK}(t,T). \tag{3.10}$$

$\mathbf{Q}$ has the property $\mathbf{Q}(\tau > t+1|\tau > t)\mathbf{Q}(\tau > t) = \mathbf{Q}(\tau > t+1)$ from conditional probability.

Transition matrices are based on empirical data and therefore cannot be used directly to compute $\mathbf{Q}$. Jarrow, Lando, and Turnbull (1997) separate the transition generator matrix into an empirical part and risk adjustments. The risk-neutral matrix is then given as $\Lambda = \hat{\Lambda} U$ where U is a diagonal matrix of risk premia with elements μ_i such that $q_{ij} = p_{ij}\mu_i$, where p_{ij} is the empirical transition probability. U can be recovered from traded bond prices by an iterative procedure. First, solve (3.9) for $\mathbf{Q}$. It can be shown that

$$\mathbf{Q}(\tau \leq T) = 1 - \mathbf{Q}(\tau > T) = \frac{P(0,T) - P_i^d(0,T)}{P(0,T)(1-\delta)}.$$

We have $\mathbf{Q}^i(\tau \leq 1) = \mu_i(0) p_{iK}$ and therefore

$$\mu_i(0) = \frac{P(0,1) - P_i^d(0,1)}{(P(0,1)(1-\delta)p_{iK})}.$$

Any $q_{ij}(0,1)$ can now be calculated as $q_{ij}(0,1) = p_{ij}(0,1)\mu_i(0)$. From (3.10) and conditional probability,

$$\mathbf{Q}^i(\tau \leq t+1) = \sum_{1}^{K-1} q_{ij}(0,t)\mu_i(t)p_{iK},$$

and therefore

$$\mu_i(t) = \frac{P(0, t+1) - P_i^d(0, t+1)}{\left(\sum_{j=1}^{K-1} q_{ij}(0, t)\right)(P(0, t+1)(1-\delta)p_{iK})}$$

for any $t \leq T$.

A historical approach will use transition probabilities as they are available from credit rating agencies. Recovery rate can also be estimated from historical values. Because transition probabilities are empirical, risk premia have to be estimated. Jarrow, Lando, and Turnbull (1997) propose a way of estimating risk premia from traded zero bond prices such that model prices computed according to (3.9) match observed market prices. Another implementation with estimates is Henn (1997). He also consideres the extended model of Lando (1998).

The model of Jarrow, Lando, and Turnbull (1997) is an extension of Jarrow and Turnbull (1995) that eliminates the assumption that default intensities are constant over time. The independence assumption and the assumption of exogenously specified and constant recovery rates, however, are maintained. Das and Tufano (1996) extend the model to alleviate the restriction that default intensities and interest rates are independent.

A problem specific to the transition matrix approach based on credit ratings is that the model implies the same yield spread for all bonds belonging to the same rating class. Longstaff and Schwartz (1995a), however, show that bonds of equal rating exhibit quite different yield spreads depending on industry sector. Another problem arises from the fact that in the model, credit spreads can only change whenever a rating transition occurs. This may not reflect reality, as changing credit quality tends to be reflected in ratings later than in market prices. Hand, Holthausen, and Leftwich (1992) showed that the explanatory power of rating changes with respect to observed yield changes is generally small. Spreads of corporate bonds change even if the rating remains unchanged, as can be seen in Table 1.1 on page 2. Because of these problems, Lando (1998) proposes an extension to Jarrow, Lando, and Turnbull (1997) which allows for credit spread changes without simultaneously changing credit rating. Of course, these restrictions do not apply if the payoff of the instrument is specifically tied to the rating class of the underlying bond. In this case a rating-based model is the canonical choice.

The estimation of the transition matrix is also sometimes complicated by the fact that highly rated bonds have very low default probabilities. For example, over a time horizon of one year, it is not uncommon for some bonds to have zero default probability. To avoid the numerical problems associated with zero-probabilities, Jarrow, Lando, and Turnbull (1997) replace zero-probabilities with very small non-zero numbers. Because this solution is somewhat ad-hoc, Kijima and Komoribayashi (1998) propose a different premium-adjustment to ensure non-zero probabilities.

3.1.4.3 Other Intensity Models. Madan and Unal (1998) propose an intensity model with stochastic intensity,

$$\lambda(t) = \frac{c}{\ln^2 \frac{s}{d}},$$

where c, d are constants and s is the deflated firm value, i.e.,

$$ds = \sigma_s s dW(t).$$

d can be interpreted as the firm's liability. It can be seen that as s approaches d, default intensity rises to infinity. d is therefore a default boundary, i.e., s cannot decrease below d without default occurring. The recovery rate is also stochastic, but independent of interest rates, and modeled as a beta distribution.

Lando (1998) showed that defaultable bonds can be priced in the same way as default-free bonds if an adjusted short rate is applied. Assuming that the recovery rate is zero in default, the price of the bond can be written

$$\begin{aligned} P^d(t,T) &= \mathbf{E_Q}[\exp - \left(\int_t^T r^d(s)\, ds \right) |\mathcal{F}_t] \\ &= \mathbf{E_Q}[\exp \left(- \int_t^T r(s) + \lambda(s)\, ds \right) |\mathcal{F}_t], \end{aligned}$$

given that default has not occurred at time t. The intensity is a stochastic process and need not be independent of the short rate process. This means that adding the arrival rate of default to the riskless rate gives a default-adjusted rate $r^d(t)$ such that the bond can be priced as a default-free bond. The modeling of defaultable term structures is therefore no different from that of default-free term structures.

Duffie, Schroder, and Skiadas (1996), Duffie and Singleton (1997), and Duffie and Singleton (1999) also take advantage of this fact. They generalize the above result to include a fixed fractional recovery rates. Under the assumption that the recovery made upon default is a fraction of the corresponding no-default value of the bond, it is possible to characterize the price of the bond again as

$$P^d(t,T) = \mathbf{E_Q}[\exp - \left(\int_t^T r^d(s)\, ds \right) |\mathcal{F}_t],$$

where $r^d(t) = r(t) + \lambda(t)(1 - \delta(t))$. Note that this definition of δ is different than that used by Jarrow and Turnbull (1995). Jarrow and Turnbull assume that a creditor makes a fixed recovery of the value of a risk-free bond whereas here fractional recovery of the pre-default value of the bond is considered.

Schönbucher (1998) develops a methodology for modelling prices of defaultable bonds in the tradition of the Heath, Jarrow, and Morton (1992) approach. He specifies conditions as drift restrictions under which the evolution of the term structure of credit-risky interest rates is arbitrage-free. He further shows that while the spread between the short rates of defaultable and default-free bonds is always non-negative under the no-arbitrage conditions, the spread of the respective forward rates can become negative in some circumstances.

3.2 Pricing Derivatives with Counterparty Risk

Financial derivatives with counterparty default risk can be priced using several of the models presented in the previous section, namely firm value models and intensity models. Because of the complicated nature of their cash-flows and bilateral credit risk, swaps are treated separately in the last subsection.

3.2.1 Firm Value Models

The first valuation model that addresses the issue of counterparty risk is that of Johnson and Stulz (1987). They also introduce the term *vulnerable* option for an option that contains counterparty default risk. Their approach is an extension of the corporate bond model by Merton (1974). Specifically, they assume that the payoff of a vulnerable call option is given by

$$\Lambda(T) = \min(V(T), \max(S(T) - K)),$$

where $V(T)$ is the value of the counterparty firm's assets, S_T the value of the underlying security, and K the strike price of the option. Johnson and Stulz (1987) identify a number of situations that can be modeled as a vulnerable option aside from the standard OTC option. For example, insurance and, in particular, loan and debt insurance where the insuring party is itself subject to default are applications of vulnerable options.

Although Johnson and Stulz do not derive a analytical closed-form pricing formula for vulnerable options, they do obtain prices by numerical integration. They also establish a number of properties of vulnerable options. They show that in the face of counterparty bankruptcy, it may be optimal to exercise vulnerable options of the American type when default-free options with equivalent contract specifications would not be exercised. For example, it may pay to exercise vulnerable call options on a stock that does not have any payouts during the lifetime of the option before expiration. Similarly, the optimal exercise time for put options with counterparty risk is earlier than for default-free put options. They also showed that in some circumstances the value of a call option without dividends may be a decreasing function of the interest rate and time, contrary to default-free options. More surprisingly, they observe that even the volatility of the underlying asset can reduce the price of the option, a result totally contradictory to traditional option-pricing theory. Johnson and Stulz (1987) also notice that the correlation between the underlying asset and the firm value affects the price of the option.

Since Johnson and Stulz (1987) use the model of Merton (1974), they also subject themselves to those assumptions. In particular, Merton assumes that there is only one class of debt outstanding. Translated to the vulnerable option model, this means that the vulnerable option that is valued is the only claim against the counterparty. The main effect of this assumption is that the price of the the underlying via the payoff of the option has a potentially large impact on the ability of the firm to pay off the claim at maturity. For example, if the only outstanding claim is a call option and this contingent liability is substantial relative to the firm's assets, the high volatility in the underlying asset increases the value of the option, but at the same time decreases the credit quality of the counterparty. Depending on which effect prevails, the vulnerable option increases or decreases with increasing volatility in the underlying asset.

Although the assumption of a single option claim on the liability side of the firm's balance sheet may be a realistic assumption in some cases, for most situations it is not. It turns out, however, that the model of Johnson and Stulz (1987) can easily be extended to allow for many small claims. By specifying the following payoff for a call option,

$$\Lambda(T) = \max(S(T) - K, 0) \min\left(\frac{V(T)}{D}, 1\right).$$

Klein (1996) implicitly assumes that each outstanding claim is insignificant relative to the entirety of all claims. In case of bankruptcy at maturity, each claim holder receives a share of the remaining assets according to the value of the claim relative to the sum of the values of all outstanding claims. Klein (1996) developes closed-form solutions for Black-Scholes option prices. Klein's model and extensions thereof are discussed in further detail in Chapter 4.

Another pricing model designed for vulnerable options is given in Hull and White (1992) and Hull and White (1995). This model is related to the first passage time bond pricing models discussed in Section 3.1.3. The firm defaults on its obligations when a exogenously specified boundary is reached by the firm value. The recovery rate is also exogenous and assumed to be zero in their implementation. Hull and White (1995) also consider correlation between firm value and underlying asset value. They obtain prices for vulnerable options by constructing a binomial lattice.

3.2.2 Intensity Models

Some of the intensity models of the previous section can also be used to value vulnerable options. By Theorem 2.2.2, the price of a contingent claim is

$$X_t = B(t)\mathbf{E}_{\mathbf{Q}}[B^{-1}(T)X(T)|\mathfrak{F}_t], \tag{3.11}$$

where $B(t)$ is the risk-free money market account. This fact pertains also to vulnerable claims. The only difference in the case of vulnerable claims is that the payoff function $X(t)$ is different.

By using their foreign currency analogy,[9] Jarrow and Turnbull (1995) arrive at a simple representation for credit-risky options. $X(T)$ in (3.11) becomes $X(T)e(T)$ for the risky claim. Because $e(T)$ is independent of interest rates, it follows that

$$X^d(t) = X(t)\mathbf{E_Q}[e(T)|\mathcal{F}_t].$$

The notation is unchanged from Section 3.1.4. Evaluating the expectation gives

$$\begin{aligned} X^d(t) &= X(t)\mathbf{E_Q}[\mathbf{1}_{\{\tau>T\}} + \delta\mathbf{1}_{\{\tau\leq T\}}|\mathcal{F}_t] \\ &= X(t)\left(\mathbf{Q}(\tau > T|\tau > t) + \delta(1 - \mathbf{Q}(\tau > T|\tau > t))\right) \\ &= X(t)\left(e^{-\lambda\gamma^d(T-t)} + \delta(1 - e^{-\lambda\gamma^d(T-t)})\right). \end{aligned}$$

This result is strictly analogous to their pricing formula for credit-risky bonds and can be applied to any option derivatives. Jarrow and Turnbull (1996b) apply it to Eurocurrency derivatives. A similarly straightforward extension to the bond pricing model can also be used to apply the rating-based model of Jarrow, Lando, and Turnbull (1997) to vulnerable derivative instruments.

Other approaches that readily extend to vulnerable option derivatives are Lando (1998), Duffie and Singleton (1995) and Duffie, Schroder, and Skiadas (1996). The main result of these intensity models is that credit risk can be taken into account by changing the discounting rate accordingly, i.e.,

$$r^d(t) = r(t) - \lambda(t)(1 - \delta(t)),$$

where $\lambda(t)$ is the default intensity and $1 - \delta(t)$ the write-down on the claim when default occurs, as defined in Section 3.1.4. The general representation of the price of a contingent claim in expression (3.11) then becomes

$$X^d(t) = \mathbf{E_Q}[\exp\left(-\int_t^T r(s) - \lambda(s)(1 - \delta(s))\,ds\right) X(T)|\mathcal{F}_t].$$

This is the general form given in Duffie and Singleton (1995) and Duffie, Schroder, and Skiadas (1996). Lando (1998) is a special case with $\delta(t) = 0$, for all t.

3.2.3 Swaps

So far, only the valuation of option derivatives with default risk has been considered. Swaps pose two special problems for credit risk pricing models.

[9] The foreign currency analogy is introduced in Jarrow and Turnbull (1992a) and Jarrow and Turnbull (1992b).

First, swaps involve a series of cash-flows. This makes default time modeling critical to any analysis. Second, default risk is two-sided, i.e., both firms party to a swap may default. Furthermore, swap contracts often contain implicit netting agreements with the effect that only payoff differences are paid instead of an exchange of gross payoffs.

The valuation of swaps with default risk was first considered by Sundaresan (1991). Sundaresan used an idea of Ramaswamy and Sundaresan (1986) to incorporate default risk as an instantaneous premium on the short rate. They assume the short rate to evolve according to Cox, Ingersoll, and Ross (1985). Accordingly, they specify a default premium x governed by a Cox-Ingersoll-Ross process $dx = b(a - x)dt + \sigma\sqrt{x}dW(t)$, where $W(t)$ is uncorrelated with the Brownian motion of the short rate process. Sundaresan obtains an equilibrium structure of swap prices across different swaps. No attempt is made, however, to determine the magnitude of the default premium. Similarly, Nielsen and Ronn (1997) also model the yield spread directly. Parameters are estimated from empirical data.

Cooper and Mello (1991) address the credit risk effects on swaps within the firm value framework of Merton (1974). They consider the situation in which a risky firm enters into a swap agreement with a riskless counterparty. If the firm does not default, the single swap payoff to the riskless party is $\Lambda(T) = F(T) - X(T)$, where F and X are the values of fixed and variable parts of the swap, respectively. Cooper and Mello assume that there is only one final settlement without netting. Moreover, the counterparty must make the payment even if the risky firm is already bankrupt. They show that under those settlement conditions, the swap value can be characterized as

$$S_t = F_t - X_t - P_t(V, X, F). \tag{3.12}$$

S_t is the value of the swap, F_t and X_t the values of the fixed and variable legs of the swap at a time t, prior to settlement, and $P_t(V, X, F)$ the value of an option on the maximum of firm value and variable payment with a strike price of the value of the fixed payment. This result only pertains if swap claims are junior to other debt claims. Cooper and Mello (1991) apply this analysis to both interest rate and currency swaps, and extend their analysis to cover alternative settlement rules, namely to the case where payments are made only if both parties are solvent and to the case where swap claims are settled before bond claims, i.e., where swap claims are effectively senior to bond claims. In the latter case, the expression for the swap price from (3.12) simplifies to

$$S_t = F_t - X_t - P_t(V + X, F),$$

where $P_t(V + X, F)$ is a standard put option struck at the value of the fixed leg of the swap at the settlement date. This option is the equivalent of the put option that appears in the price of a defaultable bond in the Merton (1974) model. Refer to expressions (3.1) and (3.2) for comparison.

By assuming only one final settlement, Cooper and Mello (1991) circumvent the first difficulty that arises when analyzing the default risk of swaps. In fact, this assumption reduces the problem to the forward contract case. Their second assumption, that default risk is one-sided, i.e., that one party cannot default, further reduces the problem to the standard problem of contingent claims with counterparty default risk, albeit with more complicated settlement rules. This is why they are able to use the model of Merton (1974) for their analysis.

Sorensen and Bollier (1994) analyze the issue of bilateral default risk. They use a traditional approach, as discussed in Section 3.1.1, based on empirical probabilities of default, not a credit risk model. Other approaches include Solnik (1990), Abken (1993), Hull and White (1995) and Li (1998). Hull and White (1995) use their credit risk model, presented in Section 3.2.1, to price swaps subject to default by one counterparty.

The first to address the valuation problem of two-sided default risk within a credit risk model are Duffie and Huang (1996). They consider the valuation problem within the credit risk model of Duffie, Schroder, and Skiadas (1996), and solve a PDE numerically to obtain swap spreads. More recently, Hübner (2001) addresses the problem of bilateral default risk in swap contracts. Using a Gaussian bond pricing model, he derives swap prices for a reduced-form credit risk model.

Jarrow and Turnbull (1996a) extend the discrete-time version of the approach of Jarrow and Turnbull (1995) to bilateral swap default risk. Instead of only one bankruptcy process, there are two such processes, one for each counterparty. If the credit risks of the parties differ, so will the intensities of the default processes. As in Jarrow and Turnbull (1995), the intensities are exogenous and constant. The procedure can be implemented in a tree structure similar to that presented in Figure 3.1.

3.3 Pricing Credit Derivatives

Credit derivatives are derivatives on pure credit risk where the payoff is dependent on a credit variable only and is unaffected by other variables such as interest rates. We distinguish between debt insurance and spread derivatives.

3.3.1 Debt Insurance

The simplest form of credit derivatives are debt insurance contracts. Debt insurance or debt guarantees have existed for a long time are designed to cover the entire financial loss resulting from a default event. All credit risk models that provide prices for credit-risky bond can also price such simple credit derivatives. The price of the insurance is determined by the difference of the prices of default-free and defaultable bonds.

Merton (1977) gives an analytic solution for the price of loan guarantees based on the credit risk model of Merton (1974). The price of deposit insurance or loan guarantee is given by the put option used to derive expression (3.3). Since insured deposits are assumed to be free of default risk, the value of the deposits is $P(t,T)$. The insurance cost $G(t,T)$ per unit currency of deposits is then

$$G(t,T) = \frac{P(t,T)N(-d+\sigma_V\sqrt{T-t}) - VN(-d)}{P(t,T)}.$$

This simplifies to

$$G(t,T) = N(-d+\sigma_V\sqrt{T-t}) - \Gamma N(-d),$$

with

$$d = \frac{\ln\Gamma + \frac{1}{2}\sigma_V^2(T-t)}{\sigma_V\sqrt{T-t}},$$

where Γ in this case is the assets-to-deposits ratio, $VP(t,T)^{-1}$. The cost of insuring deposits is dependent on the assets-to-deposits ratio and the volatility of the assets backing the deposits. Increasing asset value with constant deposits reduces the cost of insurance while increasing asset volatility increases the cost of insurance. Among insured institutions, depositors are indifferent regarding where to deposit their funds. Therefore, the insurer has to bear the cost of surveillance. Merton (1978) extends the model presented to include surveillance costs of institutions with insured deposits.

Since this deposit insurance model is derived from the credit risk model in Merton (1974), all the assumptions and restrictions of that model apply, as pointed out on page 50 in this monograph.

3.3.2 Spread Derivatives

Recently more complex credit derivatives have been introduced. These derivatives distinguish themselves from traditional credit derivatives in that they do not insure the loss of default, but a change of credit quality. This change of credit quality is usually measured in the form of yields spreads. However, derivatives on credit ratings are also conceivable. Simple credit models such as the ones presented above and generally credit risk models for risky bonds cannot be used to value spread credit derivatives, although sometimes extensions are possible.

Das (1995) extends the bond credit model by Merton (1974) to credit spread derivatives. The idea is that a spread option is an option on credit risk which is in turn on option on the firm value. Das (1995) values options with payoff

$$\Lambda(T) = P^{d*}(T,U) - P^{d}(T,U),$$

where $P^{d*}(T,U) = P(T,U)\exp(-r^{d*}(U-T))$. $r^{d*}(T,U)$ is the strike yield spread. Unless interest rates are deterministic, $P(T,U)$ is stochastic. Therefore, the option has a stochastic strike price. The defaultable bond $P^d(T,U)$ is modeled as

$$\begin{aligned} P^d(T,U) &= P(T,U) - p_T(V(T), P^{d*}(T,U)) \\ &= V(T) + c_T(V(T), P^{d*}(T,U)). \end{aligned}$$

$p_T(V(T), P^{d*}(T,U))$ is a put option on the firm value struck at $P^{d*}(T,U)$. Similarly, $c_T(V(T), P^{d*}(T,U))$ is a call option with the same strike price. This way of modeling the defaultable bond is proposed in Merton (1974). For constant interest rates, Das (1995) obtains prices for the compound option by solving an integral numerically. For stochastic interest rates, a binomial approach is taken.

Alternatively, the yield spread could be modeled directly. This is the approach taken by Longstaff and Schwartz (1995b). They assume the following spread dynamics under the risk-neutral measure,

$$dX = (a - bX)\,dt + s dW_1(t).$$

Interest rates are assumed to evolve according to Vasicek (1977), i.e.,

$$dr = (\alpha - \beta r)\,dt + \sigma dW_2(t).$$

The correlation between $W_1(t)$ and $W_2(t)$ is denoted by ρ. Under these dynamics, they derive a closed-form solution for European options on X. The price of a call option with payoff $\Lambda(T) = \max(\exp(X(T)) - K, 0)$ is given by

$$\Lambda(t) = P(t,T)(e^{\mu + \frac{1}{2}\eta^2} N(d) - KN(d-\eta)),$$

where

$$d = \frac{\ln K + \mu + \eta^2}{\eta}.$$

μ and η are the mean and the standard deviation of the normal distribution of $X(T)$ implied by the SDE for the spread. They can be expressed in terms of the process parameters.

Although this valuation method is tempting because of its simplicity, the exogenous spread process definition is somewhat arbitrary. Also, estimating the parameters of the credit spread process may be difficult. Moreover, the parameters a and b imply a specific initial term structure of credit spread which may not match the observed spread structure. Because this is a time-homogeneous model of the credit spreads, it may not be possible to choose the parameters such that the initial term structure is fitted to the observed term structure. This is the same problem that occurs in time-homogeneous term structure models such as Vasicek (1977). However, the model could be extended to the time-inhomogeneous case to allow the fitting of the initial term structure of yield spreads. In that case, no closed-form solution can be derived.

Pierides (1997) values credit spread options and option on risky bonds using the model of Merton (1974). As a consequence of using Merton's model, interest rates are assumed to be constant. Pierides (1997) obtains prices for spread options by numerically solving a PDE.

Another spread-modeling approach is proposed by Das and Sundaram (1998). They use a Heath, Jarrow, and Morton (1992) model for the evolution of interest rates and a similar specification of the evolution of credit spreads. It is a pragmatic approach in the tradition of Longstaff and Schwartz (1995b) but in the context of a Heath, Jarrow, and Morton (1992) framework. Consequently, similar critisisms regarding the economic foundation of the spread model apply.

3.4 Empirical Evidence

A number of empirical studies have been conducted testing the model of Merton (1974) or extensions thereof. Jones, Mason, and Rosenfeld (1984) examine firms with simple capital structures to test the explanatory power of Merton's model with respect to corporate bond prices. They find that although the model seems to explain prices of low-grade bond reasonably well, it is not able to explain the prices of investment-grade bonds. The credit risk spreads implied by the model are consistently lower than observed market spreads. Merton's model also explaines only a small part of the systematic risk in corporate bond returns, as a study by Weinstein (1983) showes.

Titman and Torous (1989), on the other hand, report fairly good explanatory power of Merton's model in a study that tests the model with commercial mortgage loans. Even in their study, only the general magnitude of the credit spread is reasonably well explained, while the change of credit spreads with changing maturity remaines unexplained. While the model indicates that the yield spread eventually decreases for long maturities, actual rate quotes do not show this behavior. This finding contrasts a study by Sarig and Warga (1989). They investigate Merton's model with respect to the term structure of credit spreads and find that the shapes of term structures implied by the model are generally consistent with empirical term structures. Anderson and Sundaresan (2000) compare the empirical performance of the Merton (1974) model with extensions that incorporate endogenous bankruptcy. For the extensions, they use the model by Leland (1994b) and an adaptation of Anderson and Sundaresan (1996) and Mella-Barral and Perraudin (1997). The models seem to be able to explain empirical spreads reasonably well. Moreover, they find evidence that the more recent structural models are somewhat superior to the original Merton (1974) model.

Longstaff and Schwartz (1995a) report good explanatory power of their model with respect to actual credit spreads. Their model implies that credit spreads increase with increasing correlation between interest rates and firm value. This effect is confirmed by their empirical analysis. Duffee (1999) assumes an intensity model based on Duffie and Singleton (1999) and a square-root interest rate process as in Pearson and Sun (1994) and estimates the term structure of corporate bonds with a Kalman-filter approach. Using a similar model setup as Duffee (1999) for the term structure model, Zheng (2000) presents a method for extracting implied credit-spread volatilities from corporate bond prices. The approach is based on a barrier-option model for credit risk.

Empirical research on the credit risk of derivative securities is rare. Such work tends to be jeopardized by the lack of reliable data on derivative transactions because derivatives subject to counterparty risk are traded over-the-counter. Because of the large market size and the availability of data, a few studies have investigated the credit risk component of interest-rate swap prices. For example, Cossin and Pirotte (1998) test the Longstaff and Schwartz (1995a) credit risk model applied to swap spreads as proposed in Longstaff and Schwartz (1994). Their tests are based on transaction data from the Swiss swap market. They find wide discrepancies between the spreads implied by the model and market spreads.

3.5 Summary

In this chapter we give an overview of existing credit risk models. There are several basic approaches to model credit risk. Roughly, they can be broken down into

- Firm value models
- First passage time models
- Intensity models

In pure firm value models, credit risk is an option written on firm value. The time of default is not explicitly modeled. It is usually assumed that default occurs at maturity of the instrument. Consequently, American claims are difficult to handle within such a framework. On the other hand, firm value models offer the advantage of easy tractability and the availability of explicit pricing solutions for credit-risky bonds.

First passage time models are also firm value models in the sense that they model the evolution of the firm value. However, the recovery rate on a claim is not necessarily dependent on the firm value. Rather, the firm value is used to model the time of default. Usually, default is assumed to occur when the firm value crosses an exogenous default boundary. It is thus based on a first passage time distribution. The recovery rate paid out in the case of default is often an exogenously specified fraction of the no-default value of the claim. First passage time models also admit the derivation of explicit pricing formulae for risky bonds in some cases.

By using an exogenous default process, intensity models make no reference to the concept of the firm value driving occurrence and severity of default. The default process is a jump process with an intensity which represents the risk of default in a given time interval. Closed-form solutions are only available for very simple instances of this class of models. On the other hand, intensity models can be relatively easily extended to value options on credit-risky bonds and credit derivatives.

The classification into firm value, first passage time, and intensity models is of course not clear-cut. For example, some models allow for a dependency between default intensity and state variable, which may be the firm value. Such an approach is proposed in Chapter 5.

4. A Firm Value Pricing Model for Derivatives with Counterparty Default Risk

In this chapter we propose a credit risk model for the valuation of derivative securities with counterparty default risk. Derivative securities that are subject to counterparty default risk are sometimes called *vulnerable* derivative securities. The credit risk model we propose is based on the stochastic evolution of the value of the firm's assets and liabilities.

We derive explicit pricing formulae for vulnerable call and put options under the following assumptions: deterministic interest rates and deterministic counterparty liabilities, deterministic interest rates and stochastic counterparty liabilities, stochastic interest rates and deterministic liabilities, and stochastic interest rates and stochastic liabilities. We also analyze a number of special cases of the four versions of the model. For example, as the credit risk model by Merton (1974) is based on Black and Scholes (1973), a credit risk model with stochastic liabilities can be constructed based on Margrabe (1978). We show that both are special cases of our model.

In addition, we give explicit pricing formulae for forward contracts with bilateral counterparty risk. It turns out that this problem can be dealt with in a straightforward way by the model for vulnerable options.

In this chapter, we begin by presenting the credit risk model which serves as the working model for the remainder of the chapter. Next, we derive closed-form pricing formulae for vulnerable options assuming deterministic interest rates and liabilities. The model is subsequently generalized to stochastic liabilities and stochastic interest rates. Closed-form solutions are also derived for the generalized versions of the model. The model is then applied to forward contracts. A section with numerical examples under various assumptions closes the chapter.

4.1 The Credit Risk Model

This chapter is based on a credit risk model similar to that used by Merton (1974). Let $X(T)$ be the promised amount paid by the counterparty firm at time T. If the firm remains solvent throughout the lifetime of the claim, the claim will pay out X at its maturity date T. In this chapter it is assumed that X is a claim of the European type, i.e., that it cannot be claimed before time T.

It the firm goes bankrupt anytime from the present day denoted by t to its maturity date T, the holder of the claim may not receive $X(T)$, but only a fraction thereof. The amount the claim holder receives is dependent on the value of the assets of the firm relative to its liabilities. Denote the value of the assets by $V(t)$ and the value of the liabilities by $D(t)$. In case of bankruptcy, the payoff of the claim will be

$$X^d(T) = X(T)\frac{V(T)}{D(T)}.$$

This expression shows that all claims of the firm have equal seniority and therefore bear an equal relative share in the loss. Because $V(T)D^{-1}(T)$ denotes the value recovered "on the dollar", it is commonly called *recovery rate*.

The payoff of a credit-risky claim can therefore be expressed as

$$X^d(T) = X(T)\mathbf{1}_{\{\tau > T\}} + \delta(T)X(T)\mathbf{1}_{\{\tau \leq T\}}.$$

τ is the time of default or bankruptcy and $\delta(t)$ the recovery rate. For now, we assume that $\tau \geq T$, i.e., that default or bankruptcy cannot occur before maturity of the claim. Later, this assumption will be relaxed. Although default may occur if $V(T) > D(T)$, this case is not relevant in this simple model because in such a case the claim will be fully paid. The value of the claim at its due date T can therefore also be expressed as

$$X^d(T) = X(T)\mathbf{1}_{\{V_T \geq D_T\}} + \delta(T)X(T)\mathbf{1}_{\{V_T < D_T\}}.$$

In a direct application of Proposition 2.2.2, if the market admits a martingale measure, the price of the risky claim at time $t < T$ is given by

$$X_t^d = B_t\,\mathbf{E_Q}[B_T^{-1}X(T)(\mathbf{1}_{\{V_T \geq D_T\}} + \delta_T\,\mathbf{1}_{\{V_T < D_T\}})|\mathcal{F}_t], \qquad (4.1)$$

where $\delta_T = \frac{V_T}{D_T}$. $\mathbf{Q}$ denotes the martingale measure. B_t serves as the deflating asset and is to be interpreted as the value of a default-free money market account that accumulates at the riskless interest rate r_t satisfying the ODE $dB_t = r_t B_t dt$.

Evaluating X_t^d requires modeling of relevant variables X_t, V_t, D_t. As will be shown in subsequent sections of this chapter, under the assumptions that X_T is deterministic, r_t and D_t are constants, and V_t follows geometric Brownian motion, then the evaluation of X_t^d according to (4.1) yields the Merton (1974) pricing formulae for risky zero-coupon bonds.

Derivative securities, however, have stochastic payoffs X_T. The main objective of this chapter is to show that pricing formulae can be derived under various assumptions for options that contain counterparty default risk. The payoff of a call option is $X_T = \max(S_T - K, 0)$, where S_T denotes the price of the underlying security at time T and K the strike price of the option. Similarly, the payoff of a put option is $X_T = \max(K - S_T, 0)$.

4.2 Deterministic Liabilities

In a framework similar to the Black-Scholes setting, we assume that both the firm value and the value of the underlying asset are processes that follow geometric Brownian motion. S_t and V_t therefore satisfy the SDEs

$$dS = S\big(\alpha_S dt + \sigma_S dW_S(t)\big),$$
$$dV = V\big(\alpha_V dt + \sigma_V dW_V(t)\big),$$

for constants $\alpha_S, \alpha_V, \sigma_S, \sigma_V$. $W_S(t)$ and $W_V(t)$ are assumed to be correlated, $\mathbb{R}^d$-valued Brownian motions[1] under measure $\mathbf{P}$ such that

$$\langle W^i, W^j \rangle = \rho_{ij} t.$$

We also assume that the short rate r is constant such that

$$B_T = B_t e^{r(T-t)}, \qquad t \leq T.$$

We introduce an equivalent probability measure $\mathbf{Q}$ such that $V_t B_t^{-1}$ and $S_t B_t^{-1}$ are martingales. The new probability measure is defined by

$$\frac{d\mathbf{Q}}{d\mathbf{P}} = \exp\left(\gamma W_T - \frac{1}{2}\gamma^2 T\right),$$

with γ_t in $\mathbb{R}^2$, having elements

$$\gamma_S = \frac{\alpha_S - r}{\sigma_S}, \qquad \gamma_V = \frac{\alpha_V - r}{\sigma_V}.$$

An application of Girsanov's theorem shows that

$$\begin{aligned} \frac{dS}{S} &= r dt + \sigma_S d\tilde{W}_S(t), \\ \frac{dV}{V} &= r dt + \sigma_V d\tilde{W}_V(t), \end{aligned} \tag{4.2}$$

or equivalently,

$$\begin{aligned} S_T &= S_t \exp\left((r - \frac{1}{2}\sigma_S^2)(T-t) + \sigma_S(\tilde{W}_S(T) - \tilde{W}_S(t))\right), \\ V_T &= V_t \exp\left((r - \frac{1}{2}\sigma_V^2)(T-t) + \sigma_V(\tilde{W}_V(T) - \tilde{W}_V(t))\right), \end{aligned} \tag{4.3}$$

where $\tilde{W}(t)$ denotes a Brownian motion under the martingale measure $\mathbf{Q}$.

[1] σ_S and σ_V can be interpreted as Euclidian norms of uncorrelated d-dimensional Brownian motion. Cf. Definition A.4.2 in Section A.4.

4.2.1 Prices for Vulnerable Options

The payoff of vulnerable standard options allows the derivation of analytical pricing solutions. The explicit formula of the following proposition first appeared in Klein (1996), though without formal derivation. We give a new, formal proof for it in Section 4.9.1.

By Proposition 2.2.2, the price of a credit-risky call option is

$$X_t^d = B_t\,\mathbf{E}_{\mathbf{Q}}[B_T^{-1}(S_T - K)^+(\mathbf{1}_{\{V_T \geq D\}} + \delta_T\,\mathbf{1}_{\{V_T < D\}})|\mathcal{F}_t], \tag{4.4}$$

where S_T and V_T are given by (4.3). Evaluating this expression yields the following proposition.

Proposition 4.2.1. *The price X_t of vulnerable call option with promised payoff $X_T = (S_T - K)^+$ and actual payoff $X_T = \delta(S_T - K)^+$ with $\delta = \frac{V_T}{D}$ in case of bankruptcy is given by*

$$\begin{aligned} X_t &= S_t N_2(a_1, a_2, \rho) - e^{-r(T-t)} K N_2(b_1, b_2, \rho) \\ &\quad + \frac{V_t}{D}\left(e^{(r+\rho\sigma_S\sigma_V)(T-t)} S_t N_2(c_1, c_2, -\rho) - K N_2(d_1, d_2, -\rho)\right), \end{aligned} \tag{4.5}$$

with parameters

$$\begin{aligned} a_1 &= \frac{\ln\frac{S_t}{K} + (r + \frac{1}{2}\sigma_S^2)(T-t)}{\sigma_S\sqrt{T-t}}, \\ a_2 &= \frac{\ln\frac{V_t}{D} + (r - \frac{1}{2}\sigma_V^2 + \rho\sigma_S\sigma_V)(T-t)}{\sigma_V\sqrt{T-t}}, \\ b_1 &= \frac{\ln\frac{S_t}{K} + (r - \frac{1}{2}\sigma_S^2)(T-t)}{\sigma_S\sqrt{T-t}} = a_1 - \sigma_S\sqrt{T-t}, \\ b_2 &= \frac{\ln\frac{V_t}{D} + (r - \frac{1}{2}\sigma_V^2)(T-t)}{\sigma_V\sqrt{T-t}} = a_2 - \rho\sigma_S\sqrt{T-t}, \\ c_1 &= \frac{\ln\frac{S_t}{K} + (r + \frac{1}{2}\sigma_V^2 + \rho\sigma_S\sigma_V)(T-t)}{\sigma_S\sqrt{T-t}} = a_1 + \rho\sigma_V\sqrt{T-t}, \\ c_2 &= -\frac{\ln\frac{V_t}{D} + (r + \frac{1}{2}\sigma_V^2 + \rho\sigma_S\sigma_V)(T-t)}{\sigma_V\sqrt{T-t}} = -a_2 - \rho\sigma_S\sqrt{T-t}, \\ d_1 &= \frac{\ln\frac{S_t}{K} + (r - \frac{1}{2}\sigma_V^2 + \rho\sigma_S\sigma_V)(T-t)}{\sigma_S\sqrt{T-t}} = a_1 + (\rho\sigma_V + \sigma_S)\sqrt{T-t}, \\ d_2 &= -\frac{\ln\frac{V_t}{D} + (r + \frac{1}{2}\sigma_V^2)(T-t)}{\sigma_V\sqrt{T-t}} = -(a_2 - (\rho\sigma_S + \sigma_V)\sqrt{T-t}). \end{aligned} \tag{4.6}$$

N_2 denotes the cumulated distribution function[2] *of a bivariate standard joint normal random variable defined by*

$$N_2(x_1, x_2, \rho) = \int_{-\infty}^{x_1} \int_{-\infty}^{x_2} f(z_1, z_2, \rho)\, dz_2 dz_1,$$

with density

$$f(z_1, z_2, \rho) = \frac{1}{2\pi\sqrt{1-\rho^2}} e^{-\frac{1}{2(1-\rho^2)}(z_1^2 - 2\rho z_1 z_2 + z_2^2)}.$$

Proof. Cf. Section 4.9.1.

Remark 4.2.1. Although the specification of interest rate r and volatility functions σ_S and σ_V suggested that these variables are constants, it is sufficient that they be deterministic. In that case $r(T-t)$ can be replaced with $\int_t^T r(s)\, ds$ and $\sigma_i^2(T-t)$ with $\int_t^T \sigma_i^2(s)\, ds$, for $i \in \{S, V\}$, in the formula above.

Proposition 4.2.2. *The price Y_t of vulnerable put option with promised payoff $X_T = (K - S_T)^+$ and payoff $Y_T = \delta(K - S_T)^+$ with $\delta = \frac{V_T}{D}$ in case of bankruptcy is given by*

$$\begin{aligned} Y_t = {} & e^{-r(T-t)} K N_2(-b_1, b_2, -\rho) - S_t N_2(-a_1, a_2, -\rho) \\ & + \frac{V_t}{D}\left(K N_2(-d_1, d_2, \rho) - e^{(r+\rho\sigma_S\sigma_V)(T-t)} S_t N_2(-c_1, c_2, \rho)\right), \end{aligned}$$

with parameters given in (4.6).

Proof. The proof is analogous to that of Proposition 4.2.1.

Instead of directly evaluating the expectation in (4.4), we could also take a differential equation approach. A vulnerable contingent claim is a function $F = f(S, V, t)$. Given the SDEs in (4.2) for S and V, respectively, the dynamics of F is obtained by an application of Itô's formula:

$$\begin{aligned} dF = {} & F^V dV + F^S dS + F^t dt \\ & + \frac{1}{2} F^{VV} V^2 \sigma_V^2\, dt + \frac{1}{2} F^{SS} S^2 \sigma_S^2\, dt + F^{VS} V S \sigma_V \sigma_S \rho\, dt\,. \end{aligned} \tag{4.7}$$

Superscripts of F denote partial derivatives. We form a dynamically riskless hedge similar to the original Black-Scholes derivation. The hedge portfolio is $-F^V V - F^S S$ such that the hedged position is given by

[2] Approximation techniques for the evaluation of the double integral can be found in Abramowitz and Stegun (1972) and Drezner (1978). A fast algorithm for exact evaluation is Dutt (1975). However, we evaluate the double integral by standard numerical integration methods as described in Press, Vetterling, Teukolsky, and Flannery (1992).

$$dH = dF - F^V dV - F^S dS.$$

Substituting for dF from (4.7) gives

$$dH = F^t dt + \frac{1}{2} F^{VV} V^2 \sigma_V^2 dt + \frac{1}{2} F^{SS} \sigma_S^2 S^2 \, dt + F^{VS} V S \sigma_V \sigma_S \rho \, dt.$$

This portfolio's dynamics are purely deterministic and therefore the portfolio earns the riskless rate r. Hence, $dH = rH dt$. Substituting for H and dropping dt gives the PDE

$$0 = -rF + rF^V V + rF^S S + F^t + \frac{1}{2} F^{SS} S^2 \sigma_S^2 + \frac{1}{2} F^{VV} V^2 \sigma_V^2 + F^{VS} V S \sigma_V \sigma_S \rho.$$

Given suitable boundary conditions, this PDE can be solved for F by Fourier transforms or similarity transformations. For example, the boundary condition for a vulnerable call option is

$$F(S, V, 0) = \max(S - K, 0) \min(VD^{-1}, 1).$$

Similarly, for a vulnerable put option,

$$F(S, V, 0) = \max(K - S, 0) \min(VD^{-1}, 1).$$

For a vulnerable call option, the S and V positions in the hedge portfolio can be shown to be

$$\delta_S = F^S = N_2(a_1, a_2, \rho) + \frac{V_t}{D} e^{(r+\rho\sigma_S\sigma_V)(T-t)} N_2(c_1, c_2, -\rho),$$
$$\delta_V = F^V = \frac{S_t}{D} e^{(r+\rho\sigma_S\sigma_V)(T-t)} N_2(c_1, c_2, -\rho) - \frac{K}{D} N_2(d_1, d_2, -\rho).$$

Similarly, for put options we have

$$\delta_S = -N_2(-a_1, a_2, -\rho) - \frac{V_t}{D} e^{(r+\rho\sigma_S\sigma_V)(T-t)} S_t N_2(-c_1, c_2, \rho),$$
$$\delta_V = \frac{K}{D} N_2(-d_1, d_2, \rho) - e^{(r+\rho\sigma_S\sigma_V)(T-t)} \frac{S_t}{D} N_2(-c_1, c_2, \rho).$$

4.2.2 Special Cases

The pricing formula given in Proposition 4.2.1 encompasses a number of simpler valuation problems as special cases. For example, a very simple special case is the Black-Scholes option pricing formula. Other special cases include claims with a fixed recovery rate in case of bankruptcy and deterministic claims, such as bonds. In case of deterministic claims, the model simplifies to the Merton (1974) credit risk model.

4.2.2.1 Fixed Recovery Rate. Another special case of the firm value model is a specification with a fixed recovery rate. In that case, the firm value only serves to determine the occurrence of default but has no influence on the recovery rate. Such a specification of the firm value model reminds of the first passage time models presented in Section 3.1.3. However, the firm value model with an exogenous, fixed recovery rate is not a first passage time model because it does not model the time of default explicitly. For the firm value model, it is irrelevant whether the path of the firm value process has crossed the default boundary during the lifetime of the claim. Only its value at maturity determines whether the claim is paid in full or not.

Proposition 4.2.3. *The price X_t of a vulnerable option as in Proposition 4.2.1, but with exogenous and constant recovery rate δ, is given by*

$$\begin{aligned} X_t &= S_t \left(N_2(a_1, a_2, \rho) + \delta N_2(a_1, -a_2, -\rho) \right) \\ &\quad - e^{-r(T-t)} K \left(N_2(b_1, b_2, \rho) + \delta N_2(b_1, -b_2, -\rho) \right), \end{aligned} \tag{4.8}$$

where parameters a_1, a_2, b_1, b_1 are defined as in expression (4.6).

Proof. This proof very closely follows that of Proposition 4.2.1. The value of the claim can also be expressed as $X_t = E_1 - E_2 + E_3 - E_4$. The four component terms are defined in expression (4.35). δ, however, is an exogenous constant in this case. E_1 and E_2 remain unchanged from expressions (4.44) and (4.45). The evaluation of E_3 is similar to that of E_1 starting on page 115. S_t, however, is multiplied by the recovery rate and the second indicator function is evaluated differently.

$$\begin{aligned} \mathbf{E}_{\dot{\mathbf{Q}}}[\mathbf{1}_{\{V_T<D\}}] &= \dot{\mathbf{Q}}\left(\sigma_V(\dot{W}_T - \dot{W}_t) < \ln D - \ln V_t - (r + \frac{1}{2}\sigma_V^2)(T-t) \right) \\ &= \dot{\mathbf{Q}}\left(\dot{z}_2 < -\frac{\ln V_t - \ln D + (r - \frac{1}{2}\sigma_V^2 + \rho\sigma_S\sigma_V)(T-t)}{\sigma_V\sqrt{T-t}} \right) \\ &= \dot{\mathbf{Q}}\left(\dot{z}_2 < -a_2 \right). \end{aligned}$$

Thus, $E_3 = \delta S_t N_2(a_1, -a_2, -\rho)$. The evaluation of E_4 closely corresponds to that of E_2 on page 117. Similar to E_3, the second indicator function is different. $E_4 = \delta e^{-r(T-t)} K N_2(b_1, -b_2, -\rho)$. Rearranging terms yields the proposition.

Remark 4.2.2. Consider $\delta = 1$. In this case there is no credit risk since the full payout is made even in bankruptcy. Therefore, the price of the vulnerable claim must coincide with the the price for a claim free of credit risk. To see this, we apply the following property of the bivariate normal distribution to expression (4.8):

$$N_2(x, y, \rho) = N(x) - N_2(x, -y, -\rho),$$

where $N(x)$ is the distribution function of a univariate standard normal. Expression (4.8) can therefore be rewritten as

$$\begin{aligned} X_t = &S_t\left(\delta N(a_1) + (1-\delta)N_2(a_1, a_2, \rho)\right) \\ &- e^{-r(T-t)}K\left(\delta N(b_1) + (1-\delta)N_2(b_1, b_2, \rho)\right). \end{aligned}$$

For $\delta = 1$, this expression clearly simplifies to the Black-Scholes formula for standard options.

4.2.2.2 Deterministic Claims. Consider the case when the underlying asset is deterministic, i.e., $\sigma_S \to 0$. Since $N_2(x_1, \infty, \rho) = N(x_1)$, we have

$$X_t = \left(S_t - e^{-r(T-t)}K\right)N(a_2) + \frac{V_t}{D}\left(e^{r(T-t)}S_t - K\right)N(c_2).$$

Define $d_1 = -c_2$, $d_2 = b_2$, and note that $N(x) = 1 - N(-x)$. Consequently,

$$X_t = \left(S_t - e^{-r(T-t)}K\right)N(a_2) + \frac{V_t}{D}\left(e^{r(T-t)}S_t - K\right)(1 - N(c_2)).$$

Now set $S = S_t \exp(r(T-t))$. It follows that

$$X_t = (S-K)e^{-r(T-t)}N(d_2) - (S-K)\frac{V_t}{D}N(d_1) + (S-K)\frac{V_t}{D}.$$

Simplifying gives

$$X_t = (S-K)\left(\frac{V_t}{D} - \left(\frac{V_t}{D}N(d_1) - e^{-r(T-t)}N(d_2)\right)\right). \tag{4.9}$$

In other words, we obtain the situation where the credit-risky deterministic claim $S - K$ is equivalent to having, on the one hand, a share in the firm's assets, V_t, corresponding to the size of the claim relative to the sum of all liabilities, and on the other hand, a call option written to the equity holders of the firm.

We already know that, by the put-call parity, the share in the firm's assets plus a short call option on those assets must be equal to the default-free value of the claim plus a short put option on the firm assets struck at the liability value. The put-call parity can easily be demonstrated within this example.

Expression (4.9) can be rewritten

$$X_t = (S-K)\left(\frac{V_t}{D}(1 - N(d_1)) + e^{-r(T-t)}N(d_2)\right).$$

Using $N(x) = 1 - N(-x)$, we obtain

$$\begin{aligned} X_t &= (S-K)\left(\frac{V_t}{D}N(-d_1) + e^{-r(T-t)}(1 - N(-d_2))\right) \\ &= (S-K)\left(e^{-r(T-t)} - \left(e^{-r(T-t)}N(-d_2) - \frac{V_t}{D}N(-d_1)\right)\right). \end{aligned} \tag{4.10}$$

The last equation shows that the value of the deterministic claim can also be interpreted as the value of the default-free claim less a put option. Implicit in (4.9) and (4.10) is the assumption that the claim is only one among many on the balance sheet of the firm. We can further simplify things by assuming that the firm has only one claim outstanding, i.e., $S - K = D$. We immediately find

$$\begin{aligned} X_t &= V_t - (V_t N(d_1) - e^{-r(T-t)} D N(d_2)) \\ &= D e^{-r(T-t)} - (e^{-r(T-t)} D N(-d_2) - V_t N(-d_1)), \end{aligned}$$

which represents the insight by Black and Scholes (1973) and Merton (1974).

4.3 Stochastic Liabilities

The following proposition is a generalization of Proposition 4.2.1. Counterparty liabilities are no longer assumed constant, but are specified to follow geometric Brownian motion given by the process

$$D_T = D_t \exp\left((\alpha_D - \frac{1}{2}\sigma_D^2)(T-t) + \sigma_D W_t\right), \tag{4.11}$$

with σ_D constant. Under the martingale measure $\mathbf{Q}$, we have

$$D_T = D_t \exp\left((r - \frac{1}{2}\sigma_D^2)(T-t) + \sigma_D \tilde{W}_t\right).$$

The processes for the underlying security, S, and the firm asset value, V, remain unchanged from the previous section. Expression (4.3) gives their definitions.

The covariance matrix A of S, V, and D is given by

$$A = \begin{pmatrix} \sigma_S^2 & \rho_{SV}\sigma_S\sigma_V & \rho_{SD}\sigma_S\sigma_D \\ \rho_{SV}\sigma_S\sigma_V & \sigma_V^2 & \rho_{VD}\sigma_V\sigma_D \\ \rho_{SD}\sigma_S\sigma_D & \rho_{VD}\sigma_V\sigma_D & \sigma_D^2 \end{pmatrix}.$$

The correlation coefficients can be interpreted in the sense of Definition A.4.2. As before, a volatility coefficient σ can be thought of as a Euclidian norm of a vector of sensitivities to sources of uncertainty represented by independent Brownian motions. Therefore, we have, for example, $\langle \sigma_S \tilde{W}_S(t), \sigma_V \tilde{W}_V(t) \rangle = \rho_{SV}\sigma_S\sigma_V t$. In a slight abuse of terminology, we refer to ρ_{SV} as the "correlation between S and V."

From one of the examples accompanying Theorem A.5.2, the dynamics of $\delta_t = V_t D_t^{-1}$ is given by

$$\frac{d\delta}{\delta} = \frac{dV}{V} - \frac{dD}{D} + \left(\frac{dD}{D}\right)^2 - \frac{dV}{V}\frac{dD}{D}. \tag{4.12}$$

Substituting for $\frac{dV}{V}$ and $\frac{dD}{D}$ under the measure $\mathbf{Q}$ and applying the usual multiplication rules (cf. remark on Definition A.4.1) gives

$$\frac{d\delta}{\delta} = \sigma_V dW_t^V - \sigma_D dW_t^D + \sigma_D^2 dt - \rho_{VD}\sigma_V\sigma_D dt. \tag{4.13}$$

Define $\sigma_\delta W_t^\delta = \sigma_V W_t^V - \sigma_D W_t^D$. Since the variation of a linear combination of processes can be expressed as $\langle ax + by\rangle = a^2\langle x\rangle + b^2\langle y\rangle + 2ab\langle x, y\rangle$, the instantaneous volatility of δ is given by

$$\sigma_\delta = \sqrt{\sigma_V^2 + \sigma_D^2 - 2\rho_{VD}\sigma_V\sigma_D}. \tag{4.14}$$

The Doléans-Dade solution of (4.13) is

$$\delta_T = \delta_t e^{(\sigma_D^2 - \rho_{VD}\sigma_V\sigma_D - \frac{1}{2}\sigma_\delta^2)(T-t) + \sigma_\delta(\tilde{W}_T^\delta - \tilde{W}_t^\delta)}. \tag{4.15}$$

Using (4.14), we can simplify (4.15) to give

$$\delta_T = \delta_t e^{(\frac{1}{2}\sigma_D^2 - \frac{1}{2}\sigma_V^2)(T-t) + \sigma_\delta(\tilde{W}_T^\delta - \tilde{W}_t^\delta)}. \tag{4.16}$$

Solutions (4.15) and (4.16) can be verified easily by applying Itô's formula.[3]

By the linearity of variation, $\langle S, V - D\rangle = \langle S, V\rangle - \langle S, D\rangle$, we have

$$\rho_{S\delta}\sigma_S\sigma_\delta = \rho_{SV}\sigma_S\sigma_V - \rho_{SD}\sigma_S\sigma_D$$

and therefore

$$\rho_{S\delta} = \frac{\rho_{SV}\sigma_V - \rho_{SD}\sigma_D}{\sigma_\delta}.$$

Remark 4.3.1. $\delta = f(V, D)$ is a function of two traded assets, i.e., processes that are martingales under the measure $\mathbf{Q}$ for a money market numéraire. Equation 4.13 shows that δ itself is not a martingale under the measure $\mathbf{Q}$. Therefore, δ cannot be a traded asset if arbitrage is to be ruled out. This, however, does not mean that we cannot price a derivative contract on δ. We can show that δ is a deterministic function of traded assets and, as such, replicable by a portfolio of traded assets.

Consider a portfolio ψ consisting of one unit of V and short one unit of D. The SDE for this portfolio is given by $d\psi/\psi = \sigma_V dW_t^V - \sigma_D dW_t^D$. The difference to 4.13 is the drift term. Therefore, for $\delta_0 = \psi_0$, we have $\delta_t = e^{(\sigma_D^2 - \rho_{VD}\sigma_V\sigma_D)t}\psi_t$. In other words, we can build a self-financing portfolio that replicates the non-traded asset δ.

[3] Let $\delta = f(x) = e^x$ and $x = (\frac{1}{2}\sigma_D^2 - \frac{1}{2}\sigma_V^2)(T-t) + \sigma_\delta(\tilde{W}_T^\delta - \tilde{W}_t^\delta)$. Therefore, by Itô's formula (cf. A.5.1), $d\delta = \delta(\frac{1}{2}\sigma_D^2 dt - \frac{1}{2}\sigma_V^2 dt + \sigma_V d\tilde{W}_V - \sigma_D d\tilde{W}_D + \frac{1}{2}(dx)^2)$ with $(dx)^2 = (\sigma_V^2 + \sigma_D^2 - 2\rho_{VD}\sigma_V\sigma_D)\, dt$. Resultant is the SDE in (4.13).

4.3.1 Prices of Vulnerable Options

By Proposition 2.2.2, the price of a credit-risky call option is

$$X_t^d = B_t \, \mathbf{E}_{\mathbf{Q}}[B_T^{-1}(S_T - K)^+(\mathbf{1}_{\{V_T \geq D_T\}} + \delta_T \, \mathbf{1}_{\{V_T < D_T\}})|\mathcal{F}_t], \tag{4.17}$$

where S_T, V_T, and D_T are given by (4.3) and (4.11), respectively. B_t denotes the money-market account with $B_t = \exp(rt)$. Evaluating (4.17) yields the following proposition.

Proposition 4.3.1. *The price X_t of a vulnerable call option with promised payoff $X_T = (S_T - K)^+$ and actual payoff $X_T = \delta(S_T - K)^+$ with $\delta = \frac{V_T}{D_T}$ in case of bankruptcy is given by*

$$\begin{aligned} X_t = {} & S_t \, N_2(a_1, a_2, \rho) - K e^{-r(T-t)} \, N_2(b_1, b_2, \rho) \\ & + S_t \frac{V_t}{D_t} e^{(\rho_{SV}\sigma_S\sigma_V - \rho_{SD}\sigma_S\sigma_D - \rho_{VD}\sigma_V\sigma_D + \sigma_D^2)(T-t)} \, N_2(c_1, c_2, -\rho) \\ & - e^{(-r+\sigma_D^2 - \rho_{VD}\sigma_V\sigma_D)(T-t)} K \frac{V_t}{D_t} \, N_2(d_1, d_2, -\rho), \end{aligned} \tag{4.18}$$

with parameters

$$\begin{aligned} a_1 &= \frac{\ln \frac{S_t}{K} + (r + \frac{1}{2}\sigma_S^2)(T-t)}{\sigma_S\sqrt{T-t}}, \\ a_2 &= \frac{\ln \frac{V_t}{D_t} - (\frac{1}{2}\sigma_V^2 - \frac{1}{2}\sigma_D^2 - \theta)(T-t)}{\sigma_\delta\sqrt{T-t}}, \\ b_1 &= \frac{\ln \frac{S_t}{K} + (r - \frac{1}{2}\sigma_S^2)(T-t)}{\sigma_S\sqrt{T-t}}, \\ b_2 &= \frac{\ln \frac{V_t}{D} - \frac{1}{2}(\sigma_V^2 - \sigma_D^2)(T-t)}{\sigma_\delta\sqrt{T-t}}, \\ c_1 &= \frac{\ln S_t - \ln K + (r + \frac{1}{2}\sigma_S^2 + \theta)(T-t)}{\sigma_S\sqrt{T-t}}, \\ c_2 &= -\frac{\ln \delta_t + (\frac{3}{2}\sigma_D^2 + \frac{1}{2}\sigma_V^2 + \theta - 2\rho_{VD}\sigma_V\sigma_D)(T-t)}{\sigma_\delta\sqrt{T-t}}, \\ d_1 &= \frac{\ln \frac{S_t}{K} + (r - \frac{1}{2}\sigma_S^2 + \theta)(T-t)}{\sigma_S\sqrt{T-t}}, \\ d_2 &= -\frac{\ln \delta_t + (\frac{1}{2}\sigma_V^2 + \frac{3}{2}\sigma_D^2 - 2\rho_{VD}\sigma_V\sigma_D)(T-t)}{\sigma_\delta\sqrt{T-t}}, \end{aligned} \tag{4.19}$$

and

$$\begin{aligned} \rho &= \frac{\rho_{SV}\sigma_V - \rho_{SD}\sigma_D}{\sigma_\delta}, \\ \sigma_\delta &= \sqrt{\sigma_V^2 + \sigma_D^2 - 2\rho_{VD}\sigma_V\sigma_D}, \\ \theta &= \rho_{SV}\sigma_S\sigma_V - \rho_{SD}\sigma_S\sigma_D. \end{aligned}$$

N_2 denotes the cumulated distribution function of a bivariate standard joint normal random variable defined in Proposition 4.2.1.

Proof. Cf. Section 4.9.2.

The question arises whether Proposition 4.3.1 could have been reached in a less cumbersome way than given by the proof in Section 4.9.2 by means of a numéraire change similar to the one employed in the proof to Proposition 2.3.5. Unfortunately, this question has to be negated since the problem solved by Proposition 4.3.1 is multi-factor. The standard Margrabe model is two-factor, but can be reduced to one factor by the numéraire change. This reduction is not possible for Proposition 4.3.1.

Proposition 4.3.2. *The price Y_t of a vulnerable put option with promised payoff $Y_T = (K - S_T)^+$ and actual payoff $Y_T = \delta(K - S_T)^+$ with $\delta = \frac{V_T}{D_T}$ in case of bankruptcy is given by*

$$\begin{aligned} X_t &= Ke^{-r(T-t)}\, N_2(-b_1, b_2, -\rho) - S_t\, N_2(-a_1, a_2, -\rho) \\ &\quad + e^{(-r+\sigma_D^2-\rho_{VD}\sigma_V\sigma_D)(T-t)} K \frac{V_t}{D_t}\, N_2(-d_1, d_2, \rho) \\ &\quad - S_t \frac{V_t}{D_t} e^{(\rho_{SV}\sigma_S\sigma_V - \rho_{SD}\sigma_S\sigma_D - \rho_{VD}\sigma_V\sigma_D + \sigma_D^2)(T-t)}\, N_2(-c_1, c_2, \rho), \end{aligned}$$

with parameters given in (4.19).

Proof. The proof is analogous to that of Proposition 4.3.1.

4.3.2 Special Cases

The general valuation problem that led to Proposition 4.3.1 has the payoff function

$$\begin{aligned} X_T &= \max(S_T - K, 0)\, \min(\delta_T, 1) \\ &= \max(S_T - K, 0)\,(1 - \max(1 - \delta_T, 0)), \end{aligned}$$

which translates to

$$X_T = ((S_T - K)^+ (\mathbf{1}_{\{\delta_T \geq 1\}} + \delta_T \mathbf{1}_{\{\delta_T < 1\}}).$$

An obvious special case is $\sigma_D \to 0$ and $D = D_t \exp(-r(T-t))$. It can easily be seen that the original formula in (4.5) for constant liabilities emerges.

Another simple special case is impossible bankruptcy. If $D_t \to \infty$, then the last two terms of (4.18) disappear as $c_2 \to -\infty$ and $d_2 \to -\infty$. At the same time $a_2 \to \infty$ and $b_2 \to \infty$. Since $N_2(x_1, -\infty, \rho) = 0$ and $N_2(x_1, \infty, \rho) = N(x_1)$, the Black-Scholes formula emerges.

4.3.2.1 Asset Claims. Consider the special case when $K \to 0$. The second and fourth terms of (4.18) disappear and since $\lim_{K\to 0}(-\ln K) = \infty$, we have $N_2(x_1, x_2, \rho) = N(x_2)$ for $x \in \{a, c\}$. Thus, expression (4.18) simplifies to

$$X_t = S_t\, N(a_2) + S_t \frac{V_t}{D_t} e^{(\rho_{SV}\sigma_S\sigma_V - \rho_{SD}\sigma_S\sigma_D - \rho_{VD}\sigma_V\sigma_D + \sigma_D^2)(T-t)}\, N(c_2). \tag{4.20}$$

This is the price of a vulnerable asset claim. In practice, asset claims often occur in the form of baskets or index units[4] issued by financial institutions. For credit risk purposes, such OTC units possess all the properties of a vulnerable derivative claim, albeit with a strike price of zero.

4.3.2.2 Debt Claims. We can simplify the asset claim model by assuming that the underlying of the claim is the firm's debt, i.e., that $S_s = D_s$ for all $s \in [t, T]$. Consequently, $\rho_{SD}\sigma_S\sigma_D = \sigma_D^2$ and $\rho_{SV}\sigma_S\sigma_V = \rho_{VD}\sigma_V\sigma_D$. Therefore, (4.20) simplifies to

$$X_t = D_t\, N(a_2) + V_t\, N(c_2),$$

with

$$a_2 = \frac{\ln \frac{V_t}{D_t} - (\frac{1}{2}\sigma_V^2 + \frac{1}{2}\sigma_D^2 - \rho_{VD}\sigma_V\sigma_D)(T-t)}{\sigma_\delta\sqrt{T-t}},$$

$$c_2 = -\frac{\ln \delta_t + (\frac{1}{2}\sigma_V^2 + \frac{1}{2}\sigma_D^2 - \rho_{VD}\sigma_V\sigma_D)(T-t)}{\sigma_\delta\sqrt{T-t}}.$$

a_2 and $d = -c_2$ are the integration parameters used for the Margrabe formula in Proposition 2.3.5. Since $N(-x) = 1 - N(x)$, we have $X_t = D_t\, N(a_2) + V_t\,(1 - N(d))$. $a_2 = d - v$ where

$$v = \sqrt{(\sigma_V^2 + \sigma_D^2 - 2\rho_{VD}\sigma_V\sigma_D)(T-t)}.$$

Hence,

$$X_t = V_t - (V_t\, N(d) - D_t\, N(d - v)). \tag{4.21}$$

[4] Of course, using the formula above to price baskets and index units is not without problems, at least from a theoretical point of view, since it implies the assumption of the value of a basket following geometric Brownian motion. This assumption is inconsistent with the usual assumption that a single asset follows geometric Brownian motion. However, for practical purposes, it is no more wrong to use this formula for vulnerable index units than it is to use an extension of the Black-Scholes formula for default-free index options, which is standard practice. Cf. Hull (1997), Ch. 12. It has also been argued that geometric Brownian motion may in fact be a better approximation for baskets and indices than for single stocks. Cf. Brenner (1990).

According to expression (4.21), a vulnerable claim on the liabilities (i.e., on the debt) of a firm is composed of the value of the firm's assets and a Margrabe exchange option. The option gives the equity holders the right to release the asset collateral which was bespoken to the creditors by exchanging the liabilities for the assets, i.e., by paying off the debt holders. It is clear that $V_t - x_t(D, V)$, where $x_t(D, V)$ is the option to obtain V for the price of D, must be equal to $D_t - x_t(V, D)$. Therefore, the vulnerable debt contract can also be interpreted to consist of the default-free value plus a written exchange option. This exchange option given to the equity holders allows them to default, i.e., to avoid paying off the bond holders by obtaining the liabilities in exchange for the assets of the firm.

Of course, we can further simplify our model by setting

$$D = D_t \exp(r(T - t))$$

constant and $\sigma_D \to 0$. This means that the liabilities of the firm are deterministic. In this case, the model of Merton (1974) emerges and we see that the risky bond is equivalent either to a risk-free bond less a put option on the firm's assets struck at the face value of the bond or to the firm's assets less a call option on the assets struck at the face value of the bond.

4.4 Gaussian Interest Rates and Deterministic Liabilities

In this section we propose a generalized credit risk model that allows for stochastic interest rates. The short interest rate is assumed to be of the Gaussian type, i.e., to have a deterministic volatility function. The model presented in this and the following section first appeared in Ammann (1998) and Ammann (1999). In parallel, similar work (although limited to constant liabilities) was conducted by Klein and Inglis (1999).

The derivation of explicit option pricing formulae for standard stock and bond options under stochastic interest rates was pioneered by Merton (1973) and generalized by Amin and Jarrow (1992), Jamshidian (1989), Jamshidian (1991a), Jamshidian (1993), among others. In the following, explicit pricing formulae are derived for options with counterparty default risk within a general Gaussian interest rate framework.

In Section 2.3.3 on the Heath-Jarrow-Morton methodology we note that under the martingale measure $\mathbf{Q}$, the family of instantaneous forward rate processes can be written

$$f(t,T) = f(0,T) + \int_0^t \sigma(s,T) \cdot \sigma'(s,T)\, ds + \int_0^t \sigma(s,T) \cdot d\tilde{W}(s),$$

$\forall t$ s.t. $0 \leq t \leq T$. This is expression (2.23). Recall that $d\tilde{W}(t)$ is Brownian motion in $\mathbb{R}^d$. d denotes the number of sources of uncertainty, also called factors, that drive the evolution of the forward rates. Note that we do not impose the restriction that $d = 1$, but allow d to be an arbitrary positive integer.

In this section we require that the volatility function $\sigma(t,T)$ be deterministic. This implies that forward rates $f(t,T)$ have Gaussian law. Hence, interest rate process specifications with such a restriction on the volatility function are often called Gaussian interest rate models.

Short rates r_t are also normally distributed. Because bond prices can be written in terms of forward rates such that

$$P(t,T) = e^{-\int_t^T f(t,s)ds}.$$

bond prices have a lognormal law. Bond prices and short rates under $\mathbf{Q}$ are given in expressions (2.25) and (2.24).

4.4.1 Forward Measure

The arbitrage-free bond price evolution for an arbitrary bond with maturity $T \in [0, \mathcal{T}]$ under the risk-neutral measure $\mathbf{Q}$ is given by

$$P(t,T) = P(0,T)\exp\left(\int_0^t r(s)ds - \frac{1}{2}\int_0^t \sigma'(s,T)\,ds - \int_0^t \sigma'(s,T)\cdot d\tilde{W}(s)\right).$$

This is expression (2.25) and was derived on page 38. For notational simplicity, we sometimes write $b(t,T) = -\sigma'(t,T)$ for the bond price volatility.

In this section it is convenient to work with the forward neutral measure $\mathbf{F}$ introduced in Section 2.3.4. By Definition 2.3.2, the forward price is the bond price in terms of a numéraire bond, i.e.,

$$F(t,T,\tau) = \frac{P(t,\tau)}{P(t,T)}.$$

Note that if bond prices have a deterministic volatility function, so do forward bond prices. Proposition 2.3.9 establishes that, under a forward martingale measure defined by (2.27), the forward bond price follows a martingale. By (2.28), the SDE of $F(t,T,\tau)$ is given by

$$\frac{dF(t,T,\tau)}{F(t,T,\tau)} = -\int_T^\tau \sigma(t,u)\,du\cdot dW^{\mathbf{F}}(t).$$

For easier reference, we define

$$\beta(t,T,\tau) = -\int_T^\tau \sigma(t,u)\,du. \tag{4.22}$$

We require β to be a bounded deterministic function.

This forward measure approach is not restricted to bond price processes, but can be applied to arbitrary price processes. Because we shall later price stock options and need to model the value of the firm's assets, we have to find the stock price process under the forward neutral measure. The price process of a stock under measure **Q** is

$$S_t = S_0 \exp\left(\int_0^t r(s)\,ds - \frac{1}{2}\int_0^t \|\sigma_S(s)\|^2\,ds + \int_0^t \sigma_S(s)\cdot d\tilde{W}(s)\right).$$

The forward price of a stock is given by

$$S_F(t,T) = \frac{S_t}{P(t,T)}.$$

To find the SDE of S_F under **F** we proceed as in the proof to Proposition 2.3.9. Under **Q** the dynamics of $S_F(t,T)$ are

$$\begin{aligned}\frac{dS_F(t,T)}{S_F(t,T)} &= \left(\sigma_S(t) + \int_t^T \sigma(t,u)\,du\right)\cdot d\tilde{W}_t \\ &\quad + \int_t^T \|\sigma(t,u)\|^2\,du\,dt + \sigma_t \cdot \int_t^T \sigma(t,u)\,du\,dt.\end{aligned}$$

Changing measure from **Q** to **F** implies, by Girsanov's theorem,

$$\begin{aligned}\frac{dS_F(t,T)}{S_F(t,T)} &= \left(\sigma_S(t) + \int_t^T \sigma(t,u)\,du\right)\cdot\left(dW^{\mathbf{F}}(t) - \int_t^T \sigma(t,u)\,du\right) \\ &\quad + \int_t^T \|\sigma(t,u)\|^2\,du\,dt + \sigma_S(t)\cdot\int_t^T \sigma(t,u)\,du\,dt \\ &= \left(\sigma_S(t) + \int_t^T \sigma(t,u)\,du\right)\cdot dW^{\mathbf{F}}(t).\end{aligned}$$

As in (4.22), we denote the asset price volatility vector relative to the numéraire $P(t,T)$ by $\beta(t,T)$, i.e.,

$$\beta_S(t,T) = \sigma_S(t) + \int_t^T \sigma(t,u)\,du \tag{4.23}$$

such that

$$\frac{dS_F(t,T)}{S_F(t,T)} = \beta_S(t,T)\cdot dW_t^{\mathbf{F}}. \tag{4.24}$$

$S_F(T,T)$ can be written

$$S_F(T,T) = S_F(t,T)\exp\left(\int_t^T \beta_S(s,T)\cdot dW^{\mathbf{F}}(s) - \frac{1}{2}\int_t^T \|\beta_S(s,T)\|^2\,ds\right). \tag{4.25}$$

We also have $S_F(T,T) = S_T$ and $S_F(t,T) = S_t P(t,T)^{-1}$, by the definition of the forward price. The process V_t is defined analogously, with volatility parameter β_V under the forward measure $\mathbf{F}$.

Expressions (4.22) and (4.23) are strictly analogous, as can be seen by separating the integral in (4.22), i.e.,

$$\beta(t,T,\tau) = -\int_t^{\tau} \sigma(t,u)\, du + \int_t^{T} \sigma(t,u)\, du\,. \tag{4.26}$$

In fact, β denotes the difference between asset price and numéraire volatility. Recall that bond volatility has a negative sign because σ refers to forward rate volatility. The interpretation of (4.23) as a difference between volatilities is immediate. We therefore have the forward volatility functions, for bonds and stocks, respectively,

$$\begin{aligned} \beta(t,T,\tau) &= b(t,\tau) - b(t,T), \\ \beta_S(t,T) &= \sigma_S(t) - b(t,T), \end{aligned} \tag{4.27}$$

with $b(t,T) = -\sigma'(t,T) = \int_t^T \sigma(t,s)\, ds$.

Expressions (4.23), (4.24), and (4.25) apply analogously to the firm value process. The forward firm value is written $V_F(t,T)$.

The Brownian motion $W^{\mathbf{F}}(t)$ is a process in $\mathbb{R}^d$. Because $W^{\mathbf{F}}(t)$ must be interpreted as a d-dimensional vector of independent Brownian motions, the variances of the forward stock and firm values are given by

$$\nu_i^2(t,T) = \int_t^T \|\beta_i(s,T)\|^2\, ds, \qquad \forall i \in \{S,V\}.$$

Similarly, covariance and correlation are defined as

$$\begin{aligned} \phi(t,T) &= \int_t^T \beta_S(s,T) \cdot \beta_V(s,T)\, ds, \\ \rho(t,T) &= \int_t^T \frac{\beta_S(s,T) \cdot \beta_V(s,T)}{\|\beta_S(s,T)\|\, \|\beta_V(s,T)\|}\, ds\,. \end{aligned}$$

4.4.2 Prices of Vulnerable Stock Options

By Proposition 2.2.2, the price of a credit-risky call option for a numéraire $P(t,T)$ and the forward neutral measure $\mathbf{F}$ admits the representation

$$X_t^d = P(t,T)\mathbf{E}_{\mathbf{F}}[g(T,T)(\mathbf{1}_{\{V_F(T,T)\geq K\}} + \delta_F(T,T)\mathbf{1}_{\{V_F(T,T)<D\}})|\mathcal{F}_t],$$

where $g(T,T) = (S_F(T,T) - K)^+$. $S_F(T,T)$ and $V_F(T,T)$ are given by (4.25). Evaluating the expectation expression yields the following proposition.

Proposition 4.4.1. *The price X_t of vulnerable call option with promised payoff $X_T = (S_T - K)^+$ and actual payoff $X_T = \delta(S_T - K)^+$ with $\delta = \frac{V_T}{D}$ in case of bankruptcy is given by*

$$X_t = S_t N_2(a_1, a_2, \rho(t,T)) - P(t,T) K N_2(b_1, b_2, \rho(t,T)) + \frac{S_t V_t}{D P(t,T)} e^{\phi(t,T)} N_2(c_1, c_2, -\rho(t,T)) - \frac{V_t K}{D} N_2(d_1, d_2, -\rho(t,T)),$$

with parameters

$$\begin{aligned}
a_1 &= \frac{\ln S_t - \ln K - \ln P(t,T) + \frac{1}{2}\nu_S^2(t,T)}{\nu_S(t,T)}, \\
a_2 &= \frac{\ln V_t - \ln D - \ln P(t,T) + \phi(t,T) - \frac{1}{2}\nu_V^2(t,T)}{\nu_V(t,T)}, \\
b_1 &= \frac{\ln S_t - \ln K - \ln P(t,T) - \frac{1}{2}\nu_S^2(t,T)}{\nu_S(t,T)}, \\
b_2 &= \frac{\ln V_t - \ln D - \ln P(t,T) - \frac{1}{2}\nu_V^2(t,T)}{\nu_V(t,T)}, \\
c_1 &= \frac{\ln S_t - \ln K - \ln P(t,T) + \frac{1}{2}\nu_S^2(t,T) + \phi(t,T)}{\nu_S(t,T)}, \\
c_2 &= -\frac{\ln V_t - \ln D - \ln P(t,T) + \frac{1}{2}\nu_V^2(t,T) + \phi(t,T)}{\nu_V(t,T)}, \\
d_1 &= \frac{\ln S_t - \ln K - \ln P(t,T) + \phi(t,T) - \frac{1}{2}\nu_S^2(t,T)}{\nu_S(t,T)}, \\
d_2 &= -\frac{\ln V_t - \ln D - \ln P(t,T) + \frac{1}{2}\nu_V^2(t,T)}{\nu_V(t,T)},
\end{aligned} \tag{4.28}$$

and

$$\begin{aligned}
\nu_i^2(t,T) &= \int_t^T \|\beta_i(s,T)\|^2 \, ds, \qquad \forall i \in \{S, V\}, \\
\phi(t,T) &= \int_t^T \beta_S(s,T) \cdot \beta_V(s,T) \, ds, \\
\rho(t,T) &= \int_t^T \frac{\beta_S(s,T) \cdot \beta_V(s,T)}{\|\beta_S(s,T)\| \, \|\beta_V(s,T)\|} \, ds.
\end{aligned} \tag{4.29}$$

β_i, $\forall i \in \{S, V\}$, is a bounded and deterministic function.

Proof. Cf. Section 4.9.3.

Remark 4.4.1. The pricing formula can also be written in terms of forward prices. In this case, we have

$$\begin{aligned}X_t &= P(t,T)\big(S_F(t,T)N_2(a_1,a_2,\rho(t,T)) - KN_2(b_1,b_2,\rho(t,T))\\&+ \frac{S_F(t,T)V_F(t,T)}{D}e^{\phi(t,T)}N_2(c_1,c_2,-\rho(t,T))\\&- \frac{V_F(t,T)K}{D}N_2(d_1,d_2,-\rho(t,T))\big).\end{aligned}$$

In the parameter functions, $\ln S_F(t,T)$ replaces $\ln S_t - \ln P(t,T)$, $\forall S \in \{S,V\}$.

Proposition 4.4.2. *The price Y_t of vulnerable put option with promised payoff $Y_T = (K - S_T)^+$ and actual payoff $Y_T = \delta(K - S_T)^+$ with $\delta = \frac{V_T}{D}$ in case of bankruptcy is given by*

$$\begin{aligned}Y_t &= P(t,T)KN_2(-b_1,b_2,-\rho(t,T)) - S_tN_2(-a_1,a_2,-\rho(t,T)) -\\&+ \frac{V_tK}{D}N_2(-d_1,d_2,\rho(t,T)) - \frac{S_tV_t}{DP(t,T)}e^{\phi(t,T)}N_2(-c_1,c_2,\rho(t,T)),\end{aligned}$$

with parameters given in (4.28) *and* (4.29).

Proof. The proof is analogous to that of Proposition 4.4.1.

4.4.3 Prices of Vulnerable Bond Options

Proposition 4.4.1 immediately extends to zero-coupon bond options. We simply need to replace S_t with $P(t,\tau)$ and $\beta_S(t,T)$ with $\beta_P(t,T,\tau)$. If desired, $\ln P(t,\tau) - \ln P(t,T)$ can be replaced with $\ln F(t,T,\tau)$ in the option price formula, by the definition of the forward price (cf. remark on Proposition 4.4.1.) Note that the underlying bond $P(t,T)$ is free of credit risk.

For a one-factor model, Jamshidian (1989) showed that, by adjusting the strike prices, options on coupon bonds can be priced by computing the prices of zero-coupon bond options for each coupon of the bond. For example, a three-year European call option on a five-year bond that pays annual coupons of $c\%$ of its face value would be separated into a three-year option of a four-year zero bond with face value $c\%$ and a three-year option on a five-year zero bond with face value $100\% + c\%$. The strike prices of the options can be computed by a simple numerical procedure. For vulnerable coupon bond options, each of the coupon options is itself a vulnerable option. As they all have the same maturity, there is no bias with respect to the recovery rate obtained in case of bankruptcy for different payoff dates.

4.4.4 Special Cases

Proposition 4.4.1 is more general than Proposition 4.2.1 in that it allows for stochastic interest rates. Consequently, the special cases of Proposition 4.2.1 also extend to the Gaussian interest framework. In particular, the Black-Scholes formula for Gaussian rates emerges as a special case:

$$X_t = S_t N(d) - P(t,T) K N(d - \nu_S(t,T)),$$

with

$$d = \frac{\ln S_t - \ln K - \ln P(t,T) + \frac{1}{2}\nu_S^2(t,T)}{\nu_S(t,T)}.$$

The Merton (1974) model, but extended to Gaussian interest rates, can be derived by the same steps as in Section 4.2.2.2.

4.5 Gaussian Interest Rates and Stochastic Liabilities

In this section, we generalize the Gaussian interest rate model to stochastic counterparty liabilities. A similar generalization of the constant liability result in Proposition 4.2.1 was obtained in Proposition 4.3.1 for deterministic interest rates.

The process for the underlying security is defined in expression (4.25). The processes for the value of the assets and liabilities of the firm are assumed to follow the same type of process as S. They are given by

$$V_F(T,T) = V_F(t,T) \exp\left(\int_t^T \beta_V(s,T) \cdot dW^{\mathbf{F}}(s) - \frac{1}{2} \int_t^T \|\beta_V(s,T)\|^2 \, ds \right),$$

$$D_F(T,T) = D_F(t,T) \exp\left(\int_t^T \beta_D(s,T) \cdot dW^{\mathbf{F}}(s) - \frac{1}{2} \int_t^T \|\beta_D(s,T)\|^2 \, ds \right).$$

V and D denote the values of assets and liabilities of the firm, respectively. $\beta_V(t,T)$ and $\beta_D(t,T)$ are defined analogously to $\beta_S(t,T)$ in (4.23).

Under the forward neutral measure, the SDE of the underlying security S is given by expression (4.24). V and D have analogous SDEs. The SDE of the recovery rate δ is computed according to (4.12), giving

$$\frac{d\delta_F(t,T)}{\delta_F(t,T)} = (\beta_V(t,T) - \beta_D(t,T)) \cdot dW_t^{\mathbf{F}} + \|\beta_D\|^2 \, dt - \beta_V \cdot \beta_D \, dt.$$

$\beta_V(t,T)$, and $\beta_D(t,T)$ are defined according to (4.23). We also define $\phi_{S\delta}(t,T)$ such that

$$\begin{aligned}
\phi_{S\delta}(t,T) &= \int_t^T (\beta_V(s,T) - \beta_D(s,T)) \cdot \beta_S ds \\
&= \int_t^T \beta_S(s,T) \cdot \beta_V(s,T) - \beta_S(s,T) \cdot \beta_D(s,T) \, ds \\
&= \phi_{SV}(t,T) - \phi_{SD}(t,T).
\end{aligned}$$

Furthermore, we have

$$
\begin{aligned}
\nu_\delta^2(t,T) &= \int_t^T \|\beta_V(s,T) - \beta_D(s,T)\|^2\, ds \\
&= \int_t^T \|\beta_V(s,T)\|^2 + \|\beta_D(s,T)\|^2 - 2\beta_V(s,T)\cdot\beta_D(s,T)\, ds \\
&= \nu_V^2(t,T) + \nu_D^2(t,T) - 2\phi_{VD}(t,T).
\end{aligned}
$$

$S_F(T,T)$ and $\delta_F(T,T)$ can be written

$$
\begin{aligned}
S_F(T,T) &= S_F(t,T)\exp\left(\eta_S(t,T) - \frac{1}{2}\nu_S(t,T)\right), \\
\delta_F(T,T) &= \delta_F(t,T)\exp\left(\eta_\delta(t,T) - \frac{1}{2}\nu_\delta^2(t,T) + \nu_D^2(t,T) - \phi_{VD}(t,T)\right) \\
&= \delta_F(t,T)\exp\left(\eta_\delta(t,T) - \frac{1}{2}\nu_V^2(t,T) + \frac{1}{2}\nu_D^2(t,T)\right),
\end{aligned}
\tag{4.30}
$$

where

$$
\begin{aligned}
\eta_\delta(t,T) &= \eta_V(t,T) - \eta_D(t,T) \\
\nu_\delta(t,T) &= \nu_V(t,T) - \nu_D(t,T),
\end{aligned}
$$

and

$$
\eta_i(t,T) = \int_t^T \beta_i(s,T)\cdot dW^{\mathbf{F}}(s), \qquad \forall i \in \{V, D\}.
$$

We also have $S_F(T,T) = S_T$, $S_F(t,T) = S_t P(t,T)^{-1}$, and $\delta_F(t,T) = \delta_t$, by the definition of the forward price in (2.3.2).

4.5.1 Prices of Vulnerable Stock Options

The price of the vulnerable call option X_t^d has the representation

$$
X_t^d = P(t,T)\mathbf{E_F}[\Lambda(T)(\mathbf{1}_{\{\delta_F(T,T)\geq 1\}} + \delta_F(T,T)\mathbf{1}_{\{\delta_F(T,T)<1\}})|\mathcal{F}_t],
$$

where

$$
\begin{aligned}
\Lambda(T) &= (S_F(T) - K)^+, \\
\delta_F(T,T) &= \frac{V_F(T,T)}{D_F(T,T)}.
\end{aligned}
$$

The process definitions of $\delta_F(T,T)$ and $S_F(T,T)$ are given by (4.30). Evaluating this expectation yields the following proposition.

Proposition 4.5.1. *The price X_t of a vulnerable call option with promised payoff $X_T = (S_T - K)^+$ and actual payoff $X_T = \delta_T(S_T - K)^+$ with $\delta_T = \frac{V_T}{D_T}$ in case of bankruptcy is given by*

$$
\begin{aligned}
X_t = {} & S_t N_2(a_1, a_2, \rho) - P(t,T) K N_2(b_1, b_2, \rho) \\
& + \frac{S_t V_t}{D_t} e^{\nu_D^2(t,T) + \phi_{SV}(t,T) - \phi_{SD}(t,T) - \phi_{VD}(t,T)} N_2(c_1, c_2, -\rho) \\
& - P(t,T) K \frac{V_t}{D_t} e^{\nu_D^2(t,T) - \phi_{VD}(t,T)} N_2(d_1, d_2, -\rho),
\end{aligned}
$$

with parameters

$$
\begin{aligned}
a_1 &= \frac{\ln S_t - \ln K - \ln P(t,T) + \frac{1}{2}\nu_S^2(t,T)}{\nu_S(t,T)}, \\
a_2 &= \frac{\ln V_t - \ln D_t - \frac{1}{2}\nu_V^2(t,T) + \frac{1}{2}\nu_D^2(t,T) + \theta(t,T)}{\nu_\delta(t,T)}, \\
b_1 &= \frac{\ln S_t - \ln K - \ln P(t,T) - \frac{1}{2}\nu_S^2(t,T)}{\nu_S(t,T)}, \\
b_2 &= \frac{\ln V_t - \ln D_t - \frac{1}{2}\nu_V^2(t,T) + \frac{1}{2}\nu_D^2(t,T)}{\nu_\delta(t,T)}, \\
c_1 &= \frac{\ln S_t - \ln K - \ln P(t,T) + \frac{1}{2}\nu_S^2(t,T) + \theta(t,T)}{\nu_S(t,T)}, \\
c_2 &= -\frac{\ln V_t - \ln D_t + \frac{3}{2}\nu_D^2(t,T) + \frac{1}{2}\nu_V^2(t,T) + \theta(t,T) - 2\phi_{VD}(t,T)}{\nu_\delta(t,T)}, \\
d_1 &= \frac{\ln S_t - \ln K - \ln P(t,T) - \frac{1}{2}\nu_S^2(t,T) + \theta(t,T)}{\nu_S(t,T)}, \\
d_2 &= -\frac{\ln V_t - \ln D_t + \frac{1}{2}\nu_V^2(t,T) + \frac{3}{2}\nu_D^2(t,T) - 2\phi_{VD}(t,T)}{\nu_\delta(t,T)},
\end{aligned}
\tag{4.31}
$$

and

$$
\begin{aligned}
\theta(t,T) &= \phi_{SV}(t,T) - \phi_{SD}(t,T), \\
\nu_i^2(t,T) &= \int_t^T \|\beta_i(s,T)\|^2 \, ds, \\
\phi_{ij}(t,T) &= \int_t^T \beta_i(s,T) \cdot \beta_j(s,T) \, ds, \\
\rho_{ij}(t,T) &= \int_t^T \frac{\beta_i(s,T) \cdot \beta_j(s,T)}{\|\beta_i(s,T)\| \, \|\beta_j(s,T)\|} \, ds, \\
\nu_\delta^2(t,T) &= \int_t^T \|\beta_V(s,T) - \beta_D(s,T)\|^2 \, ds,
\end{aligned}
\tag{4.32}
$$

$\forall i \in \{S, V, D\}$ with β_i a bounded deterministic function. N_2 denotes the bivariate distribution function with correlation coefficient $\rho = \rho_{SV}(t,T) - \rho_{SD}(t,T)$.

Proof. Cf. Section 4.9.4.

Proposition 4.5.2. *The price Y_t of vulnerable put option with promised payoff $Y_T = (K - S_T)^+$ and actual payoff $Y_T = \delta_T (K - S_T)^+$ with $\delta_T = \frac{V_T}{D}$ in case of bankruptcy is given by*

$$\begin{aligned} Y_t = & P(t,T) K N_2\big(-b_1, b_2, -\rho\big) - S_t N_2\big(-a_1, a_2, -\rho\big) \\ & + P(t,T) K \frac{V_t}{D_t} e^{\nu_D^2(t,T) - \phi_{VD}(t,T)} N_2\big(-d_1, d_2, \rho\big) \\ & - S_t \frac{V_t}{D_t} e^{\nu_D^2(t,T) + \phi_{SV}(t,T) - \phi_{SD}(t,T) - \phi_{VD}(t,T)} N_2\big(-c_1, c_2, \rho\big), \end{aligned}$$

with parameters given in (4.31) *and* (4.32).

Proof. The proof is analogous to that of Proposition 4.5.1.

4.5.2 Prices of Vulnerable Bond Options

Proposition 4.5.1 immediately extends to bond options. We simply need to replace S_t with $P(t,\tau)$ and $\beta_S(t,T)$ with $\beta_P(t,T,\tau)$. We also might replace $\ln P(t,\tau) - \ln P(t,T)$ with $\ln F(t,T,\tau)$ in the option price formula, by the definition of the forward price (cf. remark on Proposition 4.4.1).

4.5.3 Special Cases

It is evident that Propositions 4.2.1, 4.3.1, and 4.4.1 are all special cases of Proposition 4.5.1. Consequently, their special cases are also special cases of Proposition 4.5.1. For debt claims, in particular, a generalization of Proposition 2.3.5 (the Margrabe formula) emerges that allows for Gaussian interest rates.

4.6 Vulnerable Forward Contracts

Forward contracts, as typical OTC instruments, are also subject to counterparty default risk. The price of the counterparty risk in a forward contract is, however, more difficult to determine than for options because both counterparties can default. In the case of options, only one party of the contract faces counterparty risk. The option writer never bears any credit risk because the buyer has fulfilled all contractual obligations by paying the price of the option. After this point, the option holder has only rights, whereas the option writer has only contingent liabilities. This is different in the case of forward contracts. As symmetrical derivative instruments, forward contracts usually cost nothing at inception and represent a contingent liability for both parties of the contract.

It is well known that a forward contract can be represented by two options: a call option and a short put option. The party that is long in the contract is long in the call option and short in the put option and vice versa for the party that is short. The two options must have the same strike price. At inception of the contract, the strike price is usually chosen such that both options are of equal value. If the options do not have the same value at the outset, the forward contract would require a payment from one party.

The fact that it is possible to represent a forward contract in terms of options allows us to compute an adjusted forward price that accounts for the two-sided credit risk involved. The party that is long in the forward is exposed to the counterparty risk of the call while the party that is short is exposed to the counterparty risk of the put. Consider the case where party A is long and party B is short in a forward contract that expires at time T on a bond $P(t,U)$. Denote a call option vulnerable to default by firm B by $c_t(V_t^B, X)$ and a put option vulnerable to default by firm A by $p_t(V_t^A, X)$. The price of the contract is given by

$$f(t,T,P(t,U)) = c_t(V_t^B, X) - p_t(V_t^A, X).$$

The strike price X is usually chosen such that $f(t,T,P(t,U)) = 0$. In this case X is called *forward price*.

4.7 Numerical Examples

In this section some numerical examples of the models proposed above are presented. First, examples are given for vulnerable stock options assuming deterministic interest rates. These examples correspond to Sections 4.2 and 4.3. Second, examples for vulnerable stock and bond options are provided assuming stochastic interest rates. These examples correspond to Sections 4.4 and 4.5.

4.7.1 Deterministic Interest Rates

Table 4.1 shows the effect of credit risk on Black-Scholes option prices. The numbers in the table are percentage values of price reductions relative to the Black-Scholes price. The prices displayed refer to a two-year vulnerable option with an annual volatility of 25%. The price reductions in the table are consistent with a counterparty credit spread of 1.76% over the riskless rate of 4.88%. Rates are continuously compounded. Such a credit spread results from a bond price $P^d(0,2) = 87.56$ compared to the price of riskless bonds of $P(0,2) = 90.70$. The price reductions in the table are consistent with this credit spread in the sense that the parameters of the credit risk model, $K/V = 0.8$, $\sigma_V = 0.15$, $r = \ln(1.05)$, imply such a bond credit spread. The correlation between underlying and firm value has no effect on the bond credit spread. The price reduction on the risky bond price is 3.46%. As can be seen in the table, this is equal to the price reduction on the vulnerable options if the correlation between the underlying security of the option and the counterparties firm assets is zero. This observation will later be confirmed by Proposition 5.1.1.

Positive correlation reduces the price reduction while negative correlation increases it for call options. The opposite applies to put options. It can also be seen that the strike price affects the extent of the price reduction for correlations different from zero. Out-of-the-money options are generally more sensitive to correlation effects than in-the-money options. The further out-of-the-money the option is, the more the underlying has to change for the option to expire in-the-money. Consider the case of a call option. Assuming a high positive correlation, the more the price of the underlying rises, the more the firm value tends to rise. In other words, an out-of-the-money call option tends to expire in-the-money only if the firm value has also risen, thus making default less likely if the option is in-the-money. On the other hand, if the firm value drops, the underlying tends to drop as well if correlation is positive. A decreasing value of the underlying results in the option expiring worthless for an option that is initially out-of-the-money. An out-of-the-money and thus worthless option at expiry is not subject to credit risk. If the option is deeply in-the-money already at the outset, high positive correlation may very well result in both firm value and underlying decreasing in value. However, since the option was in-the-money at the outset, it may still be in the money after the drop in price and firm value and thus still be subject to credit risk while credit risk has increased. A analogous line of reasoning applies to put options and to negative correlation.

Table 4.2 shows another correlation effect. In this table the correlation between underlying and firm value is assumed to be zero, but the correlation between the firm's assets and its liabilities is varied. It can be seen that a higher correlation between assets and liabilities reduces the risk of the asset value falling below the value of the liabilities and therefore demands a smaller price reduction. Note that while in Table 4.1 all price reductions were consistent with a credit spread of 1.75% a year, the price reductions displayed in this table imply different credit spreads. The correlation between assets and underlying has no effect upon the prices of the firm's zero-coupon bonds. The correlation between the firm's assets and its liabilities, however, does affect the prices of the firm's outstanding bonds. For example, a correlation of 0.5 implies a yield spread of only 0.74% and a bond price of $P^d(0,2) = 89.38$, which is 1.46% less than the price of the riskless bond. The correlation between assets and liabilities affects all option prices equally, regardless if they are call or put options, in-the-money or out-of-the-money.

Table 4.3 shows yet another correlation effect. In this case the default correlations are $\sigma_{SV} = 0.5$ $\sigma_{VD} = 0.5$, and $\sigma_{SD} = 0.5$. Otherwise, the parameters are unchanged from Tables 4.1 and 4.2. These correlation values give the price reductions displayed in the first column. The price reduction is 1.46% for call and put options and for any strike. Coincidentally, this price

Table 4.1. Price reductions for different ρ_{SV}

in %		ρ_{SV}				
		-1	-0.5	0	0.5	1
X/S=1.1025	Call	12.37	7.13	3.46	1.17	0.13
	Put	0.29	1.53	3.46	5.96	8.94
X/S=0.9	Call	9.76	6.20	3.46	1.50	0.31
	Put	0.10	1.14	3.46	7.07	12.15
X/S=1.25	Call	14.59	7.86	3.46	0.98	0.06
	Put	0.49	1.78	3.46	5.42	7.56

Percentage price reductions of vulnerable options with respect to the corresponding Black-Scholes option price. Parameters: $K/V = 0.8$, $\sigma_S = 0.25$, $\sigma_V = 0.15$, $\sigma_D = 0.15$, $\rho_{VD} = 0$, $\rho_{SD} = 0$, $r = \ln 1.05$, $t = 2$. X strike, S price of underlying.

Table 4.2. Price reductions for different ρ_{VD}

in %		ρ_{VD}				
		-1	-0.5	0	0.5	1
X/S=1.1025	Call	6.56	5.14	3.46	1.46	0.00
	Put	6.56	5.14	3.46	1.46	0.00
X/S=0.9	Call	6.56	5.14	3.46	1.46	0.00
	Put	6.56	5.14	3.46	1.46	0.00

Percentage price reductions of vulnerable options with respect to the corresponding Black-Scholes option price. Parameters: $K/V = 0.8$, $\sigma_S = 0.25$, $\sigma_V = 0.15$, $\sigma_D = 0.15$, $\rho_{SV} = 0$, $\rho_{SD} = 0$, $r = \ln 1.05$, $t = 2$. X strike, S price of underlying.

Table 4.3. Price reductions for various correlations

in %		ρ	ρ_{SD}	ρ_{SD}	ρ_{SV}	ρ_{SV}
		0.5	0.8	1	0.8	1
X/S=1.1025	Call	1.46	2.97	4.33	0.53	0.19
	Put	1.46	0.68	0.31	2.47	3.26
X/S=0.9	Call	1.46	2.58	3.52	0.66	0.31
	Put	1.46	0.52	0.17	2.94	4.24

Percentage price reductions of vulnerable options with respect to the corresponding Black-Scholes option price. The first column shows reductions for the default parameters. In the following columns, one correlation parameter is modified. Default parameters: $K/V = 0.8$, $\sigma_S = 0.25$, $\sigma_V = 0.15$, $\sigma_D = 0.15$, $\rho_{SV} = 0.5$, $\rho_{SD} = 0.5$, $\rho_{VD} = 0.5$, $r = \ln 1.05$, $t = 2$. X strike, S price of underlying.

reduction exactly corresponds to the price reduction showed in Table 4.2 for a correlation between firm value and liabilities of 0.5. The reason for this equal number is the fact that volatilities of firm assets and liabilities have been assumed to be equal at 0.15 in this example. In the other columns, either correlation ρ_{SV} or ρ_{SD} is increased to 0.8 or 1. The reduction numbers confirm the effect of ρ_{SV} from Table 4.1 and demonstrate the opposite effect of ρ_{SD}. It can be seen that a higher correlation between liabilities and underlying increases the effect of credit risk on call options and decreases it for put options.

Given these results, it may be overly simplistic to work with a model that only considers correlation between firm asset value and underlying but neglects liabilities. If the effect of credit risk on option prices is measured by the simple model that does not take stochastic liabilities into account, it is conceivable that important credit risk effects are missed. Moreover, compare the reductions for $\rho_{SV} = 1$ to the corresponding values in Table 4.1. While the out-of-the money call shows a reduction of 0.31% is both cases, the reduction of the at-the-money call is 0.19% compared to 0.13%. This is an extreme case with a correspondingly small price reduction. However, the striking observation is not the absolute amount, but the fact that we have a higher reduction in Table 4.3 for a bond credit spread of 1.46% than in Table 4.1 for a bond credit spread of 3.46%. The same observation is made when comparing the reduction for $\rho_{VS} = 0.5$, which is 1.46% in Table 4.3 and only 1.17% in Table 4.1. The difference is not so pronounced for the out-of-the-money call option.

For put options, we find a similar effect. Recall that, unlike call options, a high correlation between firm value and underlying implies high reductions for put options. While the bond price spread is a little more than twice as high in Table 4.1 than in Table 4.3, price reductions for put options can be as much as four times higher in Table 4.1, e.g., 5.96% vs. 1.46%. This example demonstrates that it is the difference in the correlation between assets and underlying and the correlation between liabilities and underlying which determines the correlation effect.

4.7.2 Stochastic Interest Rates

To compute prices for vulnerable options in a stochastic interest rate framework, it is necessary to specify a Gaussian interest rate model. In principle, any Gaussian model could be selected such as the models proposed by Vasicek (1977), Ho and Lee (1986), Hull and White (1990), Hull and White (1993b), or Heath, Jarrow, and Morton (1992).

Without loss of generality, we specify a time-homogeneous Vasicek short rate model for the numerical examples in this section. This corresponds to the model originally proposed by Vasicek (1977). It has subsequently been extended to the time-inhomogeneous case such that an exact fit to the initial term structure is made possible. Moreover, it can be extended to several sources of uncertainty. Although the pricing formulae presented above do not require a time-homogeneous model nor are they restricted to one factor, we choose, for simplicity, the original one-factor Vasicek model for the subsequent examples.

The short rate model proposed by Vasicek (1977) assumes that the short rate satisfies the SDE

$$dr = a(b - r)dt + \sigma d\tilde{W}(t), \tag{4.33}$$

where $\tilde{W}(t)$ is a one-dimensional Brownian motion under measure $\mathbf{Q}$ and a, b are constants. Such a process is called an Ornstein-Uhlenbeck process and its distinctive property is mean-reversion, meaning that the process always tends to the long-run mean of b. Parameter a governs the speed of mean-reversion and is called the rate of mean-reversion. The solution to SDE (4.33) can be shown to be

$$r(t) = b + \exp(-at)(r(0) - b) + \sigma \int_0^t \exp(-a(t-s))\, d\tilde{W}(s)$$

The process $r(t)$ has Gaussian law with mean $b + \exp(-at)(r(0) - b)$ and variance $\sigma^2(1 - \exp(-2at))/2a$.[5] As $t \to \infty$, the distribution converges. In the limit, it is Gaussian with mean b and $\sigma^2/2a$.

Vasicek showed that the behavior of the short rates as defined in (4.33) implies bond prices given by

$$P(t,T) = A(t,T)e^{-B(t,T)r(t)},$$

with

$$A(t,T) = \exp\left(\frac{(B(t,T) - (T-t))(a^2 b - \sigma^2/2)}{a^2} - \frac{\sigma^2 B^2(t,T)}{4a}\right),$$
$$B(t,T) = a^{-1}(1 - e^{-a(T-t)}).$$

It can be seen that the term structure is fully determined by the values of $r(t)$ and parameters a, b, and σ. The SDE for the bond price can be shown to be

$$dP(t,T) = P(t,T)(r(t)dt + b(t,T)d\tilde{W}(t)), \tag{4.34}$$

with bond volatility function $b(t,T) = \sigma B(t,T)$.[6]

[5] Cf. Karatzas and Shreve (1991), Section 5.6, for a more detailed description of linear SDEs, including solutions and moments.

[6] Compare the SDE (4.34) with (2.26) on page 38. These expressions are equal if $\sigma'(t,T) = b(t,T)$. Hence, the Vasicek short rate model is easily expressed in terms of Heath-Jarrow-Morton forward rates. By computing $\sigma(t,T) = \frac{\partial b(t,T)}{\partial T}$, we obtain the well-known Vasicek forward rate volatility function $\sigma(t,T) = \sigma \exp(-a(T-t))$.

For the numerical examples below, the following term structure parameters have been chosen: $a = 0.1$, $b = \ln(1.08)$, $\sigma = 0.02$, $r = \ln(1.05)$. Figure 4.1 illustrates the volatility function implied by those parameters. For comparison, a volatility function with no mean-reversion (i.e., a=0) is also given in the figure.

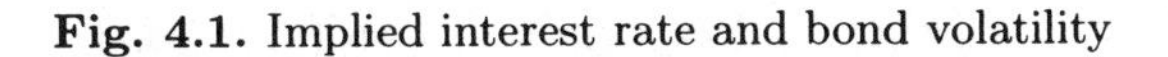

Fig. 4.1. Implied interest rate and bond volatility

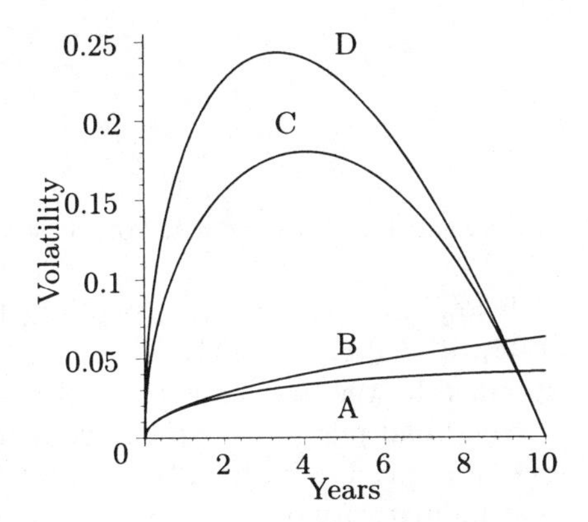

Implied interest rate and bond volatility for parameters: $r(0) = \ln(1.05)$, $a = 0.1$, $b = \ln(1.08)$, $\sigma = 0.02$. A: Interest rate volatility for indicated parameters, B: Volatility function of A, but without mean reversion, C: Volatility of a 10-year zero-coupon bond for varying holding time horizons for volatility function A, D: Volatility of 10-year bond for volatility function B.

For the stock price we have the usual SDE

$$dS = S(r(t)dt + \sigma_S d\tilde{W}_S(t)).$$

$\tilde{W}_S(t)$ is a one-dimensional Brownian motion under $\mathbf{Q}$. σ from (4.33) and σ_S are interpreted as norms such that $\langle \tilde{W}(t), \tilde{W}_S(t) \rangle = \rho t$. To price a vulnerable bond option, however, we need the forward volatility functions of bond and stock. The forward volatility function of a bond is given in (4.27) and is $\beta(t,T) = b(t,\tau) - b(t,T)$. Hence,

$$\nu^2(t,T,\tau) = \int_t^T \|b(s,\tau) - b(s,T)\|^2 \, ds.$$

For the Vasicek bond volatility function $b(t,T) = \sigma B(t,T)$, this gives

$$\nu^2(t,T,\tau) = \int_t^T (b(s,\tau) - b(s,T))^2\, ds$$
$$= \frac{\sigma^2}{2a^3}(1 - e^{-2a(T-t)})(1 - e^{-a(\tau-T)}).$$

To obtain $\nu_S^2(t,T)$, we integrate $\|\sigma_S(t) - b(t,T)\|^2$ and obtain

$$\nu_S^2(t,T) = \int_t^T \left(\sigma_S^2(s) - 2\rho\sigma_S(s)b(s,T) + b^2(s,T)\right) ds$$
$$= \sigma_S^2(T-t) - \frac{\sigma^2}{2a}B^2(t,T) - \frac{\sigma^2}{a^2}B(t,T) + \frac{2\rho\sigma\sigma_S}{a}B(t,T)$$
$$+ \frac{\sigma^2(T-t)}{a^2} - \frac{2\rho\sigma\sigma_S(T-t)}{a}.$$

ρ denotes the correlation between bond and stock. For ν_V we have a similar expression.

The prices computed below are based on the following term structure information. $r(0) = \ln(1.05)$, $a = 0.1$, $b = \ln(1.08)$, $\sigma = 0.02$. The credit-risky term structure is specified by the parameters $D/V(0) = 0.9$ and $\sigma_V = 0.15$. The correlation between bond prices and firm value is irrelevant when determining the term structure since the value of the bond if default has not occurred is deterministic at its maturity date.

Figure 4.2 displays the prices of risk-free and risky bonds with respect to time to maturity. Figure 4.3 computes the corresponding zero-coupon bond yield term structures. The default-free term structure is upward-sloping with a mean reversion level of ln(1.08). The risky term structures are based on an asset volatility of 15% and forward liability values of 70% and 90% of asset value, respectively. Compared to the firm with a liability level of 70%, the firm with 90% is more likely to default and therefore commands a higher spread.

The term structure for the firm of lesser credit quality is hump-shaped. It shows sharply increased yields for increasing maturities at the short-term end of the figure and decreasing yields for maturities in excess of one year. The yield seems to level off around 6%, which is, for a 10-year time horizon an approximate yield spread of 100 basis points. The term structure of the high-grade firm is upward-sloping. It has been observed in practice that lower-grade firms often exhibit hump-shaped term structures while the credit yield spread monotonically increases for higher-grade firms. Firms close to bankruptcy often even exhibit a downward-sloping term structure.[7]

[7] This is consistent with empirical findings by Johnson (1967), Rodriguez (1988), and Sarig and Warga (1989). Moreover, Fons (1994) shows that such spreads are consistent with default experience during 1970-1993. He observes that holders of investment grade debt initially face low default risk. However, credit quality may deteriorate over time. Speculative-grade issuers are subject to high initial default risk that tends to decrease if the issuer survives the first few years, having by then overcome the problems which caused the low initial rating.

Fig. 4.2. Implied bond prices

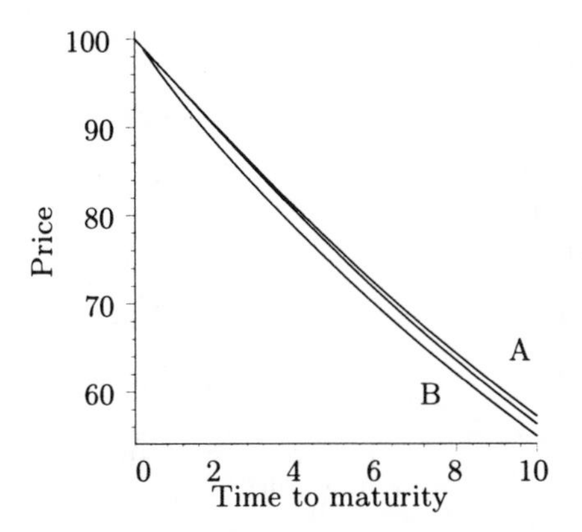

Term structures with and without credit risk. Parameters: $r(0) = \ln(1.05)$, $a = 0.1$, $b = \ln(1.08)$, $\sigma = 0.02$, $\sigma_V = 0.15$. A: riskless. B: $D_1/V = 0.9$. Not annotated: $D_2/V = 0.7$.

Fig. 4.3. Implied term structures

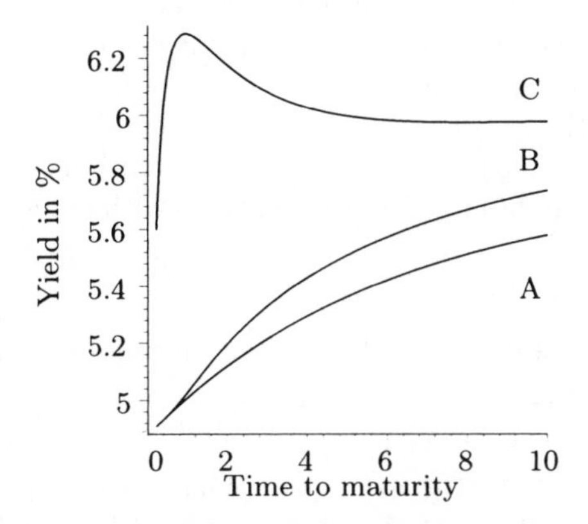

Two credit-risky term structures compared to the riskless term structure. Parameters: $r(0) = \ln(1.05)$, $a = 0.1$, $b = \ln(1.08)$, $\sigma = 0.02$, $\sigma_V = 0.15$. A: riskless. B: $D_2/V = 0.7$. C: $D_1/V = 0.9$.

Fig. 4.4. Implied yield spreads

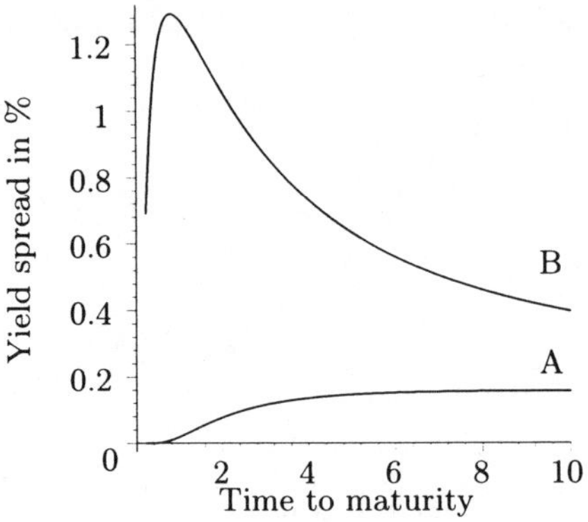

Yield credit spreads for the term structures of Figure 4.3. Parameters: $r(0) = \ln(1.05)$, $a = 0.1$, $b = \ln(1.08)$, $\sigma = 0.02$, $\sigma_V = 0.15$. A: $D_2/V = 0.7$. B: $D_1/V = 0.9$.

Figure 4.5 shows the effect of the firm's liabilities and their volatility on price reductions of bonds and options depending on the time horizon of the claim. The solid lines denote the volatility $\sigma_V = 0.15$, the dashed lines $\sigma_V = 0.20$. The liability levels are given by $D_1/V = 0.9$ and $D_2/V = 0.7$.

Table 4.4 shows the effect of credit risk on bond options. The values depicted are for two-year options on a five-year zero bond. The correlation effects are similar to those found for stock options with deterministic interest rates.

Table 4.4. Price reductions for different ρ_{VP}

in %		ρ_{VP}				
		- 0.9	-0.5	0	0.5	0.9
K=84.73	Call	10.91	5.96	2.09	0.30	0.00
	Put	0.00	0.52	2.09	4.18	5.86
K=80.00	Call	7.05	4.50	2.09	0.57	0.03
	Put	0.00	0.22	2.09	6.10	11.03
K=90.00	Call	19.10	8.62	2.09	0.12	0.00
	Put	0.15	0.97	2.09	3.06	3.55

Percentage price reductions of vulnerable zero bond options with respect to the corresponding default-free option price. Parameters: $\sigma_V = 0.15$, $r = \ln 1.05$, $t = 2$, $D/V = 0.9$. K strike.

Fig. 4.5. Price reductions for different maturities

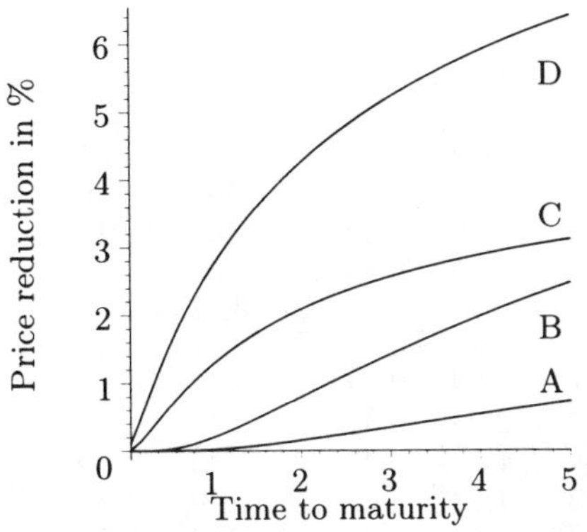

Option price reductions over time with varying liabilities and volatility of liabilities. Parameters: $r(0) = \ln(1.05)$, $a = 0.1$, $b = \ln(1.08)$. A: $\sigma_1^V = 0.15$, $D_2/V = 0.7$. B: $\sigma_2^V = 0.2$, $D_2/V = 0.7$. C: $\sigma_1^V = 0.15$, $D_1/V = 0.9$. D: $\sigma_2^V = 0.2$, $D_1/V = 0.9$.

Table 4.5 displays the price reductions of a vulnerable stock option within the stochastic interest framework presented above. Reductions for three different correlations are shown: correlations between bond and stock prices, ρ_{SP}, bond and firm asset values, ρ_{VP}, and stock and firm asset values, ρ_{SV}. In this example the values are chosen such that the resulting 3×3 matrix is a valid covariance matrix, i.e., it is positive definite. In the rightmost column, the bond price spreads corresponding to the parameter values are displayed. It can be seen that the correlation between bonds and firm value affects the credit spread. In fact, high correlation between bond prices and firm value can reduce the credit spread by a substantial amount, as is evident from Table 4.5. When calibrating a credit risk model to market data, it therefore seems essential to take into consideration the correlation between firm asset and bond prices. If this correlation is neglected in the calibration process, estimated model parameters will be unlikely to reflect the real credit risk situation.

It can also be observed that the effect of the correlation between bonds and underlying is minimal. This may seem surprising given the influence of correlation between bonds and firm assets which was noted in the previous paragraph. Recall, however, that the values showed in the table are relative price reductions due to credit risk effects. Correlation between bonds and underlying assets will likely have a noticeable effect on option prices, but not on credit-risk induced reductions thereof. This is different from correlation between firm value and bonds since firm value immediately affects the credit standing of the firm.

Table 4.5. Price reductions for different ρ_{VP}, ρ_{SV}, ρ_{SP}

in %		Calls			Puts			Bond
ρ_{VP}	ρ_{SP}	−0.5	0.0	0.5	−0.5	0.0	0.5	
	−0.5	6.65	2.45	0.39	0.63	2.45	5.12	
−0.5	0.0	6.62	2.45	0.40	0.62	2.45	5.14	2.45
	0.5	6.58	2.45	0.40	0.62	2.45	5.17	
	−0.5	5.81	2.09	0.32	0.51	2.09	4.44	
0.0	0.0	5.78	2.09	0.32	0.51	2.09	4.46	2.09
	0.5	5.75	2.09	0.32	0.50	2.09	4.48	
	−0.5	4.94	1.73	0.25	0.40	1.73	3.74	
0.5	0.0	4.92	1.73	0.25	0.40	1.73	3.76	1.73
	0.5	4.89	1.73	0.25	0.39	1.73	3.77	

Percentage price reductions of vulnerable stock options relative to the corresponding default-free option price. Parameters: $\sigma_V = 0.15$, $r = \ln 1.05$, $t = 2$, $K/S = 1.1025$.

Going back to Table 4.4, we find that we can identify a similar bond-firm asset correlation effect. We noted before that the differences in price reductions were similar to the reductions encountered in the case of deterministic interest rates. Additionally, it can also be seen that negative correlations between bonds and firm asset value affect call options more than positive correlations affect put options. For example, the reduction for call options can be as high as 19.10% in the example while put options are not subject to reductions higher than 11.03%. Of course, the strike price does influence the extent of the reduction, as we noted already in the section on deterministic interest rates. There is, however, an additional factor involved in this case. Namely, the correlation between assets has the usual influence, which was observed to be large before, but also incorporates the bond-firm asset correlation effect observed in Table 4.5. The two effects are therefore combined. While the latter impact affects call and put options alike, the former has the opposite effect on call and put options. Hence, while the credit risk effect is mitigated for put options, it is exacerbated for call options.

4.7.3 Forward Contracts

Table 4.6 shows deviations from the default-free forward price of a bond if credit risk is taken into account. It is assumed that party A is long in the forward contract, party B is short. This means that party A is exposed to the credit risk of party B with a long call option and party B is exposed to party A credit risk through a long put option. A and B have equal credit characteristics with $D_A/V_A = D_B/V_B = 0.9$ and $\sigma_{V_A} = \sigma_{V_B} = 0.15$. Since the parties have equal credit parameter, their bonds would have the same price. In this situation the credit-adjusted forward price is equal to the default-free forward price if correlation between interest rates and both A's and B's assets is zero. In this case both parties are exposed to an equal amount of credit risk from each other and therefore there is no credit effect in the forward price.

If the firm's assets are correlated with interest rates, however, the forward price will deviate from the risk-free forward price. The term structure parameters are set to $\sigma = 0.02$, $a = 0.01$, $b = \ln 1.08$ and the forward contract considered matures in 2 years, the underlying bond in 5 years from present time. Note that credit and term structure parameters are unchanged from previous numerical examples in Table 4.4 and Figures 4.3 and 4.4.

It can be seen that the deviation from the riskless forward price is highest for high positive or negative correlations and if both firms exhibit similar correlations. On the other hand, opposite correlations tend to offset each other. Consider, for example, the realistic case where two firms the assets of which are both highly positively correlated with bond prices enter into a bond forward contract. Such a setting results in the credit-adjusted forward price being farthest from its no-default value because correlation affects credit risk in opposite directions for call and put options. While positive correlation mitigates the credit risk effect on call options, it exacerbates it for put options. Since a forward contract includes both a call and a put option, the credit risk effect is increased for correlations with equal sign.

Correlations between interest rates and firm values also affect the credit spread of the bond, as shown in Table 4.4. It may therefore be argued that the deviations from the risk-free forward price in Table 4.6 are biased because different correlations imply different credit spreads for the counterparty's debt. Although it is true that not all deviations in Table 4.6 are based on the same implied credit spread for the counterparties' debt, this effect does not invalidate the main observation that deviations are highest for high correlations with equal sign. It has to be taken into account that, in this example, both firms were assumed to have equal credit parameters. This means that if they exhibit the same correlation with respect to interest rates, they will also have the same credit spread. The same credit spread implies that credit risk is symmetrical and that the two credit risk effects exactly offset each other. However, we observe that this is not the case. Quite to the contrary, this is the situation where the forward price adjusted for credit risk deviates most from the risk-free forward price.

Table 4.7 shows absolute deviations (in basis points) from the default-free forward price for stock options with a constant interest rate of $r = \ln 1.05$. In general, we make similar observations as in Table 4.6. In the stock forward case, however, we find that deviations increase more with high equal correlation. Consider the case where the firm value of both parties exhibits a correlation of 0.9 with the underlying asset. While, in the bond option case, an increase of correlation from 0.5 to 0.9 only increased the deviation

Table 4.6. Deviations from forward bond price

84.73			Call	$\rho_{V_B P}$		
		-0.9	-0.5	0.0	0.5	0.9
	-0.9	-25.2				
	-0.5		-12.5	-4.4	0	
Put	0.0		-7.2	0.0	3.8	
$\rho_{V_A P}$	0.5		0.0	5.6	8.8	
	0.9					13.4

Absolute deviations from the unadjusted forward price of 84.727 in basis points. Parameters: $D/V = 0.9$, $\sigma_V = 0.15$, $r = \ln 1.05$, $\sigma_r = 0.02$, $a = 0.01$, $b = \ln(1.08)$, $t = 2$, $T = 5$.

Table 4.7. Deviations from forward stock price

44.10			Call	$\rho_{V_B S}$		
		-0.9	-0.5	0.0	0.5	0.9
	-0.9	-67				
	-0.5		-34	-10	1	
Put	0.0		-23	0	11	
$\rho_{V_A S}$	0.5		-8	15	26	
	0.9					43

Absolute deviations from the unadjusted forward price of 44.10 in basis points. Parameters: $\sigma_V = 0.15$, $r = \ln 1.05$, $t = 2$, $D/V = 0.9$.

from 8.8 to 13.2 basis points, we find that the same increase in correlation increases the deviation for stocks from 26 to 43 basis points. Although those numbers are not directly comparable because they are absolute values and are based on different parametric assumptions, it is apparent that deviations do not increase quite as rapidly for bond options as for stock options. The opposite effect holds for high negative correlations. In this case we find that bond options are affected more than stock options. For example, while the absolute deviation in the case of a correlation of -0.9 is more than twice the value for a correlation of 0.9 for bond options, it is only about one- half more for stock options, i.e. 67 as opposed to 43 basis points. This effect occurs because correlation affects the credit spread of counterparty debt in the bond option case but not in the stock option case. The correlation with the underlying stock has no effect on the overall credit risk of the firm while the correlation with interest rates does have an effect. High correlation with bond prices implies a smaller credit spread, as Table 4.4 has shown. Hence, the increase in deviation for increased correlation is somewhat mitigated by decreasing overall credit risk of the firm. Similarly, the effect is somewhat increased for high negative correlations.

4.8 Summary

In this chapter we proposed a firm value model to determine the values of derivative securities which are subject to counterparty default risk. We derived explicit pricing formula for vulnerable call and put options under various assumptions. The base version of the model used the Black-Scholes assumptions. In particular, the firm value and the underlying of the derivative contract were assumed to follow geometric Brownian motion. The counterparties' liabilities were assumed to be deterministic. The base model was then generalized to allow stochastic liabilities of the counterparty and stochastic interest rates with a deterministic volatility function. Closed-form solutions for option prices were derived for

- Deterministic interest rates and deterministic liabilities
- Deterministic interest rates and stochastic liabilities
- Gaussian interest rates and deterministic liabilities
- Gaussian interest rates and stochastic liabilities

We showed that a number of other models can be expressed as special cases of our model. In particular, the option pricing formulae of Black and Scholes (1973), Merton (1973), Margrabe (1978), and Jamshidian (1989) for various types of options can all be derived as special cases. Moreover, the credit risk model by Merton (1974) and a similar credit risk model where credit risk is priced by the Margrabe exchange option pricing model are also special cases of our model.

We also showed that forward contracts can be analyzed easily within our model. Forward contracts are usually subject to bilateral credit risk because a forward contract is a symmetrical instrument and can become a liability to either firm party to the contract. We showed that the price of a forward contract can be modeled in terms of call and put options, in a similar fashion as the default-free case.

Numerical examples were provided for a number of different options and forward contracts under constant and Vasicek (1977) interest rates. The main insight gained was that correlation can greatly influence the amount of credit risk inherent in a derivative security. For example, an OTC stock call option contains little credit risk if the stock process exhibits high positive correlation with the asset value process of the counterparty. On the other hand, the relative price reduction caused by credit risk can be a multiple of the price reduction on the counterparty's bond issues if this correlation is negative. These findings are reversed for put options.

A forward contract can be constructed from a call option and a put option with equal strike price. We showed that the forward price adjusted by credit risk effects can deviate substantially from the theoretical risk-free value. Whether credit risk has an increasing or decreasing effect on the forward price depends on the credit quality of the counterparties. However, even if credit risk is symmetrical, i.e., if both parties are of equal credit quality, the forward price is likely to deviate from the risk-free price. This deviation is particularly pronounced if the correlation between firm value and underlying security is similar for both parties.

Interest rate risk can also have noticeable effects on the prices of credit-risky derivatives. Our results reveal that positive correlation between interest rates and firm value increases the credit spread if the other credit risk parameters are held constant. This result is consistent with the theoretical and empirical findings of Longstaff and Schwartz (1995a). Consequently, the effect of credit risk on bond option prices and stock option prices with stochastic interest rates is more pronounced for positive correlations between interest rates and firm value.

In sum, this chapter provided a simple credit risk model based on firm value. As one of its main advantages, the model facilitates closed-form pricing solutions for options and forward contracts on various security classes. Moreover, the model captures correlation and interest-rate risk effects on the prices of credit-risky securities.

4.9 Proofs of Propositions

This section contains the proofs of the propositions in this chapter.

4.9.1 Proof of Proposition 4.2.1

Proposition 4.2.1 gives a closed-form pricing formula for the price of stock or bond options subject to counterparty default risk. Credit risk is modeled with a firm-value approach. Interest rates and the liabilities of the counterparty are assumed to be deterministic.

Proof. The price of a vulnerable claim can be expressed as

$$X_t = B_t \, \mathbf{E_Q}[B_T^{-1}(S_T - K)^+(\mathbf{1}_{\{V_T \geq D\}} + \delta_T \, \mathbf{1}_{\{V_T < D\}})|\mathcal{F}_t],$$

where $\delta_T = \frac{V_T}{D}$. $\mathbf{Q}$ denotes the risk-neutral measure, B the money market account. The expectation can be separated into four terms such that

$$X_t = E_1 - E_2 + E_3 - E_4,$$

where

$$\begin{aligned} E_1 &= B_t \, \mathbf{E_Q}[B_T^{-1} \, S_T \, \mathbf{1}_{\{S_T > K\}} \mathbf{1}_{\{V_T \geq D\}}|\mathcal{F}_t], \\ E_2 &= B_t \, \mathbf{E_Q}[B_T^{-1} \, K \, \mathbf{1}_{\{S_T > K\}} \mathbf{1}_{\{V_T \geq D\}}|\mathcal{F}_t], \\ E_3 &= B_t \, \mathbf{E_Q}[B_T^{-1} \, S_T \delta_T \, \mathbf{1}_{\{S_T > K\}} \mathbf{1}_{\{V_T < D\}}|\mathcal{F}_t], \\ E_4 &= B_t \, \mathbf{E_Q}[B_T^{-1} \, K \delta_T \, \mathbf{1}_{\{S_T > K\}} \mathbf{1}_{\{V_T < D\}}|\mathcal{F}_t]. \end{aligned} \tag{4.35}$$

Each of the four terms can be evaluated separately. To obtain closed-form solutions, we assume that $B_t \, \mathbf{E_Q}[B_T^{-1}|\mathcal{F}_t] = e^{-r(T-t)}$.

Evaluation of term E_1. Substituting for S_T gives

$$E_1 = \mathbf{E_Q}[S_t e^{-\frac{1}{2}\sigma_S^2(T-t) + \sigma_S(\tilde{W}_T - \tilde{W}_t)} \mathbf{1}_{\{S_T > K\}} \mathbf{1}_{\{V_T \geq D\}}|\mathcal{F}_t].$$

$\tilde{z} = \frac{\tilde{W}_T - \tilde{W}_t}{\sqrt{T-t}}$ has law $N(0,1)$. Therefore, E_1 can be expressed in terms of a bivariate normal distribution.

$$E_1 = \int_{-\infty}^{\infty} \int_{-\infty}^{\infty} S_t e^{-\frac{1}{2}\sigma_S^2(T-t) + \sigma_S \tilde{z}_1 \sqrt{T-t}} \mathbf{1}_{\{S_T > K\}} \mathbf{1}_{\{V_T \geq D\}} \frac{1}{2\pi\sqrt{1-\rho^2}} e^{-\frac{1}{2(1-\rho^2)}(\tilde{z}_1^2 - 2\rho\tilde{z}_1\tilde{z}_2 + \tilde{z}_2^2)} \, d\tilde{z}_1 d\tilde{z}_2 \tag{4.36}$$

To rearrange terms, we use the following equality, which is easily verified.

$$
\begin{aligned}
&-\frac{z_1^2 - 2\rho z_1 z_2 + z_2^2}{2(1-\rho^2)} + a z_1 + b z_2 + c \\
&= -\frac{(z_1 - a - \rho b)^2 - 2\rho(z_1 - a - \rho b)(z_2 - b - \rho a) + (z_2 - b - \rho a)^2}{2(1-\rho^2)} \\
&\qquad + \frac{1}{2}a^2 + \rho ab + \frac{1}{2}b^2 + c. \qquad (4.37)
\end{aligned}
$$

By (4.37), E_1 can be written

$$
\begin{aligned}
E_1 = \int_{-\infty}^{\infty}\int_{-\infty}^{\infty} & S_t \mathbf{1}_{\{S_T > K\}} \mathbf{1}_{\{V_T \geq D\}} \\
& \frac{1}{2\pi\sqrt{1-\rho^2}} e^{-\frac{1}{2(1-\rho^2)}\left(v_1^2 - 2\rho v_1 v_2 + v_2^2\right)} \, d\tilde{z}_1 d\tilde{z}_2, \qquad (4.38)
\end{aligned}
$$

with $v_1 = \tilde{z}_1 - \sigma_S\sqrt{T-t})^2$ and $v_2 = \tilde{z}_2 - \rho\sigma_S\sqrt{T-t}$. This bivariate random variable has law $N_2(\sigma_S\sqrt{T-t}, \rho\sigma_S\sqrt{T-t}, 1, 1, \rho)$.

We introduce an equivalent probability measure defined by

$$
\frac{d\dot{\mathbf{Q}}}{d\mathbf{Q}} = \exp(\gamma \tilde{W}_T - \frac{1}{2}\sigma^2 T), \qquad (4.39)
$$

with γ and $\tilde{W}$ vectors in $\mathbb{R}^2$. γ is defined to have elements $\gamma_S = \sigma_S$ and $\gamma_V = \rho\sigma_S$. For notational simplicity, subscripts of the scalars W_S and W_V are dropped if it is clear from context whether W is a vector or a scalar.

By Girsanov's theorem, $\dot{W}_t = \tilde{W}_t - \gamma t$ is a $\mathbb{R}^2$-valued standard Brownian motion under $\dot{\mathbf{Q}}$. Therefore,

$$
\tilde{z} = \frac{\tilde{W}_T - \tilde{W}_t}{\sqrt{T-t}} = \frac{\dot{W}_T - \dot{W}_t + \gamma(T-t)}{\sqrt{T-t}} = \dot{z} + \gamma\sqrt{T-t}. \qquad (4.40)
$$

This implies that the density in expression (4.36) is that of a standard bivariate normal distribution under measure $\dot{\mathbf{Q}}$. Thus, E_1 is given by $E_1 = S_t N_2(a_1, a_2, \rho)$ with the parameters a_1 and a_2 of the bivariate joint distribution function N_2 to be determined by the evaluation of the indicator functions. The indicator functions can be evaluated as

$$
\begin{aligned}
\mathbf{E}_{\dot{\mathbf{Q}}}[\mathbf{1}_{\{S_T > K\}}] &= \dot{\mathbf{Q}}(S_T > K) \\
&= \dot{\mathbf{Q}}\left(S_t e^{(r - \frac{1}{2}\sigma_S^2)(T-t) + \sigma_S(\dot{W}_T - \dot{W}_t + \sigma_S(T-t))} > K\right) \\
&= \dot{\mathbf{Q}}\left(\sigma_S(\dot{W}_T - \dot{W}_t) > \ln K - \ln S_t - (r + \frac{1}{2}\sigma_S^2)(T-t)\right) \qquad (4.41) \\
&= \dot{\mathbf{Q}}\left(\dot{z}_1 < \frac{\ln S_t - \ln K + (r + \frac{1}{2}\sigma_S^2)(T-t)}{\sigma_S\sqrt{T-t}}\right),
\end{aligned}
$$

$$
\begin{aligned}
\mathbf{E}_{\dot{\mathbf{Q}}}[\mathbf{1}_{\{V_T \geq D\}}] &= \dot{\mathbf{Q}}(V_T > D) \\
&= \dot{\mathbf{Q}}\left(V_t e^{(r-\frac{1}{2}\sigma_V^2)(T-t)+\sigma_V(\dot{W}_T-\dot{W}_t+\rho\sigma_S(T-t))} > D\right) \\
&= \dot{\mathbf{Q}}\left(\sigma_V(\dot{W}_T-\dot{W}_t) > \ln D - \ln V_t - (r - \frac{1}{2}\sigma_V^2 + \rho\sigma_S\sigma_V)(T-t)\right) \\
&= \dot{\mathbf{Q}}\left(\dot{z}_2 < \frac{\ln V_t - \ln D + (r-\frac{1}{2}\sigma_V^2 + \rho\sigma_S\sigma_V)(T-t)}{\sigma_V\sqrt{T-t}}\right).
\end{aligned}
\tag{4.42}
$$

From the properties of the joint normal distribution (cf. Abramowitz and Stegun (1972)), we have

$$
\begin{aligned}
&\int_{-\infty}^{\infty}\int_{-\infty}^{\infty} \mathbf{1}_{\{z_1 \geq a_1\}}\mathbf{1}_{\{z_2 \geq a_2\}} f(z_1, z_2, \rho)\, dz_2\, dz_1 \\
&\quad = \int_{a_1}^{\infty}\int_{a_2}^{\infty} f(z_1, z_2, \rho)\, dz_2\, dz_1 \\
&\quad = \int_{a_1}^{\infty}\int_{-\infty}^{-a_2} f(z_1, z_2, -\rho)\, dz_2\, dz_1 \\
&\quad = \int_{-\infty}^{-a_1}\int_{-\infty}^{-a_2} f(z_1, z_2, \rho)\, dz_2\, dz_1,
\end{aligned}
\tag{4.43}
$$

where $f(z_1, z_2, \rho)$ denotes the joint density function.

Consequently, from the evaluation of the indicator functions in expressions (4.41) and (4.42) and from the equalities in expression (4.43), it follows that

$$
E_1 = S_1 N_2(a_1, a_2, \rho), \tag{4.44}
$$

where the parameters of the distribution function are given by

$$
\begin{aligned}
a_1 &= \frac{\ln \frac{S_t}{K} + (r + \frac{1}{2}\sigma_S^2)(T-t)}{\sigma_S\sqrt{T-t}}, \\
a_2 &= \frac{\ln \frac{V_t}{D} + (r - \frac{1}{2}\sigma_V^2 + \rho\sigma_S\sigma_V)(T-t)}{\sigma_V\sqrt{T-t}}.
\end{aligned}
$$

Evaluation of term E_2. The second term can be written as

$$
E_2 = \int_{-\infty}^{\infty}\int_{-\infty}^{\infty} K e^{-r(T-t)} \mathbf{1}_{\{S_T \geq K\}}\mathbf{1}_{\{V_T \geq D\}} \frac{1}{2\pi\sqrt{1-\rho^2}} e^{-\frac{1}{2(1-\rho^2)}(\tilde{z}_1^2 - 2\rho\tilde{z}_1\tilde{z}_2 + \tilde{z}_2^2)}\, d\tilde{z}_1 d\tilde{z}_2\,.
$$

From (4.43) and since $e^{-r(T-t)}K$ is a constant, it becomes apparent that

$$E_2 = Ke^{-r(T-t)}N_2(b_1, b_2, \rho). \tag{4.45}$$

b_1 and b_2 are again determined by the evaluation of the indicator functions. In this case the indicator functions can be evaluated without change of measure.

$$\begin{aligned} \mathbf{E_Q}[\mathbf{1}_{\{S_T>K\}}] &= \mathbf{Q}(S_T > K) \\ &= \mathbf{Q}\left(S_t e^{(r-\frac{1}{2}\sigma_S^2)(T-t)+\sigma_S(\tilde{W}_T-\tilde{W}_t)} > K\right) \\ &= \mathbf{Q}\left(\sigma_S(\tilde{W}_T - \tilde{W}_t) > \ln K - \ln S_t - (r - \frac{1}{2}\sigma_S^2)(T-t)\right) \\ &= \mathbf{Q}\left(\tilde{z}_1 < \frac{\ln S_t - \ln K + (r - \frac{1}{2}\sigma_S^2)(T-t)}{\sigma_S\sqrt{T-t}}\right). \end{aligned} \tag{4.46}$$

Therefore, by expressions (4.46) and (4.43),

$$b_1 = \frac{\ln \frac{S_t}{K} + (r - \frac{1}{2}\sigma_S^2)(T-t)}{\sigma_S\sqrt{T-t}}.$$

The second indicator function, $\mathbf{E_Q}[\mathbf{1}_{\{V_T>D\}}]$, is evaluated analogously, giving

$$b_2 = \frac{\ln \frac{V_t}{D} + (r - \frac{1}{2}\sigma_V^2)(T-t)}{\sigma_V\sqrt{T-t}}.$$

Evaluation of term E_3.

$$\begin{aligned} E_3 = \int_{-\infty}^{\infty}\int_{-\infty}^{\infty} & S_t e^{-\frac{1}{2}\sigma_S^2(T-t)+\sigma_S\tilde{z}_1\sqrt{T-t}}\frac{V_t}{D}e^{(r-\frac{1}{2}\sigma_V^2)(T-t)+\sigma_V\tilde{z}_2\sqrt{T-t}} \\ & \mathbf{1}_{\{S_T\geq K\}}\mathbf{1}_{\{V_T\geq D\}}\frac{1}{2\pi\sqrt{1-\rho^2}}e^{-\frac{1}{2(1-\rho^2)}(\tilde{z}_1^2-2\rho\tilde{z}_1\tilde{z}_2+\tilde{z}_2^2)}\,d\tilde{z}_1 d\tilde{z}_2. \end{aligned} \tag{4.47}$$

Using (4.37) and setting $v_1 = \tilde{z}_1 - \sigma_S\sqrt{T-t} - \rho\sigma_V\sqrt{T-t}$ and $v_2 = \tilde{z}_2 - \sigma_V\sqrt{T-t} - \rho\sigma_S\sqrt{T-t}$, we have

$$\begin{aligned} E_3 = \int_{-\infty}^{\infty}\int_{-\infty}^{\infty} & S_t\frac{V_t}{D}e^{r(T-t)}\mathbf{1}_{\{S_T>K\}}\mathbf{1}_{\{V_T<D\}} \\ & \frac{1}{2\pi\sqrt{1-\rho^2}}e^{-\frac{1}{2(1-\rho^2)}(\tilde{v}_1^2-2\rho\tilde{v}_1\tilde{v}_2+v_2^2)}\,d\tilde{z}_1 d\tilde{z}_2. \end{aligned}$$

To transform the density in (4.47) into that of a standard bivariate random variable, we change the measure using expression (4.39) with

$$\gamma_S = \sigma_S + \rho\sigma_V\,,$$

By expression (4.40), we have

$$E_3 = e^{r(T-t)} S_t \frac{V_t}{D} e^{\rho\sigma_S\sigma_V(T-t)} N_2(c_1, c_2, -\rho).$$

The parameters of the distribution function are again determined by the evaluation of the indicator functions. The sign of the correlation parameter of the distribution function is negative because of the direction of the inequality sign of the indicator function, in accordance with the properties of the bivariate normal distribution shown in (4.43).

Evaluating the indicator functions gives

$$\begin{aligned}
\mathbf{E}_{\dot{\mathbf{Q}}}[\mathbf{1}_{\{S_T>K\}}] &= \dot{\mathbf{Q}}(S_T > K) \\
&= \dot{\mathbf{Q}}\left(S_t e^{(r-\frac{1}{2}\sigma_S^2)(T-t)+\sigma_S(\dot{W}_T-\dot{W}_t+(\sigma_S+\rho\sigma_V)(T-t))} > K\right) \\
&= \dot{\mathbf{Q}}\left(\sigma_S(\dot{W}_T - \dot{W}_t) > \ln K - \ln S_t - (r + \frac{1}{2}\sigma_S^2 + \rho\sigma_S\sigma_V)(T-t)\right) \\
&= \dot{\mathbf{Q}}\left(\dot{z}_1 < \frac{\ln S_t - \ln K + (r + \frac{1}{2}\sigma_S^2 + \rho\sigma_S\sigma_V)(T-t)}{\sigma_S\sqrt{T-t}}\right),
\end{aligned} \tag{4.48}$$

$$\begin{aligned}
\mathbf{E}_{\dot{\mathbf{Q}}}[\mathbf{1}_{\{V_T<D\}}] &= \dot{\mathbf{Q}}(V_T < D) \\
&= \dot{\mathbf{Q}}\left(V_t e^{(r-\frac{1}{2}\sigma_V^2)(T-t)+\sigma_V(\dot{W}_T-\dot{W}_t+(\sigma_V+\rho\sigma_S)(T-t))} < D\right) \\
&= \dot{\mathbf{Q}}\left(\sigma_V(\dot{W}_T - \dot{W}_t) < \ln D - \ln V_t - (r + \frac{1}{2}\sigma_V^2 + \rho\sigma_S\sigma_V)(T-t)\right) \\
&= \dot{\mathbf{Q}}\left(\dot{z}_2 < -\frac{\ln V_t - \ln D + (r + \frac{1}{2}\sigma_V^2 + \rho\sigma_S\sigma_V)(T-t)}{\sigma_V\sqrt{T-t}}\right).
\end{aligned} \tag{4.49}$$

From the second equality in expression (4.43) and from expressions (4.48) and (4.49), it follows that

$$E_3 = e^{r(T-t)} S_t \frac{V_t}{D} e^{\rho\sigma_S\sigma_V(T-t)} N_2(c_1, c_2, -\rho),$$

where

$$\begin{aligned}
c_1 &= \frac{\ln \frac{S_t}{K} + (r + \frac{1}{2}\sigma_S^2 + \rho\sigma_S\sigma_V)(T-t)}{\sigma_S\sqrt{T-t}}, \\
c_2 &= -\frac{\ln \frac{V_t}{D} + (r + \frac{1}{2}\sigma_V^2 + \rho\sigma_S\sigma_V)(T-t)}{\sigma_V\sqrt{T-t}}.
\end{aligned}$$

Evaluation of term E_4.

$$E_4 = \int_{-\infty}^{\infty}\int_{-\infty}^{\infty} K\frac{V_t}{D}e^{-\frac{1}{2}\sigma_V^2(T-t)+\sigma_V\tilde{z}_2\sqrt{T-t}}\mathbf{1}_{\{S_T\geq K\}}\mathbf{1}_{\{V_T<D\}}$$
$$\frac{1}{2\pi\sqrt{1-\rho^2}}e^{-\frac{1}{2(1-\rho^2)}(\tilde{z}_1^2-2\rho\tilde{z}_1\tilde{z}_2+\tilde{z}_2^2)}\,d\tilde{z}_1 d\tilde{z}_2. \quad (4.50)$$

As in (4.39), we define an equivalent measure, this time with $\gamma_S = \rho\sigma_V$ and $\gamma_V = \sigma_V$. The indicator functions are then evaluated under the new measure.

$$\begin{aligned}
\mathbf{E}_{\dot{\mathbf{Q}}}[\mathbf{1}_{\{S_T>K\}}] &= \dot{\mathbf{Q}}\left(S_t e^{(r-\frac{1}{2}\sigma_S^2)(T-t)+\sigma_S(\dot{W}_T-\dot{W}_t+\rho\sigma_V)(T-t)} > K\right)\\
&= \dot{\mathbf{Q}}\left(\dot{z}_1 < \frac{\ln S_t - \ln K + (r-\frac{1}{2}\sigma_S^2+\rho\sigma_S\sigma_V)(T-t)}{\sigma_S\sqrt{T-t}}\right),\\
\mathbf{E}_{\dot{\mathbf{Q}}}[\mathbf{1}_{\{V_T<D\}}] &= \dot{\mathbf{Q}}\left(V_t e^{(r-\frac{1}{2}\sigma_V^2)(T-t)+\sigma_V(\dot{W}_T-\dot{W}_t+\sigma_V(T-t))} < D\right)\\
&= \dot{\mathbf{Q}}\left(\dot{z}_2 < -\frac{\ln V_t - \ln D + (r+\frac{1}{2}\sigma_V^2)(T-t)}{\sigma_V\sqrt{T-t}}\right).
\end{aligned}$$

It follows that

$$E_4 = K\frac{V_t}{D}N_2(d_1, d_2, -\rho),$$

where

$$\begin{aligned}
d_1 &= \frac{\ln\frac{S_t}{K} + (r-\frac{1}{2}\sigma_S^2+\rho\sigma_S\sigma_V)(T-t)}{\sigma_S\sqrt{T-t}},\\
d_2 &= -\frac{\ln\frac{V_t}{D} + (r+\frac{1}{2}\sigma_V^2)(T-t)}{\sigma_V\sqrt{T-t}}.
\end{aligned}$$

This completes the proof of Proposition 4.2.1.

4.9.2 Proof of Proposition 4.3.1

This section contains the proof of the pricing formula for vulnerable stock or bond options assuming stochastic liabilities of the counterparty firm.

Proof. The price of a vulnerable claim can be expressed as

$$X_t = B_t\,\mathbf{E}_{\mathbf{Q}}[B_T^{-1}(S_T-K)^+(\mathbf{1}_{\{\delta_T\geq 1\}}+\delta_T\,\mathbf{1}_{\{\delta_T<1\}})|\mathcal{F}_t], \quad (4.51)$$

where $\delta_T = \frac{V_T}{D_T}$ is the recovery rate.

As in the proof of Proposition 4.2.1, the expectation in (4.51) can be separated into four terms such that

$$X_t = E_1 - E_2 + E_3 - E_4,$$

where

$$E_1 = B_t\, \mathbf{E}_{\mathbf{Q}}[B_T^{-1}\, S_T\, \mathbf{1}_{\{S_T>K\}}\mathbf{1}_{\{\delta_T\geq 1\}}|\mathcal{F}_t],$$
$$E_2 = B_t\, \mathbf{E}_{\mathbf{Q}}[B_T^{-1}\, K\, \mathbf{1}_{\{S_T>K\}}\mathbf{1}_{\{\delta_T\geq 1\}}|\mathcal{F}_t],$$
$$E_3 = B_t\, \mathbf{E}_{\mathbf{Q}}[B_T^{-1}\, S_T\delta_T\, \mathbf{1}_{\{S_T>K\}}\mathbf{1}_{\{\delta_T< 1\}}|\mathcal{F}_t],$$
$$E_4 = B_t\, \mathbf{E}_{\mathbf{Q}}[B_T^{-1}\, K\delta_T\, \mathbf{1}_{\{S_T>K\}}\mathbf{1}_{\{\delta_T< 1\}}|\mathcal{F}_t].$$

Each of the four terms can then be evaluated separately. To obtain closed-form solutions, we assume that $B_t\, \mathbf{E}_{\mathbf{Q}}[B_T^{-1}|\mathcal{F}_t] = e^{-r(T-t)}$.

Evaluation of term E_1. Substituting for S_T gives

$$E_1 = \mathbf{E}_{\mathbf{Q}}[S_t e^{-\frac{1}{2}\sigma_S^2(T-t)+\sigma_S(\tilde{W}_T-\tilde{W}_t)}\mathbf{1}_{\{S_T>K\}}\mathbf{1}_{\{\delta_T\geq 1\}}|\mathcal{F}_t].$$

$\tilde{z} = \frac{\tilde{W}_T-\tilde{W}_t}{\sqrt{T-t}}$ has law $N(0,1)$. Therefore, E_1 can be expressed in terms of a bivariate normal distribution.

$$E_1 = \int_{-\infty}^{\infty}\int_{-\infty}^{\infty} S_t e^{-\frac{1}{2}\sigma_S^2(T-t)+\sigma_S\tilde{z}_1\sqrt{T-t}}\mathbf{1}_{\{S_T>K\}}\mathbf{1}_{\{\delta_T\geq 1\}}$$
$$\frac{1}{2\pi\sqrt{1-\rho_{12}^2}} e^{-\frac{1}{2(1-\rho_{12}^2)}(\tilde{z}_1^2-2\rho_{12}\tilde{z}_1\tilde{z}_2+\tilde{z}_2^2)}\, d\tilde{z}_1 d\tilde{z}_2. \quad (4.52)$$

By (4.37), E_1 can be written

$$E_1 = \int_{-\infty}^{\infty}\int_{-\infty}^{\infty} S_t \mathbf{1}_{\{S_T>K\}}\mathbf{1}_{\{\delta_T\geq 1\}}$$
$$\frac{1}{2\pi\sqrt{1-\rho_{12}^2}} e^{-\frac{1}{2(1-\rho_{12}^2)}\left(v_1^2-2\rho_{12}(\tilde{z}_1-\sigma_S\sqrt{T-t})v_2+v_2^2\right)}\, d\tilde{z}_1 d\tilde{z}_2,$$

with $v_1 = \tilde{z}_1 - \sigma_S\sqrt{T-t}$ and $v_2 = \tilde{z}_2 - \rho_{12}\sigma_S\sqrt{T-t}$.

For changing measure, we set $\gamma_\delta = \rho_{S\delta}\sigma_S$ in expression (4.39) and make a Girsanov transformation as shown in expression (4.40). This implies that the density in expression (4.52) is that of a standard bivariate normal distribution under measure $\dot{\mathbf{Q}}$. Thus, E_1 is given by $E_1 = S_t N_2(a_1, a_2, \rho)$ with parameters a_1 and a_2 of the bivariate joint distribution function N_2 to be determined by the evaluation of the indicator functions and $\rho = \frac{\rho_{SV}\sigma_V - \rho_{SD}\sigma_D}{\sigma_\delta}$.

The evaluation of the indicator function $\mathbf{1}_{\{S_T>K\}}$ is given in expression (4.41). The second indicator function is evaluated as

$$\begin{aligned}
\mathbf{E}_{\dot{\mathbf{Q}}}[\mathbf{1}_{\{\delta_T\geq 1\}}] &= \dot{\mathbf{Q}}(\delta_T > 1)\\
&= \dot{\mathbf{Q}}\left(\delta_t e^{(-\frac{1}{2}\sigma_V^2+\frac{1}{2}\sigma_D^2)(T-t)+\sigma_\delta(\dot{W}_T-\dot{W}_t+\rho_{S\delta}\sigma_S(T-t))} > 1\right)\\
&= \dot{\mathbf{Q}}\left(\sigma_\delta(\dot{W}_T-\dot{W}_t) > -\ln\delta_t + (\frac{1}{2}\sigma_V^2 - \frac{1}{2}\sigma_D^2 - \rho_{S\delta}\sigma_S\sigma_\delta)(T-t)\right)\\
&= \dot{\mathbf{Q}}\left(\dot{z}_2 < \frac{\ln\delta_t - (\frac{1}{2}\sigma_V^2 - \frac{1}{2}\sigma_D^2 - \rho_{SV}\sigma_S\sigma_V + \rho_{SD}\sigma_S\sigma_D)(T-t)}{\sigma_\delta\sqrt{T-t}}\right).
\end{aligned}$$

By expression (4.43) and the evaluation of the indicator functions, it follows that

$$E_1 = S_1\, N_2\left(a_1, a_2, \frac{\rho_{SV}\sigma_V - \rho_{SD}\sigma_D}{\sigma_\delta}\right), \tag{4.53}$$

where

$$a_1 = \frac{\ln\frac{S_t}{K} + (r + \frac{1}{2}\sigma_S^2)(T-t)}{\sigma_S\sqrt{T-t}},$$
$$a_2 = \frac{\ln\frac{V_t}{D_t} - (\frac{1}{2}\sigma_V^2 - \frac{1}{2}\sigma_D^2 - \rho_{SV}\sigma_S\sigma_V + \rho_{SD}\sigma_S\sigma_D)(T-t)}{\sigma_\delta\sqrt{T-t}}.$$

Evaluation of term E_2. The second term can be written as

$$E_2 = \int_{-\infty}^{\infty}\int_{-\infty}^{\infty} Ke^{-r(T-t)}\mathbf{1}_{\{S_T\geq K\}}\mathbf{1}_{\{\delta_T\geq 1\}} \frac{1}{2\pi\sqrt{1-\rho_{12}^2}} e^{-\frac{1}{2(1-\rho_{12}^2)}(\tilde{z}_1^2 - 2\rho_{12}\tilde{z}_1\tilde{z}_2 + \tilde{z}_2^2)}\, d\tilde{z}_1 d\tilde{z}_2.$$

From (4.43) and since $e^{-r(T-t)}K$ is a constant, it immediately follows that

$$E_2 = Ke^{-r(T-t)}\, N_2\left(b_1, b_2, \frac{\rho_{SV}\sigma_V - \rho_{SD}\sigma_D}{\sigma_\delta}\right).$$

b_1 and b_2 are again determined by the evaluation of the indicator functions. In this case, the indicator functions can be evaluated without change of measure. The evaluation of $\mathbf{E_Q}[\mathbf{1}_{\{S_T>K\}}]$ is given in expression (4.46) while $\mathbf{E_Q}[\mathbf{1}_{\{\delta_T>1\}}]$ is evaluated as

$$\begin{aligned}
\mathbf{E_Q}[\mathbf{1}_{\{\delta_T\geq 1\}}] &= \mathbf{Q}(\delta_T > 1)\\
&= \mathbf{Q}\left(\delta_t e^{\frac{1}{2}\sigma_D^2 - \frac{1}{2}\sigma_V^2(T-t) + \sigma_\delta(\tilde{W}_T - \tilde{W}_t)} > 1\right)\\
&= \mathbf{Q}\left(\sigma_\delta(\tilde{W}_T - \tilde{W}_t) > -\ln\delta_t + \frac{1}{2}(\sigma_V^2 - \sigma_D^2)(T-t)\right)\\
&= \mathbf{Q}\left(\dot{z}_2 < \frac{\ln\delta_t - \frac{1}{2}(\sigma_V^2 - \sigma_D^2)(T-t)}{\sigma_\delta\sqrt{T-t}}\right).
\end{aligned}$$

Therefore,

$$b_1 = \frac{\ln\frac{S_t}{K} + (r - \frac{1}{2}\sigma_S^2)(T-t)}{\sigma_S\sqrt{T-t}},$$
$$b_2 = \frac{\ln\frac{V_t}{D} - \frac{1}{2}(\sigma_V^2 - \sigma_D^2)(T-t)}{\sigma_\delta\sqrt{T-t}}.$$

Evaluation of term E_3. We evaluate

$$E_3 = \int_{-\infty}^{\infty}\int_{-\infty}^{\infty} S_t e^{-\frac{1}{2}\sigma_S^2(T-t)+\sigma_S \tilde{z}_1\sqrt{T-t}}$$
$$\frac{V_t}{D_t} e^{(-\frac{1}{2}\sigma_\delta^2+\sigma_D^2-\rho_{VD}\sigma_V\sigma_D)(T-t)+\sigma_\delta \tilde{z}_2\sqrt{T-t}} \mathbf{1}_{\{S_T \geq K\}} \mathbf{1}_{\{\delta_T \geq 1\}}$$
$$\frac{1}{2\pi\sqrt{1-\rho_{12}^2}} e^{-\frac{1}{2(1-\rho_{12}^2)}(\tilde{z}_1^2 - 2\rho_{12}\tilde{z}_1\tilde{z}_2+\tilde{z}_2^2)} d\tilde{z}_1 d\tilde{z}_2. \quad (4.54)$$

Using (4.37) and setting

$$v_1 = \tilde{z}_1 - \sigma_S\sqrt{T-t} - \rho_{12}\sigma_\delta\sqrt{T-t},$$
$$v_2 = \tilde{z}_2 - \sigma_\delta\sqrt{T-t} - \rho_{12}\sigma_S\sqrt{T-t},$$

we have

$$E_3 = \int_{-\infty}^{\infty}\int_{-\infty}^{\infty} S_t \frac{V_t}{D_t} e^{r(T-t)} \mathbf{1}_{\{S_T > K\}} \mathbf{1}_{\{\delta_T < 1\}}$$
$$\frac{1}{2\pi\sqrt{1-\rho_{12}^2}} e^{-\frac{1}{2(1-\rho_{12}^2)}(\tilde{v}_1^2 - 2\rho_{12}\tilde{v}_1\tilde{v}_2+v_2^2)} d\tilde{z}_1 d\tilde{z}_2.$$

To transform the density in (4.54) into that of a standard bivariate random variable, we change the measure using expression (4.39) with $\gamma_S = \sigma_S + \rho_{S\delta}\sigma_\delta$ and $\gamma_\delta = \sigma_\delta + \rho_{S\delta}\sigma_S$. By expressions (4.40) and (4.13), we have

$$E_3 = S_t \frac{V_t}{D_t} e^{(\rho_{SV}\sigma_S\sigma_V - \rho_{SD}\sigma_S\sigma_D - \rho_{VD}\sigma_V\sigma_D + \sigma_D^2)(T-t)} N_2(c_1, c_2, -\rho).$$

The parameters c_1 and c_2 of the distribution function are again determined by the evaluation of the indicator functions and $\rho = \frac{\rho_{SV}\sigma_V - \rho_{SD}\sigma_D}{\sigma_\delta}$.

Evaluating the indicator functions gives

$$\begin{aligned}
\mathbf{E}_{\dot{\mathbf{Q}}}[\mathbf{1}_{\{S_T > K\}}] &= \dot{\mathbf{Q}}(S_T > K) \\
&= \dot{\mathbf{Q}}\left(S_t e^{(r-\frac{1}{2}\sigma_S^2)(T-t)+\sigma_S(\dot{W}_T - \dot{W}_t + (\sigma_S+\rho_{S\delta}\sigma_\delta)(T-t))} > K\right) \\
&= \dot{\mathbf{Q}}\left(\sigma_S(\dot{W}_T - \dot{W}_t) > \ln K - \ln S_t - (r + \frac{1}{2}\sigma_S^2 + \rho_{S\delta}\sigma_S\sigma_\delta)(T-t)\right) \\
&= \dot{\mathbf{Q}}\left(\dot{z}_1 < \frac{\ln S_t - \ln K + (r + \frac{1}{2}\sigma_S^2 + \rho_{SV}\sigma_S\sigma_V - \rho_{SD}\sigma_S\sigma_D)(T-t)}{\sigma_S\sqrt{T-t}}\right),
\end{aligned} \quad (4.55)$$

$$
\begin{aligned}
\mathbf{E}_{\dot{\mathbf{Q}}}[\mathbf{1}_{\{\delta_T<1\}}] &= \dot{\mathbf{Q}}(\delta_T < 1)\\
&= \dot{\mathbf{Q}}\left(\delta_t e^{(\sigma_D^2-\frac{1}{2}\sigma_\delta^2-\rho_{VD}\sigma_V\sigma_D)(T-t)+\sigma_\delta(\dot{W}_T-\dot{W}_t+(\sigma_\delta+\rho_{S\delta}\sigma_S)(T-t))} < 1\right)\\
&= \dot{\mathbf{Q}}\left(\delta_t e^{(\sigma_D^2+\frac{1}{2}\sigma_\delta^2+\rho_{S\delta}\sigma_S\sigma_\delta-\rho_{VD}\sigma_V\sigma_D)(T-t)+\sigma_\delta(\dot{W}_T-\dot{W}_t)} < 1\right)\\
&= \dot{\mathbf{Q}}\left(\sigma_\delta(\dot{W}_T - \dot{W}_t) < -A\right)\\
&= \dot{\mathbf{Q}}\left(\dot{z}_2 < -\frac{A}{\sigma_\delta\sqrt{T-t}}\right),
\end{aligned}
\tag{4.56}
$$

where

$$
\begin{aligned}
A &= \ln\delta_t + (\frac{3}{2}\sigma_D^2 + \frac{1}{2}\sigma_V^2 + \rho_{S\delta}\sigma_S\sigma_\delta - 2\rho_{VD}\sigma_V\sigma_D)(T-t)\\
&= \ln\delta_t + (\frac{3}{2}\sigma_D^2 + \frac{1}{2}\sigma_V^2 + \rho_{SV}\sigma_S\sigma_V - \rho_{SD}\sigma_S\sigma_D - 2\rho_{VD}\sigma_V\sigma_D)(T-t).
\end{aligned}
$$

From the second equality in expression (4.43) and from expressions (4.55), (4.56) and (4.13), it follows that

$$
E_3 = S_t\frac{V_t}{D_t}e^{(\rho_{SV}\sigma_S\sigma_V-\rho_{SD}\sigma_S\sigma_D-\rho_{VD}\sigma_V\sigma_D+\sigma_D^2)(T-t)}\, N_2\left(c_1, c_2, -\rho\right),
$$

where

$$
\begin{aligned}
c_1 &= \frac{\ln S_t - \ln K + (r + \frac{1}{2}\sigma_S^2 + \rho_{SV}\sigma_S\sigma_V - \rho_{SD}\sigma_S\sigma_D)(T-t)}{\sigma_S\sqrt{T-t}},\\
c_2 &= -\frac{\ln\delta_t + (\frac{3}{2}\sigma_D^2 + \frac{1}{2}\sigma_V^2 + \rho_{SV}\sigma_S\sigma_V - \rho_{SD}\sigma_S\sigma_D - 2\rho_{VD}\sigma_V\sigma_D)(T-t)}{\sigma_\delta\sqrt{T-t}},\\
\rho &= \frac{\rho_{SV}\sigma_V - \rho_{SD}\sigma_D}{\sigma_\delta}.
\end{aligned}
$$

Evaluation of term E_4.

$$
\begin{aligned}
E_4 = \int_{-\infty}^{\infty}\int_{-\infty}^{\infty} & K\frac{V_t}{D_t}e^{(-r-\frac{1}{2}\sigma_\delta^2+\sigma_D^2-\rho_{VD}\sigma_V\sigma_D)(T-t)+\sigma_\delta\tilde{z}_2\sqrt{T-t}}\\
& \mathbf{1}_{\{S_T\geq K\}}\mathbf{1}_{\{\delta_T<1\}}\frac{1}{2\pi\sqrt{1-\rho_{12}^2}}e^{-\frac{1}{2(1-\rho_{12}^2)}(\tilde{z}_1^2-2\rho_{12}\tilde{z}_1\tilde{z}_2+\tilde{z}_2^2)}\,d\tilde{z}_1 d\tilde{z}_2.
\end{aligned}
$$

By (4.37), this is equal to

$$
\begin{aligned}
E_4 = \int_{-\infty}^{\infty}\int_{-\infty}^{\infty} & K\frac{V_t}{D_t}e^{(-r+\sigma_D^2-\rho_{VD}\sigma_V\sigma_D)(T-t)+\sigma_\delta\tilde{z}_2\sqrt{T-t}}\\
& \mathbf{1}_{\{S_T\geq K\}}\mathbf{1}_{\{\delta_T<1\}}\frac{1}{2\pi\sqrt{1-\rho_{12}^2}}e^{-\frac{1}{2(1-\rho_{12}^2)}(\tilde{z}_1^2-2\rho_{12}\tilde{z}_1\tilde{z}_2+\tilde{z}_2^2)}\,d\tilde{z}_1 d\tilde{z}_2.
\end{aligned}
$$

The equivalent measure is defined by (4.39) with $\gamma_S = \rho_{S\delta}\sigma_\delta$ and $\gamma_\delta = \sigma_\delta$. The term will have the form of

$$E_4 = e^{(-r+\sigma_D^2-\rho_{VD}\sigma_V\sigma_D)(T-t)} K \frac{V_t}{D_t} N_2\left(d_1, d_2, -\rho\right).$$

The indicator functions are evaluated under the new measure such that

$$\begin{aligned}
\mathbf{E}_{\dot{\mathbf{Q}}}[\mathbf{1}_{\{S_T>K\}}] &= \dot{\mathbf{Q}}\left(S_t e^{(r-\frac{1}{2}\sigma_S^2)(T-t)+\sigma_S(\dot{W}_T-\dot{W}_t+\rho_{S\delta}\sigma_\delta(T-t))} > K\right) \\
&= \dot{\mathbf{Q}}\left(\dot{z}_1 < \frac{\ln S_t - \ln K + (r - \frac{1}{2}\sigma_S^2 + \rho_{S\delta}\sigma_S\sigma_\delta)(T-t)}{\sigma_S\sqrt{T-t}}\right) \\
&= \dot{\mathbf{Q}}\left(\dot{z}_1 < \frac{\ln \frac{S_t}{K} + (r - \frac{1}{2}\sigma_S^2 + \rho_{SV}\sigma_S\sigma_V - \rho_{SD}\sigma_S\sigma_D)(T-t)}{\sigma_S\sqrt{T-t}}\right),
\end{aligned}$$

$$\begin{aligned}
\mathbf{E}_{\dot{\mathbf{Q}}}[\mathbf{1}_{\{\delta_T<1\}}] &= \dot{\mathbf{Q}}\left(\delta_t e^{(-\frac{1}{2}\sigma_\delta^2+\sigma_D^2-\rho_{VD}\sigma_V\sigma_D)(T-t)+\sigma_\delta(\dot{W}_T-\dot{W}_t+\sigma_\delta(T-t))} < 1\right) \\
&= \dot{\mathbf{Q}}\left(\dot{z}_2 < -\frac{\ln \delta_t + (\frac{1}{2}\sigma_\delta^2 + \sigma_D^2 - \rho_{VD}\sigma_V\sigma_D)(T-t)}{\sigma_\delta\sqrt{T-t}}\right) \\
&= \dot{\mathbf{Q}}\left(\dot{z}_2 < -\frac{\ln \delta_t + (\frac{1}{2}\sigma_V^2 + \frac{3}{2}\sigma_D^2 - 2\rho_{VD}\sigma_V\sigma_D)(T-t)}{\sigma_\delta\sqrt{T-t}}\right).
\end{aligned}$$

It follows that

$$E_4 = e^{(-r+\sigma_D^2-\rho_{VD}\sigma_V\sigma_D)(T-t)} K \frac{V_t}{D_t} N_2\left(d_1, d_2, \frac{\rho_{SD}\sigma_D - \rho_{SV}\sigma_V}{\sigma_\delta}\right),$$

where

$$\begin{aligned}
d_1 &= \frac{\ln \frac{S_t}{K} + (r - \frac{1}{2}\sigma_S^2 + \rho_{SV}\sigma_S\sigma_V - \rho_{SD}\sigma_S\sigma_D)(T-t)}{\sigma_S\sqrt{T-t}}, \\
d_2 &= -\frac{\ln \delta_t + (\frac{1}{2}\sigma_V^2 + \frac{3}{2}\sigma_D^2 - 2\rho_{VD}\sigma_V\sigma_D)(T-t)}{\sigma_\delta\sqrt{T-t}}.
\end{aligned}$$

This completes the proof of Proposition 4.3.1.

4.9.3 Proof of Proposition 4.4.1

This section contains the proof of the pricing formula for vulnerable stock or bond options with stochastic interest rates.

Proof. For this proof, we derive the price of the vulnerable claim in a Gaussian interest rate framework by means of the forward neutral measure. The forward neutral measure is introduced in Section 2.3.4.

The price of the vulnerable claim is

$$X_t = P(t,T)\mathbf{E_F}[\Lambda(T)(\mathbf{1}_{\{V_F(T,T)\geq D\}} + \delta_F(T,T)\mathbf{1}_{\{V_F(T,T)<D\}})|\mathcal{F}_t],$$

with promised payoff $\Lambda(T) = (S_F(T) - K)^+$ and $\delta_F(T,T) = \frac{V_F(T,T)}{D}$. $\mathbf{F}$ denotes the forward neutral measure.

As in (4.35), this expression can be split into four separate terms such that

$$X_t = E_1 - E_2 + E_3 - E_4,$$

where

$$\begin{aligned}
E_1 &= P(t,T)\,\mathbf{E_F}[S_F(T,T)\,\mathbf{1}_{\{S_F(T,T)>K\}}\mathbf{1}_{\{V_F(T,T)\geq D\}}|\mathcal{F}_t],\\
E_2 &= P(t,T)\,\mathbf{E_F}[K\,\mathbf{1}_{\{S_F(T,T)>K\}}\mathbf{1}_{\{V_F(T,T)\geq D\}}|\mathcal{F}_t],\\
E_3 &= P(t,T)\,\mathbf{E_F}[S_F(T,T)\delta_F(T,T)\,\mathbf{1}_{\{S_F(T,T)>K\}}\mathbf{1}_{\{V_F(T,T)<D\}}|\mathcal{F}_t],\\
E_4 &= P(t,T)\,\mathbf{E_F}[K\delta_F(T,T)\,\mathbf{1}_{\{S_F(T,T)>K\}}\mathbf{1}_{\{V_F(T,T)<D\}}|\mathcal{F}_t].
\end{aligned}$$

Under the forward neutral measure, the dynamics of the forward asset prices S_F and V_F are given by

$$\frac{dS_F(t,T)}{S_F(t,T)} = \beta_S(t,T)\cdot dW_t^{\mathbf{F}}, \qquad \frac{dV_F(t,T)}{V_F(t,T)} = \beta_V(t,T)\cdot dW_t^{\mathbf{F}}.$$

$\beta_S(t,T)$ and $\beta_V(t,T)$ are defined in (4.23). We recall the definitions

$$\begin{aligned}
\nu_i^2(t,T) &= \int_t^T \|\beta_i(s,T)\|^2\,ds,\\
\phi(t,T) &= \int_t^T \beta_S(s,T)\cdot\beta_V(s,T)\,ds,\\
\rho(t,T) &= \int_t^T \frac{\beta_S(s,T)\cdot\beta_V(s,T)}{\|\beta_S(s,T)\|\,\|\beta_V(s,T)\|}\,ds,
\end{aligned}$$

and define

$$\eta_i(t,T) = \int_t^T \beta_i(s,T)\cdot dW^{\mathbf{F}}(s),$$

for $i \in \{S,V\}$. $S_F(T,T)$ and $V_F(T,T)$ can therefore be written

$$i_F(T,T) = i_F(t,T)\exp\left(\eta_i(t,T) - \frac{1}{2}\nu_i^2(t,T)\right), \quad \forall i\in\{S,V\}. \tag{4.57}$$

We also have $i_F(T,T) = i_T$ and $i_F(t,T) = i_t P(t,T)^{-1}$, by the definition of the forward price in 2.3.2.

Evaluation of E_1. We evaluate the expression

$$E_1 = P(t,T)\,\mathbf{E_F}[S_F(T,T)\,\mathbf{1}_{\{S_F(T,T)>K\}}\mathbf{1}_{\{V_F(T,T)\geq D\}}|\mathcal{F}_t].$$

By substituting for $S_F(T,T)$ we obtain for E_1

$$E_1 = \mathbf{E_F}[\frac{S_t}{P(t,T)} \exp\left(\int_t^T \beta_S(s,T) \cdot dW^{\mathbf{F}}(s) - \frac{1}{2}\int_t^T \|\beta_S(s,T)\|^2\, ds\right)$$
$$\mathbf{1}_{\{S_F(T,T)>K\}}\mathbf{1}_{\{V_F(T,T)\geq D\}}|\mathfrak{F}_t]P(t,T)$$
$$= S_t\mathbf{E_F}[\exp\left(\eta_S(t,T) - \frac{1}{2}\nu_S^2(t,T)\right)\mathbf{1}_{\{S_F(T,T)>K\}}\mathbf{1}_{\{V_F(T,T)\geq D\}}|\mathfrak{F}_t]. \quad (4.58)$$

We introduce a new measure $\dot{\mathbf{F}}$ defined by

$$\zeta_T = \frac{d\dot{\mathbf{F}}}{d\mathbf{F}} = \exp\left(\int_0^T \beta_S(s,T) \cdot dW^{\mathbf{F}}(s) - \frac{1}{2}\int_0^T \|\beta_S(s,T)\|^2\, ds\right), \quad (4.59)$$

with $\zeta_t = \mathbf{E_F}[\zeta_T|\mathfrak{F}_t]$, $\forall t \in [0,T]$.
Expression (4.58) can be seen to be equal to

$$E_1 = \zeta_t^{-1}\,\mathbf{E_F}[S_t\zeta_T\mathbf{1}_{\{S_F(T,T)>K\}}\mathbf{1}_{\{V_F(T,T)\geq D\}}|\mathfrak{F}_t].$$

By the Bayes rule (cf. A.6.3), this is equal to

$$E_1 = S_t\,\mathbf{E}_{\dot{\mathbf{F}}}[\mathbf{1}_{\{S_F(T,T)>K\}}\mathbf{1}_{\{V_F(T,T)\geq D\}}|\mathfrak{F}_t].$$

$S_F(T,T)$ and $V_F(T,T)$ are functions of $\eta_S(t,T)$ and $\eta_V(t,T)$, respectively. By the definition of Brownian motion, $\eta_i(t,T)$, $\forall i \in \{S,V\}$, has $\mathbf{F}$-law $N(0,\nu_i^2(t,T))$ with joint normal law $N_2(0,0,\nu_S^2(t,T),\nu_V^2(t,T),\rho(t,T))$. By the definition of conditional expectation, we have

$$\begin{aligned} E_1 &= S_t\int_t^T\int_t^T \exp\left(\eta_S - \frac{1}{2}\nu_S^2\right)\mathbf{1}_{\{S_F(T,T)>K\}}\mathbf{1}_{\{V_F(T,T)\geq D\}}\, d\eta_V\, d\eta_S \\ &= S_t\int_t^T\int_t^T \mathbf{1}_{\{S_F(T,T)>K\}}\mathbf{1}_{\{V_F(T,T)\geq D\}}\, d\dot{\eta}_V\, d\dot{\eta}_S \\ &= S_tN_2(a_1,a_2,\rho(t,T)), \end{aligned}$$

where $\dot{\eta}_i(t,T) = \int_t^T \beta_i(s,T) \cdot dW^{\dot{\mathbf{F}}}(s)$, $\forall i \in \{S,V\}$, and a_1 and a_2 are determined by the evaluation of the indicator functions.

By Girsanov's theorem, the change of measure by (4.59) implies that

$$W^{\dot{\mathbf{F}}}(t) = W^{\mathbf{F}}(t) - \int_0^t \beta_S(s,T)\, ds$$

is a standard Brownian motion under $\dot{\mathbf{F}}$. The indicator functions can therefore be evaluated as

$$\begin{aligned}
&\dot{\mathbf{F}}(S_F(T,T) > K) \\
&= \dot{\mathbf{F}}\left(S_F(t,T)\exp\left(\eta_S(t,T) - \frac{1}{2}\nu_S^2(t,T)\right) > K\right) \\
&= \dot{\mathbf{F}}\left(\frac{S_t}{P(t,T)}\exp\left(\int_t^T \beta_S(s,T)\cdot(dW^{\dot{\mathbf{F}}}(s) + \beta_S(s,T)ds)\right.\right. \\
&\qquad\qquad \left.\left. - \frac{1}{2}\nu_S^2(t,T)\right) > K\right) \qquad (4.60) \\
&= \dot{\mathbf{F}}\left(\frac{S_t}{P(t,T)}\exp\left(\eta_S(t,T) + \frac{1}{2}\nu_S^2(t,T)\right) > K\right) \\
&= \dot{\mathbf{F}}\left(\eta_S(t,T) > \ln\frac{KP(t,T)}{S_t} - \frac{1}{2}\nu_S^2(t,T)\right) \\
&= \dot{\mathbf{F}}\left(\eta_S(t,T) < \ln S_t - \ln K - \ln P(t,T) + \frac{1}{2}\nu_S^2(t,T)\right),
\end{aligned}$$

$$\begin{aligned}
&\dot{\mathbf{F}}(V_F(t,T) > D) \\
&= \dot{\mathbf{F}}\left(V_F(t,T)\exp\left(\eta_V(t,T) - \frac{1}{2}\nu_V^2(t,T)\right) > D\right) \\
&= \dot{\mathbf{F}}\left(\frac{V_t}{P(t,T)}\exp\left(\int_t^T \beta_V(s,T)\cdot(dW^{\dot{\mathbf{F}}}(s) + \beta_S(s,T)ds)\right.\right. \\
&\qquad\qquad \left.\left. - \frac{1}{2}\nu_V^2(t,T)\right) > D\right) \\
&= \dot{\mathbf{F}}\left(\frac{V_t}{P(t,T)}\exp\left(\eta_V(t,T) + \int_t^T \beta_S(s,T)\cdot\beta_V(s,T)\,ds\right.\right. \\
&\qquad\qquad \left.\left. - \frac{1}{2}\nu_V^2(t,T)\right) > D\right) \\
&= \dot{\mathbf{F}}\left(\eta_V(t,T) > \ln\frac{DP(t,T)}{V_t} - \phi(t,T) + \frac{1}{2}\nu_V^2(t,T)\right) \\
&= \dot{\mathbf{F}}\left(\eta_V(t,T) < \ln V_t - \ln D - \ln P(t,T) + \phi(t,T) - \frac{1}{2}\nu_V^2(t,T)\right). \qquad (4.61)
\end{aligned}$$

Therefore,

$$E_1 = S_t N_2(a_1, a_2, \rho(t,T)),$$

where

$$\begin{aligned}
a_1 &= \frac{\ln S_t - \ln K - \ln P(t,T) + \frac{1}{2}\nu_S^2(t,T)}{\nu_S(t,T)}, \\
a_2 &= \frac{\ln V_t - \ln D - \ln P(t,T) + \phi(t,T) - \frac{1}{2}\nu_V^2(t,T)}{\nu_V(t,T)}.
\end{aligned}$$

Evaluation of E_2. The expression to be evaluated is

$$E_2 = P(t,T)\,\mathbf{E_F}[K\,\mathbf{1}_{\{S_F(T,T)>K\}}\mathbf{1}_{\{V_F(T,T)\geq D\}}|\mathcal{F}_t].$$

Since K is a constant, we immediately have

$$\begin{aligned}
E_2 &= P(t,T)K\,\mathbf{E_F}[\mathbf{1}_{\{S_F(T,T)>K\}}\mathbf{1}_{\{V_F(T,T)\geq D\}}|\mathcal{F}_t] \\
&= P(t,T)K\int_t^T\int_t^T \mathbf{1}_{\{S_F(T,T)>K\}}\mathbf{1}_{\{V_F(T,T)<D\}}\,d\eta_V\,d\eta_S \\
&= P(t,T)KN_2(b_1,b_2,\rho(t,T)).
\end{aligned}$$

The indicator functions can be evaluated in a similar way as (4.60) and (4.61), but without change of measure. We obtain

$$E_2 = P(t,T)KN_2(b_1,b_2,\rho(t,T)),$$

where

$$\begin{aligned}
b_1 &= \frac{\ln S_t - \ln K - \ln P(t,T) - \frac{1}{2}\nu_S^2(t,T)}{\nu_S(t,T)}, \\
b_2 &= \frac{\ln V_t - \ln D - \ln P(t,T) - \frac{1}{2}\nu_V^2(t,T)}{\nu_V(t,T)}.
\end{aligned}$$

Evaluation of E_3. We evaluate the expression

$$E_3 = P(t,T)\,\mathbf{E_F}[S_F(T,T)V_F(T,T)D^{-1}\,\mathbf{1}_{\{S_F(T,T)>K\}}\mathbf{1}_{\{V_F(T,T)<D\}}|\mathcal{F}_t].$$

Substituting for $S_F(T,T)$ and $V_F(T,T)$, we obtain

$$\begin{aligned}
E_3 &= \mathbf{E_F}[\exp\left(\eta_S(t,T) - \frac{1}{2}\nu_S^2(t,T)\right)\exp\left(\eta_V(t,T) - \frac{1}{2}\nu_V^2(t,T)\right) \\
&\qquad \mathbf{1}_{\{S_F(T,T)>K\}}\mathbf{1}_{\{V_F(T,T)<D\}}|\mathcal{F}_t]\frac{S_tV_t}{DP(t,T)} \\
&= \frac{S_tV_t}{DP(t,T)}\mathbf{E_F}[\exp\left(\int_t^T (\beta_S(s,T)+\beta_V(s,T))\cdot dW^{\mathbf{F}}(s)\right. \\
&\qquad \left. - \frac{1}{2}\int_t^T \|\beta_S(s,T)\|^2 + \|\beta_V(s,T)\|^2\,ds\right) \\
&\qquad \mathbf{1}_{\{S_F(T,T)>K\}}\mathbf{1}_{\{V_F(T,T)<D\}}|\mathcal{F}_t]. \qquad (4.62)
\end{aligned}$$

We introduce a new measure $\dot{\mathbf{F}}$ defined by

$$\begin{aligned}
\frac{d\dot{\mathbf{F}}}{d\mathbf{F}} = \exp\Bigg(&\int_0^T \beta_S(s,T) + \beta_V(s,T)\cdot dW(s) \\
&- \frac{1}{2}\int_0^T \|\beta_S(s,T) + \beta_V(s,T)\|^2\,ds\Bigg),
\end{aligned}$$

with $\dot{\zeta}_t = \mathbf{E_F}[\frac{d\dot{\mathbf{F}}}{d\mathbf{F}}|\mathcal{F}_t] \quad \forall t \in [0,T]$. Since

$$\int_t^T \|\beta_S(s,T) + \beta_V(s,T)\|^2 \, ds$$
$$= \int_t^T \|\beta_S(s,T)\|^2 + \|\beta_V(s,T)\|^2 \, ds + 2\int_t^T \beta_S(s,T) \cdot \beta_V(s,T) \, ds \, ,$$

expression (4.62) can be seen to be equal to

$$\begin{aligned} E_3 &= \frac{S_t V_t}{DP(t,T)} \dot{\zeta}_t^{-1} \, \mathbf{E_F}[\dot{\zeta}_T \exp(\phi(t,T)) \mathbf{1}_{\{S_F(T,T)>K\}} \mathbf{1}_{\{V_F(T,T)<D\}} | \mathcal{F}_t] \\ &= \frac{S_t V_t}{DP(t,T)} e^{\phi(t,T)} \dot{\zeta}_t^{-1} \, \mathbf{E_F}[\dot{\zeta}_T \mathbf{1}_{\{S_F(T,T)>K\}} \mathbf{1}_{\{V_F(T,T)<D\}} | \mathcal{F}_t], \end{aligned}$$

where $\phi(t,T) = \int_t^T \beta_S(s,T) \cdot \beta_V(s,T) \, ds$. By the Bayes rule (cf. Corollary A.6.3), this is equal to

$$E_3 = \frac{S_t V_t}{DP(t,T)} e^{\phi(t,T)} \mathbf{E}_{\dot{\mathbf{F}}}[\mathbf{1}_{\{S_F(T,T)>K\}} \mathbf{1}_{\{V_F(T,T)<D\}} | \mathcal{F}_t]. \tag{4.63}$$

By the definition of Brownian motion, $\eta_i(t,T)$ has $\mathbf{F}$-law $N(0, \nu_i^2(t,T))$ with joint normal law $N_2(0, 0, \nu_S^2(t,T), \nu_V^2(t,T), \rho(t,T))$. We can therefore write

$$\begin{aligned} E_3 &= \frac{S_t V_t}{DP(t,T)} \int_t^T \int_t^T \exp\left(\eta_S - \frac{1}{2}\nu_S^2 + \eta_V - \frac{1}{2}\nu_V^2 \right) \\ &\qquad\qquad \mathbf{1}_{\{S_F(T,T)>K\}} \mathbf{1}_{\{V_F(T,T)<D\}} \, d\eta_2 \, d\eta_1 \\ &= \frac{S_t V_t}{DP(t,T)} e^{\phi(t,T)} \int_t^T \int_t^T \mathbf{1}_{\{S_F(T,T)>K\}} \mathbf{1}_{\{V_F(T,T)<D\}} \, d\dot{\eta}_2 \, d\dot{\eta}_1 \\ &= \frac{S_t V_t}{DP(t,T)} e^{\phi(t,T)} N_2(c_1, c_2, -\rho(t,T)). \end{aligned}$$

The indicator functions determine c_1 and c_2. By Girsanov's theorem, the change of measure above implies that

$$W^{\dot{\mathbf{F}}}(t) = W^{\mathbf{F}}(t) - \int_0^t \beta_S(s,T) + \beta_V(s,T) \, ds$$

is a standard Brownian motion under $\dot{\mathbf{F}}$. Therefore, the indicator functions are evaluated as

$$
\begin{aligned}
&\dot{\mathbf{F}}(S_F(t,T) > K) \\
&= \dot{\mathbf{F}}\left(S_F(t,T)\exp\left(\eta_S(t,T) - \frac{1}{2}\nu_S(t,T)^2\right) > K\right) \\
&= \dot{\mathbf{F}}\left(\frac{S_t}{P(t,T)}\exp\left(\int_t^T \beta_S(s,T)\right.\right. \\
&\qquad \left.\left.\cdot\left(dW^{\dot{\mathbf{F}}}(s) + \beta_S(s,T)\,ds + \beta_V(s,T)\,ds\right)\right) > K\right) \\
&= \dot{\mathbf{F}}\left(\frac{S_t}{P(t,T)}\exp\left(\eta_S(t,T) + \frac{1}{2}\nu_S(t,T)^2 + \phi(t,T)\right) > K\right) \\
&= \dot{\mathbf{F}}\left(\eta_S(t,T) < \ln S_t - \ln K - \ln P(t,T) + \frac{1}{2}\nu_S^2(t,T) + \phi(t,T)\right).
\end{aligned}
$$

$\dot{\mathbf{F}}(V_F(t,T) < D)$ is evaluated analogously. Thus,

$$
E_3 = \frac{S_t V_t}{DP(t,T)} e^{\phi(t,T)} N_2(c_1, c_2, -\rho(t,T)),
$$

where

$$
\begin{aligned}
c_1 &= \frac{\ln S_t - \ln K - \ln P(t,T) + \frac{1}{2}\nu_S^2(t,T) + \phi(t,T)}{\nu_S(t,T)}, \\
c_2 &= -\frac{\ln V_t - \ln D - \ln P(t,T) + \frac{1}{2}\nu_V^2(t,T) + \phi(t,T)}{\nu_V(t,T)}.
\end{aligned}
$$

Evaluation of E_4. The expression to be evaluated is

$$
\begin{aligned}
E_4 &= P(t,T)\,\mathbf{E}_{\mathbf{F}}[K\delta_F(T,T)\,\mathbf{1}_{\{S_F(T,T)>K\}}\mathbf{1}_{\{V_F(T,T)<D\}}|\mathcal{F}_t] \\
&= \frac{V_t K}{D}\,\mathbf{E}_{\mathbf{F}}[\exp\left(\eta_V(t,T) - \frac{1}{2}\nu_V^2(t,T)\right) \\
&\qquad\qquad \mathbf{1}_{\{S_F(T,T)>K\}}\mathbf{1}_{\{V_F(T,T)<D\}}|\mathcal{F}_t].
\end{aligned}
$$

In this case, the new measure is defined as

$$
\frac{d\dot{\mathbf{F}}}{d\mathbf{F}} = \exp\left(\int_0^T \beta_V(s,T)\cdot dW(s) - \frac{1}{2}\int_0^T \|\beta_V(s,T)\|^2\,ds\right).
$$

The evaluation is similar to that of E_1 but because the Radon-Nikodým derivative is defined in terms of β_V instead of β_S, $\phi(t,T)$ appears in the first parameter of the probability function. We obtain

$$
E_4 = \frac{V_t K}{D} N_2(d_1, d_2, -\rho(t,T)),
$$

where

$$d_1 = \frac{\ln S_t - \ln K - \ln P(t,T) + \phi(t,T) - \frac{1}{2}\nu_S^2(t,T)}{\nu_S(t,T)},$$
$$d_2 = -\frac{\ln V_t - \ln D - \ln P(t,T) + \frac{1}{2}\nu_V^2(t,T)}{\nu_V(t,T)}.$$

This completes the proof of Proposition 4.4.1.

4.9.4 Proof of Proposition 4.5.1

This section contains the proof of the pricing formula for vulnerable stock or bond options with stochastic interest rates and stochastic liabilities of the counterparty firm.

Proof. The price of the vulnerable claim is

$$X_t = P(t,T)\mathbf{E_F}[\Lambda(T)(\mathbf{1}_{\{\delta_F(T,T)\geq 1\}} + \delta_F(T,T)\mathbf{1}_{\{\delta_F(T,T)<1\}})|\mathcal{F}_t],$$

where

$$\Lambda(T) = (S_F(T) - K)^+,$$
$$\delta_F(T,T) = \frac{V_F(T,T)}{D_F(T,T)}.$$

$\mathbf{F}$ denotes the forward neutral measure.

As before, we evaluate the four terms below.

$$X_t = E_1 - E_2 + E_3 - E_4,$$

where

$$E_1 = P(t,T)\,\mathbf{E_F}[S_F(T,T)\,\mathbf{1}_{\{S_F(T,T)>K\}}\mathbf{1}_{\{\delta_F(T,T)\geq 1\}}|\mathcal{F}_t],$$
$$E_2 = P(t,T)\,\mathbf{E_F}[K\,\mathbf{1}_{\{S_F(T,T)>K\}}\mathbf{1}_{\{\delta_F(T,T)\geq 1\}}|\mathcal{F}_t],$$
$$E_3 = P(t,T)\,\mathbf{E_F}[S_F(T,T)\delta_F(T,T)\,\mathbf{1}_{\{S_F(T,T)>K\}}\mathbf{1}_{\{\delta_F(T,T)<1\}}|\mathcal{F}_t],$$
$$E_4 = P(t,T)\,\mathbf{E_F}[K\delta_F(T,T)\,\mathbf{1}_{\{S_F(T,T)>K\}}\mathbf{1}_{\{\delta_F(T,T)<1\}}|\mathcal{F}_t].$$

Evaluation of E_1. We evaluate expression

$$E_1 = P(t,T)\,\mathbf{E_F}[S_F(T,T)\,\mathbf{1}_{\{S_F(T,T)>K\}}\mathbf{1}_{\{\delta_F(T,T)\geq 1\}}|\mathcal{F}_t].$$

Substituting for $S_F(T,T)$, we obtain for E_1,

$$\begin{aligned} E_1 &= \mathbf{E_F}[\frac{S_t}{P(t,T)} \exp\left(\int_t^T \beta_S(s,T)\cdot dW^{\mathbf{F}}(s) - \frac{1}{2}\int_t^T \|\beta_S(s,T)\|^2\,ds\right) \\ &\qquad \mathbf{1}_{\{S_F(T,T)>K\}}\mathbf{1}_{\{\delta_F(T,T)\geq 1\}}|\mathcal{F}_t]\,P(t,T) \\ &= S_t\mathbf{E_F}[\exp\left(\eta_S(t,T) - \frac{1}{2}\nu_S^2(t,T)\right)\,\mathbf{1}_{\{S_F(T,T)>K\}}\mathbf{1}_{\{\delta_F(T,T)\geq 1\}}|\mathcal{F}_t]. \end{aligned} \tag{4.64}$$

We introduce a new measure $\dot{\mathbf{F}}$ defined by

$$\zeta_T = \frac{d\dot{\mathbf{F}}}{d\mathbf{F}} = \exp\left(\int_0^T \beta_S(s,T) \cdot dW^{\mathbf{F}}(s) - \frac{1}{2}\int_0^T \|\beta_S(s,T)\|^2\, ds\right), \quad (4.65)$$

with $\zeta_t = \mathbf{E}_{\mathbf{F}}[\zeta_T|\mathcal{F}_t] \quad \forall t \in [0,T]$.

Expression (4.64) can be seen to be equal to

$$E_1 = \zeta_t^{-1}\,\mathbf{E}_{\mathbf{F}}[S_t\zeta_T\mathbf{1}_{\{S_F(T,T)>K\}}\mathbf{1}_{\{\delta_F(T,T)\geq 1\}}|\mathcal{F}_t].$$

By the Bayes rule[8], this is equal to

$$E_1 = S_t\,\mathbf{E}_{\dot{\mathbf{F}}}[\mathbf{1}_{\{S_F(T,T)>K\}}\mathbf{1}_{\{\delta_F(T,T)\geq 1\}}|\mathcal{F}_t].$$

$S_F(T,T)$ and $\delta_F(T,T)$ are functions of $\eta_S(t,T)$ and $\eta_\delta(t,T)$, respectively. By the definition of Brownian motion, $\eta_i(t,T)$, $\forall i \in \{S,V\}$, has $\mathbf{F}$-law $N(0,\nu_i^2(t,T))$ with joint normal law $N_2(0,0,\nu_S^2(t,T),\nu_\delta^2(t,T),\rho_{SV}(t,T) - \rho_{SD}(t,T))$. By the definition of conditional expectation, we have

$$\begin{aligned} E_1 &= S_t \int_t^T\int_t^T \exp\left(\eta_S - \frac{1}{2}\nu_S^2\right)\mathbf{1}_{\{S_F(T,T)>K\}}\mathbf{1}_{\{\delta_F(T,T)\geq 1\}}\, d\eta_\delta\, d\eta_S \\ &= S_t \int_t^T\int_t^T \mathbf{1}_{\{S_F(T,T)>K\}}\mathbf{1}_{\{\delta_F(T,T)\geq 1\}}\, d\dot{\eta}_\delta\, d\dot{\eta}_S \\ &= S_t N_2(a_1,a_2,\rho_{SV}(t,T) - \rho_{SD}(t,T)), \end{aligned}$$

where, as before, $\dot{\eta}_i(t,T) = \int_t^T \beta_i(s,T) \cdot dW^{\dot{\mathbf{F}}}(s)$, $\forall i \in \{S,\delta\}$, and a_1 and a_2 are determined by the evaluation of the indicator functions. By Girsanov's theorem, the change of measure by (4.65) implies that

$$W^{\dot{\mathbf{F}}}(t) = W^{\mathbf{F}}(t) - \int_0^t \beta_S(s,T)\, ds$$

is a standard Brownian motion under $\dot{\mathbf{F}}$. The indicator functions can therefore be evaluated as

[8] $\mathbf{E}_{\mathbf{Q}}[X_t|\mathcal{F}_s] = \zeta_s^{-1}\mathbf{E}_{\mathbf{P}}[X_t\zeta_t|\mathcal{F}_s]$, where $\zeta_t = \mathbf{E}_{\mathbf{P}}[\frac{d\mathbf{Q}}{d\mathbf{P}}|\mathcal{F}_t]$.

$$
\begin{aligned}
&\dot{\mathbf{F}}(S_F(t,T) > K) \\
&\quad = \dot{\mathbf{F}}\left(S_F(t,T)\exp\left(\eta_S(t,T) - \frac{1}{2}\nu_V^2(t,T)\right) > K\right) \\
&\quad = \dot{\mathbf{F}}\Bigg(\frac{S_t}{P(t,T)}\exp\Big(\int_t^T \beta_S(s,T)\cdot(dW^{\dot{\mathbf{F}}}(s) + \beta_S(s,T)ds) \\
&\qquad\qquad - \frac{1}{2}\nu_S^2(t,T)\Big) > K\Bigg) \qquad (4.66)\\
&\quad = \dot{\mathbf{F}}\left(\frac{S_t}{P(t,T)}\exp\left(\dot{\eta}_S(t,T) + \frac{1}{2}\nu_S^2(t,T)\right) > K\right) \\
&\quad = \dot{\mathbf{F}}\left(\dot{\eta}_S(t,T) > \ln\frac{KP(t,T)}{S_t} - \frac{1}{2}\nu_S^2(t,T)\right) \\
&\quad = \dot{\mathbf{F}}\left(\dot{\eta}_S(t,T) < \ln S_t - \ln K - \ln P(t,T) + \frac{1}{2}\nu_S^2(t,T)\right),
\end{aligned}
$$

$$
\begin{aligned}
&\dot{\mathbf{F}}(\delta_F(t,T) \geq 1) \\
&\quad = \dot{\mathbf{F}}\left(\delta_F(t,T)\exp\left(\eta_\delta(t,T) - \frac{1}{2}\nu_V^2(t,T) + \frac{1}{2}\nu_D^2(t,T)\right) > 1\right) \\
&\quad = \dot{\mathbf{F}}\Bigg(\delta_t\exp\Big(\int_t^T \beta_\delta(s,T)\cdot(dW^{\dot{\mathbf{F}}}(s) + \beta_S(s,T)ds) \\
&\qquad\qquad - \frac{1}{2}\nu_V^2(t,T) + \frac{1}{2}\nu_D^2(t,T)\Big) > 1\Bigg) \\
&\quad = \dot{\mathbf{F}}\Bigg(\delta_t\exp\Big(\dot{\eta}_\delta(t,T) + \int_t^T \beta_S(s,T)\cdot\beta_\delta(s,T)\,ds \\
&\qquad\qquad - \frac{1}{2}\nu_\delta^2(t,T) + \frac{1}{2}\nu_D^2(t,T)\Big) > 1\Bigg) \qquad (4.67)\\
&\quad = \dot{\mathbf{F}}\left(\dot{\eta}_\delta(t,T) > \ln\frac{D_t}{V_t} - \phi_{S\delta}(t,T) + \frac{1}{2}\nu_V^2(t,T) - \frac{1}{2}\nu_D^2(t,T)\right) \\
&\quad = \dot{\mathbf{F}}\Bigg(\dot{\eta}_\delta(t,T) < \ln V_t - \ln D_t + \phi_{SV}(t,T) - \phi_{SD}(t,T) \\
&\qquad\qquad + \frac{1}{2}\nu_D^2(t,T) - \frac{1}{2}\nu_V^2(t,T)\Bigg).
\end{aligned}
$$

Therefore,

$$E_1 = S_t N_2(a_1, a_2, \rho_{SV}(t,T) - \rho_{SD}(t,T)),$$

where

$$a_1 = \frac{\ln S_t - \ln K - \ln P(t,T) + \frac{1}{2}\nu_S^2(t,T)}{\nu_S(t,T)},$$
$$a_2 = \frac{\ln V_t - \ln D_t + \phi_{SV}(t,T) - \phi_{SD}(t,T) + \frac{1}{2}\nu_D^2(t,T) - \frac{1}{2}\nu_V^2(t,T)}{\nu_\delta(t,T)}.$$

Evaluation of E_2. The expression to be evaluated is

$$E_2 = P(t,T)\,\mathbf{E_F}[K\,\mathbf{1}_{\{S_F(T,T)>K\}}\mathbf{1}_{\{\delta_F(T,T)\geq 1\}}|\mathcal{F}_t].$$

Since K is a constant, we immediately have

$$\begin{aligned} E_2 &= P(t,T)K\,\mathbf{E_F}[\mathbf{1}_{\{S_F(T,T)>K\}}\mathbf{1}_{\{\delta_F(T,T)\geq 1\}}|\mathcal{F}_t] \\ &= P(t,T)K\int_t^T\int_t^T \mathbf{1}_{\{S_F(T,T)>K\}}\mathbf{1}_{\{\delta_F(T,T)\geq 1\}}\,d\eta_\delta\,d\eta_S \\ &= P(t,T)KN_2(b_1,b_2,\rho_{SV}(t,T)-\rho_{SD}(t,T)). \end{aligned}$$

The indicator functions can be evaluated in a similar fashion as (4.66) and (4.67), but without change of measure. We obtain

$$E_2 = P(t,T)KN_2(b_1,b_2,\rho_{SV}(t,T)-\rho_{SD}(t,T)),$$

where

$$b_1 = \frac{\ln S_t - \ln K - \ln P(t,T) - \frac{1}{2}\nu_S^2(t,T)}{\nu_S(t,T)},$$
$$b_2 = \frac{\ln V_t - \ln D_t - \frac{1}{2}(\nu_V^2(t,T) - \nu_D^2(t,T))}{\nu_\delta(t,T)}.$$

Evaluation of E_3. We evaluate

$$E_3 = P(t,T)\,\mathbf{E_F}[S_F(T,T)\delta_F(T,T)\,\mathbf{1}_{\{S_F(T,T)>K\}}\mathbf{1}_{\{\delta_F(T,T)<1\}}|\mathcal{F}_t].$$

Substituting for $S_F(T,T)$ and $\delta_F(T,T)$, we obtain

$$\begin{aligned} E_3 &= \frac{S_tV_t}{D_t}\mathbf{E_F}[\exp\left(\eta_S(t,T) - \frac{1}{2}\nu_S^2(t,T)\right) \\ &\qquad \exp\left(\eta_\delta(t,T) - \frac{1}{2}\nu_\delta^2(t,T) - \phi_{VD}(t,T) + \nu_D^2(t,T)\right) \\ &\qquad \mathbf{1}_{\{S_F(T,T)>K\}}\mathbf{1}_{\{\delta_F(T,T)<1\}}|\mathcal{F}_t] \\ &= \frac{S_tV_t}{D_t}\mathbf{E_F}[\exp\left(\int_t^T (\beta_S(s,T)+\beta_\delta(s,T))\cdot dW^{\mathbf{F}}(s)\right. \\ &\qquad - \frac{1}{2}\int_t^T \|\beta_S(s,T)\|^2 + \|\beta_\delta(s,T)\|^2\,ds \\ &\qquad \left. - \phi_{VD}(t,T) + \nu_D^2(t,T)\right)\mathbf{1}_{\{S_F(T,T)>K\}}\mathbf{1}_{\{\delta_F(T,T)<1\}}|\mathcal{F}_t]. \end{aligned} \tag{4.68}$$

We introduce a new measure $\dot{\mathbf{F}}$ defined by

$$\frac{d\dot{\mathbf{F}}}{d\mathbf{F}} = \exp\left(\int_0^T \beta_S(s,T) + \beta_\delta(s,T) \cdot dW^{\mathbf{F}}(s) - \frac{1}{2}\int_0^T \|\beta_S(s,T) + \beta_\delta(s,T)\|^2 \, ds\right),$$

with

$$\dot{\zeta}_t = \mathbf{E_F}[\frac{d\dot{\mathbf{F}}}{d\mathbf{F}}|\mathcal{F}_t], \quad \forall t \in [0,T].$$

Since

$$\int_t^T \|\beta_S(s,T) + \beta_\delta(s,T)\|^2 \, ds = \int_t^T \|\beta_S(s,T)\|^2 + \|\beta_\delta(s,T)\|^2 \, ds + 2\int_t^T \beta_S(s,T) \cdot \beta_\delta(s,T) \, ds,$$

expression (4.68) can be seen to be equal to

$$\begin{aligned} E_3 &= \frac{S_t V_t}{D_t} \dot{\zeta}_t^{-1} \mathbf{E_F}[\dot{\zeta}_T \exp(\phi_{S\delta}(t,T) - \phi_{VD}(t,T) + \nu_D^2(t,T)) \\ &\qquad \mathbf{1}_{\{S_F(T,T)>K\}} \mathbf{1}_{\{\delta_F(T,T)<1\}} |\mathcal{F}_t] \\ &= \frac{S_t V_t}{D_t} e^{(\phi_{SV}(t,T) - \phi_{SD}(t,T) - \phi_{VD}(t,T) + \nu_D^2(t,T))} \\ &\qquad \dot{\zeta}_t^{-1} \mathbf{E_F}[\dot{\zeta}_T \mathbf{1}_{\{S_F(T,T)>K\}} \mathbf{1}_{\{\delta_F(T,T)<1\}} |\mathcal{F}_t]. \end{aligned}$$

By the Bayes rule, we have

$$E_3 = \frac{S_t V_t}{D} e^{(\phi_{SV}(t,T) - \phi_{SD}(t,T) - \phi_{VD}(t,T) + \nu_D^2(t,T))} \mathbf{E}_{\dot{\mathbf{F}}}[\mathbf{1}_{\{S_F(T,T)>K\}} \mathbf{1}_{\{\delta_F(T,T)<1\}} |\mathcal{F}_t]. \quad (4.69)$$

By the definition of Brownian motion, $\eta_i(t,T)$ has $\mathbf{F}$-law $N(0, \nu_i^2(t,T))$, $\forall i \in \{S, \delta\}$, with joint normal law $N_2(0, 0, \nu_S^2(t,T), \nu_\delta^2(t,T), \rho_{SD}(t,T) - \rho_{SV}(t,T))$. We can therefore write

$$\begin{aligned} E_3 &= \frac{S_t V_t}{D_t} \int_t^T \int_t^T \exp\left(\eta_S - \frac{1}{2}\nu_S^2 + \eta_\delta - \frac{1}{2}\nu_\delta^2 + \nu_D^2(t,T) - \phi_{VD}(t,T)\right) \\ &\qquad \mathbf{1}_{\{S_F(T,T)>K\}} \mathbf{1}_{\{\delta_F(T,T)<1\}} \, d\eta_2 \, d\eta_1 \\ &= \frac{S_t V_t}{D_t} e^{(\phi_{SV}(t,T) - \phi_{SD}(t,T) - \phi_{VD}(t,T) + \nu_D^2(t,T))} \\ &\qquad \int_t^T \int_t^T \mathbf{1}_{\{S_F(T,T)>K\}} \mathbf{1}_{\{\delta_F(T,T)<1\}} \, d\dot{\eta}_2 \, d\dot{\eta}_1 \\ &= \frac{S_t V_t}{D_t} e^{(\phi_{SV}(t,T) - \phi_{SD}(t,T) - \phi_{VD}(t,T) + \nu_D^2(t,T))} \\ &\qquad N_2(c_1, c_2, \rho_{SD}(t,T) - \rho_{SV}(t,T)). \end{aligned}$$

The indicator functions determine c_1 and c_2. By Girsanov's theorem, the change of measure above implies that

$$W^{\dot{\mathbf{F}}}(t) = W^{\mathbf{F}}(t) - \int_0^t \beta_S(s,T) + \beta_\delta(s,T)\,ds$$

is a standard Brownian motion under $\dot{\mathbf{F}}$. Therefore, the indicator functions can be evaluated as

$$\begin{aligned}
&\dot{\mathbf{F}}(S_F(t,T) > K) \\
&\quad = \dot{\mathbf{F}}\left(S_F(t,T)\exp\left(\eta_S(t,T) - \frac{1}{2}\nu_S(t,T)^2\right) > K\right) \\
&\quad = \dot{\mathbf{F}}\Bigg(\frac{S_t}{P(t,T)}\exp\Bigg(\int_t^T \beta_S(s,T)\cdot(dW^{\dot{\mathbf{F}}}(s) \\
&\qquad\qquad + \beta_S(s,T)\,ds + \beta_\delta(s,T)\,ds)\Bigg) > K\Bigg) \\
&\quad = \dot{\mathbf{F}}\left(\frac{S_t}{P(t,T)}\exp\left(\dot{\eta}_S(t,T) + \frac{1}{2}\nu_S(t,T)^2 + \phi_{s\delta}(t,T)\right) > K\right) \\
&\quad = \dot{\mathbf{F}}\Bigg(\dot{\eta}_S(t,T) < \ln S_t - \ln K - \ln P(t,T) + \frac{1}{2}\nu_S^2(t,T) \\
&\qquad\qquad + \phi_{SV}(t,T) - \phi_{SD}(t,T)\Bigg),
\end{aligned}$$

$$\begin{aligned}
&\dot{\mathbf{F}}(\delta_F(t,T) < 1) \\
&\quad = \dot{\mathbf{F}}\left(\delta_F(t,T)\exp\left(\eta_\delta(t,T) - \frac{1}{2}\nu_\delta(t,T)^2 + \nu_D^2 - \phi_{VD}(t,T)\right) < 1\right) \\
&= \dot{\mathbf{F}}\Bigg(\delta_t\exp\Bigg(\int_t^T \beta_\delta(s,T)\cdot(dW^{\dot{\mathbf{F}}}(\delta) + \beta_\delta(s,T)\,ds + \beta_S(s,T)\,ds) \\
&\qquad - \frac{1}{2}\nu_\delta(t,T)^2 + \nu_D^2 - \phi_{VD}(t,T)\Bigg) < 1\Bigg) \\
&= \dot{\mathbf{F}}\left(\delta_t\exp\left(\dot{\eta}_\delta(t,T) + \frac{1}{2}\nu_\delta(t,T)^2 + \phi_{s\delta}(t,T) + \nu_D^2 - \phi_{VD}(t,T)\right) < 1\right) \\
&= \dot{\mathbf{F}}\Bigg(\dot{\eta}_\delta(t,T) < -\ln\delta_t - (\frac{3}{2}\nu_D^2(t,T) + \frac{1}{2}\nu_V^2(t,T) + \phi_{SV}(t,T) \\
&\qquad - \phi_{SD}(t,T) - 2\phi_{VD}(t,T))\Bigg).
\end{aligned}$$

Thus,

$$E_3 = \frac{S_t V_t}{D_t} e^{(\phi_{SV}(t,T) - \phi_{SD}(t,T) - \phi_{VD}(t,T) + \nu_D^2(t,T))}$$
$$N_2(c_1, c_2, \rho_{SD}(t,T) - \rho_{SV}(t,T)),$$

where

$$c_1 = \frac{\ln S_t - \ln K - \ln P(t,T) + \frac{1}{2}\nu_S^2(t,T) + \phi_{SV}(t,T) - \phi_{SD}(t,T)}{\nu_S(t,T)},$$
$$c_2 = -\frac{\ln V_t - \ln D_t + \frac{3}{2}\nu_D^2(t,T) + \frac{1}{2}\nu_V^2(t,T)}{\nu_\delta(t,T)}$$
$$- \frac{\phi_{SV}(t,T) - \phi_{SD}(t,T) - 2\phi_{VD}(t,T)}{\nu_\delta(t,T)}.$$

Evaluation of E_4. The expression to be evaluated is

$$E_4 = P(t,T)\,\mathbf{E_F}[K\delta_F(T,T)\,\mathbf{1}_{\{S_F(T,T)>K\}}\mathbf{1}_{\{\delta_F(T,T)<1\}}|\mathcal{F}_t]$$
$$= P(t,T)K\frac{V_t}{D_t}\,\mathbf{E_F}\left[\exp\left(\eta_\delta(t,T) - \frac{1}{2}\nu_\delta^2(t,T) + \nu_D^2(t,T) - \phi_{VD}(t,T)\right)\right.$$
$$\left.\mathbf{1}_{\{S_F(T,T)>K\}}\mathbf{1}_{\{\delta_F(T,T)<1\}}|\mathcal{F}_t\right].$$

In this case the new measure is defined as

$$\frac{d\dot{\mathbf{F}}}{d\mathbf{F}} = \exp\left(\int_0^T \beta_\delta(s,T)\cdot dW^{\mathbf{F}}(s) - \frac{1}{2}\int_0^T \|\beta_\delta(s,T)\|^2\,ds\right).$$

It follows that

$$E_4 = P(t,T)\zeta_t^{-1}\,\mathbf{E_F}[K\frac{V_t}{D_t}\exp(\nu_D^2(t,T) - \phi_{VD}(t,T))\,\zeta_T$$
$$\mathbf{1}_{\{S_F(T,T)>K\}}\mathbf{1}_{\{\delta_F(T,T)<1\}}|\mathcal{F}_t].$$

By the Bayes rule, this is equal to

$$E_4 = P(t,T)K\frac{V_t}{D_t}e^{(\nu_D^2(t,T) - \phi_{VD}(t,T))}\,\mathbf{E}_{\dot{\mathbf{F}}}[\mathbf{1}_{\{S_F(T,T)>K\}}\mathbf{1}_{\{\delta_F(T,T)<1\}}|\mathcal{F}_t].$$

As in previous evaluations, we therefore have

$$E_4 = P(t,T)K\frac{V_t}{D_t}\int_t^T\int_t^T \exp\left(\eta_\delta - \frac{1}{2}\nu_\delta^2 + \nu_D^2(t,T) - \phi_{VD}(t,T)\right)$$
$$\mathbf{1}_{\{S_F(T,T)>K\}}\mathbf{1}_{\{\delta_F(T,T)<1\}}\,d\eta_\delta\,d\eta_S$$
$$= P(t,T)K\frac{V_t}{D_t}e^{(\nu_D^2(t,T) - \phi_{VD}(t,T))}$$
$$\int_t^T\int_t^T \mathbf{1}_{\{S_F(T,T)>K\}}\mathbf{1}_{\{\delta_F(T,T)<1\}}\,d\dot{\eta}_\delta\,d\dot{\eta}_S$$
$$= P(t,T)K\frac{V_t}{D_t}e^{(\nu_D^2(t,T) - \phi_{VD}(t,T))}N_2(d_1, d_2, \rho_{SD}(t,T) - \rho_{SV}(t,T)).$$

As usual, parameters d_1 and d_2 are determined by the evaluation of the indicator functions. By Girsanov's theorem, the change of measure by (4.65) implies that

$$W^{\dot{\mathbf{F}}}(t) = W^{\mathbf{F}}(t) - \int_0^t \beta_\delta(s,T)\,ds$$

is a standard Brownian motion under $\dot{\mathbf{F}}$. The indicator functions can therefore be evaluated as follows.

$$\begin{aligned}
&\dot{\mathbf{F}}(S_F(t,T) > K) \\
&\quad = \dot{\mathbf{F}}\left(S_F(t,T) \exp\left(\eta_S(t,T) - \frac{1}{2}\nu_S^2(t,T)\right) > K \right) \\
&\quad = \dot{\mathbf{F}}\Bigg(\frac{S_t}{P(t,T)} \exp\Big(\int_t^T \beta_S(s,T) \cdot (dW^{\dot{\mathbf{F}}}(s) + \beta_\delta(s,T)ds) \\
&\qquad\qquad - \frac{1}{2}\nu_S^2(t,T)\Big) > K \Bigg) \\
&\quad = \dot{\mathbf{F}}\left(\frac{S_t}{P(t,T)} \exp\left(\dot{\eta}_S(t,T) - \frac{1}{2}\nu_S^2(t,T) + \phi_{S\delta}(t,T)\right) > K \right) \\
&\quad = \dot{\mathbf{F}}\Bigg(\dot{\eta}_S(t,T) < \ln S_t - \ln K - \ln P(t,T) - \frac{1}{2}\nu_S^2(t,T) \\
&\qquad\qquad + \phi_{SV}(t,T) - \phi_{SD}(t,T) \Bigg).
\end{aligned}$$

$$\begin{aligned}
&\dot{\mathbf{F}}(\delta_F(t,T) < 1) \\
&\quad = \dot{\mathbf{F}}\left(\delta_F(t,T) \exp\left(\eta_\delta(t,T) - \frac{1}{2}\nu_\delta^2(t,T) + \nu_D^2(t,T) - \phi_{VD}(t,T)\right) < 1 \right) \\
&\quad = \dot{\mathbf{F}}\Bigg(\delta_t \exp\Big(\int_t^T \beta_\delta(s,T) \cdot (dW^{\dot{\mathbf{F}}}(s) + \beta_\delta(s,T)ds) - \frac{1}{2}\nu_\delta^2(t,T) \\
&\qquad\qquad + \nu_D^2(t,T) - \phi_{VD}(t,T)\Big) < 1 \Bigg) \\
&\quad = \dot{\mathbf{F}}\left(\delta_t \exp\left(\dot{\eta}_\delta(t,T) + \frac{1}{2}\nu_\delta^2(t,T) + \nu_D^2(t,T) - \phi_{VD}(t,T)\right) < 1 \right) \\
&\quad = \dot{\mathbf{F}}\left(\dot{\eta}_\delta(t,T) < -(\ln V_t + \ln D_t + \frac{1}{2}\nu_V^2(t,T) + \frac{3}{2}\nu_D^2(t,T) - 2\phi_{VD}(t,T)) \right).
\end{aligned}$$

We obtain

$$E_4 = P(t,T)K \frac{V_t}{D_t} e^{(\nu_D^2(t,T) - \phi_{VD}(t,T))} N_2(d_1, d_2, \rho_{SD}(t,T) - \rho_{SV}(t,T)),$$

where

$$d_1 = \frac{\ln S_t - \ln K - \ln P(t,T) - \frac{1}{2}\nu_S^2(t,T) + \phi_{SV}(t,T) - \phi_{SD}(t,T)}{\nu_S(t,T)},$$

$$d_2 = -\frac{\ln V_t + \ln D_t + \frac{1}{2}\nu_V^2(t,T) + \frac{3}{2}\nu_D^2(t,T) - 2\phi_{VD}(t,T)}{\nu_\delta(t,T)}.$$

This completes the proof of Proposition 4.5.1.

5. A Hybrid Pricing Model for Contingent Claims with Credit Risk

In this chapter we propose a hybrid pricing model for contingent claims with counterparty risk. We call the model *hybrid* because it borrows both from firm value models and from intensity models. Specifically, the firm value models the recovery rate while, at the same time, an intensity-based bankruptcy process determines the occurrence of default. We study the hybrid model under various assumptions with respect to the bankruptcy process and interest rates. In addition to pricing derivative securities which are subject to counterparty default risk, we propose a pricing approach for default-free options on credit-risky bonds.

5.1 The General Credit Risk Framework

Default is modeled by a digital process with one jump in the time interval $u \in [0, T]$. The jump process is denoted by the indicator function

$$N(u) = \mathbf{1}_{\{\tau \leq u\}}, \tag{5.1}$$

where stopping time τ is the default event. $N(u)$ is driven by a compensating *intensity*, *arrival rate*, or *hazard rate* process denoted by $\lambda(u)$ such that

$$N(u) - \int_0^u \lambda(s)\, ds$$

is a martingale. The jump process $N(u)$ can be shown to be a semimartingale.[1] If we had defined $N(u) = \sum_{n \geq 1} \mathbf{1}_{\{\tau_n \leq u\}}$ and the compensated process $N(u) - \lambda u$ with the arrival rate λ constant, then $N(u)$ would have been a standard Poisson process. However, in our case the jump process has a time-varying arrival rate and therefore the Poisson process is time-inhomogeneous. In fact, the arrival rate can depend on a random state variable process and can therefore also be random. Such a process is called a *doubly stochastic*

[1] Consequently, as most of the results used for Brownian motion extend to semimartingales, the jump process can be often be treated similarly to Brownian motion. Appendix A describes Girsanov's theorem, Itô's formula, and some other results frequently used in financial modeling, in the context of semimartingales.

Poisson process or a *Cox* process.[2] From the properties of Poisson processes, we have

$$\mathbf{P}(N(u) - N(t) = i) = \frac{(\int_t^u \lambda(s)\,ds)^i}{i!} \exp\left(-\int_t^u \lambda(s)\,ds\right), \quad \forall i \in \mathbb{N}^+,$$

for any u, $t \in [0, \mathcal{T}]$ such that $u > t$. For a bankruptcy process, only the first jump in the time interval $[t, u]$ is relevant, therefore $i = 0$. As a result,

$$\mathbf{P}(N(u) - N(t) = 0) = \exp\left(-\int_t^u \lambda(s)\,ds\right) \tag{5.2}$$

is the probability that bankruptcy will not have occurred at time u. This is called the *survival* probability. For expositional simplicity, we set $N(t) = 0$. This assumes that bankruptcy has not occurred yet at time t when the claim is valued. This assumption is trivially relaxed by making subsequent expressions conditional on $\tau > t$. However, to keep notation simple, we do not explicitly state this condition.

Let $g_u^d(S, V)$ denote the actual payoff function of a vulnerable contingent claim expiring at time u and dependent on state variables S and V. The promised payoff is $g_u(S)$ and is a function of state variable S. The payoff of the risky claim can be written

$$\begin{aligned} g_u^d(S, V) &= g_u(S)\mathbf{1}_{\{u<\tau\}} + g_u(S)\delta_u(V, S)\mathbf{1}_{\{\tau\leq u\}} \\ &= g_u(S)(1 - N_u(V, S)) + g_u(S)\delta_u(V, S)N_u(V, S) \\ &= g_u(S)\left(1 - N_u(V, S)(1 - \delta_u(V, S))\right). \end{aligned} \tag{5.3}$$

$\delta_u(V, S)$ denotes the recovery rate in case of default and is a function of state variables V and S; accordingly, $(1 - \delta_u(V, S))$ is the loss rate. If $u < \tau$ a.s., then default is certain not to occur before maturity of the claim.[3] It follows that indicator functions $\mathbf{1}_{\{\tau\leq u\}} = 0$ and $\mathbf{1}_{\{\tau>u\}} = 1$ and therefore the payoff of the claim simplifies to $g_u^d(S, V) = g_u(S)$. State variable S is the underlying instrument of the claim, such as the short rate or an equity price, V is the firm value. To simplify the notation, we write $g(u)$ for $g_u(S)$, $\delta(u)$ for $\delta_u(V, S)$, and $N(u)$ for $N_u(V, S)$.

If a martingale measure $\mathbf{Q}$ exists, the arbitrage-free value of the claim for any time $t < u$ is given, as usual, by

$$\Lambda^d(t) = B(t)\mathbf{E_Q}[B^{-1}(u)\, g^d(u)|\mathcal{F}_t],$$

[2] We do not give a rigorous treatment of jump processes and omit all technical conditions. A general theory of point processes can be found in Bremaud (1981). For Cox processes in particular, see also Grandell (1976). Refer also to these references for technical conditions.

[3] Certainty is always to be understood in the probabilistic *almost sure* sense. Refer to A.1 for details. For simplicity, the notation a.s. is usually omitted.

where $B(t)$ is the risk-free money market account. Substituting for $g^d(u)$ gives

$$\begin{aligned} \Lambda^d(t) &= B(t)\mathbf{E_Q}[B^{-1}(u)(g(u)\mathbf{1}_{\{u<\tau\}} + g(u)\delta(u)\mathbf{1}_{\{\tau\le u\}})|\mathcal{F}_t] \\ &= B(t)\mathbf{E_Q}[B^{-1}(u)g(u)\mathbf{1}_{\{u<\tau\}}|\mathcal{F}_t] \\ &\quad + B(t)\mathbf{E_Q}[B^{-1}(u)\,g(u)\delta(u)\mathbf{1}_{\{\tau\le u\}}|\mathcal{F}_t]. \end{aligned} \tag{5.4}$$

According to (5.3) we could also write

$$\begin{aligned} \Lambda^d(t) &= B(t)\mathbf{E_Q}[B^{-1}(u)\,g(u)(1-N(u))(1-\delta(u))|\mathcal{F}_t] \\ &= \Lambda(t) - B(t)\mathbf{E_Q}[B^{-1}(u)g(u)\,N(u)\,(1-\delta(u))|\mathcal{F}_t]. \end{aligned} \tag{5.5}$$

The price of a defaultable claim is therefore equal to the price of a default-free claim less the expected write-down on the position.

As an alternative to the definition of δ used in (5.4), we could define $\delta'(u)$ to be a recovery rate which is 1 if the firm has not defaulted and is $\delta(u)$ in default. Formally,

$$\delta'(t) = \mathbf{1}_{\{\tau>t\}} + \delta(t)\mathbf{1}_{\{\tau\le t\}}.$$

Then (5.4) becomes

$$\Lambda(t)^d = B(t)\,\mathbf{E_Q}[B(u)^{-1}\,g(u)\,\delta'(u)|\mathcal{F}_t], \qquad t<u. \tag{5.6}$$

As can easily be seen, $\mathbf{1}_{\{\tau\le u\}} = 0$ a.s., implies $\delta'(u) = 1$ and therefore an absence of credit risk. It reduces expressions (5.6) and (5.4) to the familiar

$$\Lambda(t) = B_t\mathbf{E_Q}[B^{-1}(u)g(u)|\mathcal{F}_t]$$

for default-free claims.

5.1.1 Independence and Constant Parameters

So far we have not made any assumptions as to the specification of the intensity $\lambda(t)$, which governs the jump process $N(t)$, the recovery rate $\delta(t)$, or the state variables S and V. Before we specify an actual working model, we will analyze some special cases.

Consider a claim $g(u)$ the payoff of which depends on state variable $S(t)$. Assume $\lambda(t)$ to be independent of the state variables $S(t)$ and $V(t)$ such that the default time is independent of recovery rate and underlying security of the claim under the martingale measure. In this case, expression (5.4) can be written

$$\begin{aligned} \Lambda^d(t) &= B_t\mathbf{E_Q}[B^{-1}(u)g(u)\mathbf{1}_{\{u<\tau\}}|\mathcal{F}_t] \\ &\quad + B_t\mathbf{E_Q}[B^{-1}(u)\,g(u)\delta(u)\mathbf{1}_{\{\tau\le u\}}|\mathcal{F}_t] \\ &= B_t\mathbf{E_Q}[B^{-1}(u)g(u)|\mathcal{F}_t]\,\mathbf{E_Q}[\mathbf{1}_{\{u<\tau\}}|\mathcal{F}_t] \\ &\quad + B_t\mathbf{E_Q}[B^{-1}(u)\,g(u)\delta(u)|\mathcal{F}_t]\,\mathbf{E_Q}[\mathbf{1}_{\{\tau\le u\}}|\mathcal{F}_t]. \end{aligned}$$

Evaluation of the indicator function expectations gives

$$\mathbf{E_Q}[\mathbf{1}_{\{\tau>u\}}|\mathcal{F}_t] = \mathbf{Q}(\tau > u|\tau > t) = \exp\left(-\int_t^u \lambda(s)\,ds\right).$$

Denote the risk-neutral survival probability by $\mathcal{S}(t,u) = \mathbf{Q}(\tau > u|\tau > t)$. Then,

$$\begin{aligned} \Lambda(t)^d = {} & B_t\mathbf{E_Q}[B^{-1}(u)g(u)|\mathcal{F}_t]\mathcal{S}(t,u) \\ & + B_t\,\mathbf{E_Q}[B^{-1}(u)\,g(u)\delta(u)|\mathcal{F}_t](1-\mathcal{S}(t,u)). \end{aligned} \tag{5.7}$$

If we make the stronger assumption that $g_u(S)$, $\lambda(u)$, and $\delta(u)$ are mutually independent, we can write

$$\begin{aligned} \Lambda^d(t) &= \Lambda(t)\mathcal{S}(t,u) + \Lambda(t)\mathbf{E_Q}[\delta(u)|\mathcal{F}_t](1-\mathcal{S}(t,u)) \\ &= \Lambda(t)\mathbf{E_Q}[\delta(u)|\mathcal{F}_t] + \mathcal{S}(t,u)\Lambda(t)(1-\mathbf{E_Q}[\delta(u)|\mathcal{F}_t]). \end{aligned} \tag{5.8}$$

The price of a vulnerable claim is a sum of the price of the claim if default is certain and the expected gain from no-default; or, in other words, the price if default is certain plus the expected gain given default does not occur multiplied by the probability that default does not occur.

This is equivalent to saying that the price of the vulnerable claim is given by the price of a riskless claim less the expected loss. To see this, consider expression (5.5). Under the independence assumption it simplifies to

$$\begin{aligned} \Lambda^d(t) &= \Lambda(t) - B(t)\mathbf{E_Q}[B^{-1}(u)g(u)\,N(u)\,(1-\delta(u))|\mathcal{F}_t] \\ &= \Lambda(t) - (1-\mathcal{S}(t,u))B(t)\mathbf{E_Q}[B^{-1}(u)g(u)\,(1-\delta(u))|\mathcal{F}_t] \\ &= \Lambda(t)\big(1-(1-\mathcal{S}(t,u))(1-\mathbf{E_Q}[\delta_u|\mathcal{F}_t])\big). \end{aligned} \tag{5.9}$$

The second equality is by independence of default time with respect to recovery rate and underlying state variable and the third equality by additional independence of state variable and recovery rate. The price of the vulnerable claim in (5.8) can easily be seen to be equal to that in (5.9).

If we make the further assumptions that the risk-neutral intensity $\lambda(u) = \lambda$ and the recovery rate $\delta(u) = \delta$ are exogenous constants. Then, by (5.7),

$$\Lambda^d(t) = \Lambda(t)\left(e^{-\lambda(u-t)} + \delta\left(1-e^{-\lambda(u-t)}\right)\right). \tag{5.10}$$

Alternatively, the price could be expressed in terms of (5.8) or (5.9).

This is the pricing solution derived by Jarrow and Turnbull (1995). Their model is therefore a special case of our approach. Other existing models can also be described in this framework. Consider, for example, the original model of Merton (1974) for valuing corporate bonds. Merton uses one state variable, the firm value V, and assumes constant interest rates. To obtain Merton's model, let $r(t) = r$ such that $B(t) = \exp(-rt)$ and $\mathbf{Q}_{\tau<u} = 0$, $\tau = u$ if $V_u < K$, and $\tau > u$, otherwise.

5.1.2 Price Reduction and Bond Prices

Using $\delta'(t)$ to represent credit risk, we arrive at an important result in a straightforward fashion. For $\delta'(t)$ independent of $g(t)$ and bankruptcy time, expression (5.6) simplifies to

$$\Lambda^d(t) = B(t)\,\mathbf{E_Q}[B(u)^{-1}g(u)|\mathcal{F}_t]\,\mathbf{E_Q}[\delta'(u)|\mathcal{F}_t], \qquad t < u. \tag{5.11}$$

In other words, the vulnerable claim is equal to the default-free claim multiplied by the expected recovery rate. If we assume that the counterparty to a vulnerable claim has a zero-coupon bond of equal maturity outstanding, then this fact leads immediately to the following proposition.

Proposition 5.1.1. *If default time and recovery rate are independent of the underlying state variable of a contingent claim, then the price of the vulnerable claim is*

$$\Lambda^d(t) = \Lambda(t)\frac{P^d(t,u)}{P(t,u)},$$

where $P^d(t,u)$ and $P(t,u)$ are the prices of defaultable and default-free zero-coupon bonds, respectively.

Proof. Expression (5.11) is valid for any contingent claim. Consequently, it also applies to zero-coupon bonds. In the case of a zero-coupon bond, it simplifies to

$$P^d(t,u) = B(t)\,\mathbf{E_Q}[B(u)^{-1}|\mathcal{F}_t]\,\mathbf{E_Q}[\delta'(u)|\mathcal{F}_t].$$

For a default-free bond,

$$P(t,u) = B(t)\,\mathbf{E_Q}[B(u)^{-1}|\mathcal{F}_t].$$

Substituting the expression for the default-free bond into that of the risky bond gives

$$\mathbf{E_Q}[\delta'(u)|\mathcal{F}_t] = \frac{P^d(t,u)}{P(t,u)},$$

which shows that the ratios between the prices of risky claims and risk-free claims and between risky bonds and risk-free bonds are equal.

Proposition 5.1.1 is a general result that is not bound to a particular pricing model. For the default model described by (5.2), but under the independence assumption, we obtain

$$\frac{P^d(t,u)}{P(t,u)} = \mathbf{E_Q}[\delta'(u)|\mathcal{F}_t] = \left(e^{-\int_t^u \lambda(s)\,ds} + \delta(u)(1 - e^{-\int_t^u \lambda(s)\,ds})\right).$$

When $\lambda(t) = \lambda$ and $\delta(t) = \delta$ are both constants, we obtain (3.8) on page 62, which is the bond price formula by Jarrow and Turnbull (1995).

5.1.3 Model Specifications

The framework outlined above is rather general and can be specified in many different ways. In the implementations in subsequent sections, we do not always use the same model specification although we usually follow one of the specifications given below.

5.1.3.1 Arrival Rate of Default. In our model, the default intensity is a function of the firm value, $\lambda(t) = f(V(t))$. We specify

$$\lambda(t) = \left(\frac{D}{V(t)}\right)^{\gamma}, \tag{5.12}$$

for some constant γ, in the numerical examples presented in Section 5.6. Although there are a number of other effects such as liquidity, it is commonly agreed that the debt to asset value is a reasonable proxy for a firm's ability to redeem debt.[4] Our specification captures the idea that the debt-to-asset ratio is a critical factor regarding the ability of the firm to retire debt, but is somewhat arbitrary, of course. The same criticism, however, applies to Madan and Unal (1998) and their default intensity specification described in Section 3.1.4. This arbitrariness of functional specifications is a drawback of these models. On the other hand, it offers a simple way of making the default intensity a stochastic variable and thus allowing default probability to be correlated with a state variable such as firm value.

The functional specification of the default intensity in expression (5.12) is extensible in a variety of ways. For example, the default intensity can be defined to depend on the price of the claim itself. This dependence is a desirable feature of a model if the contract which is valued takes up a large part of the counterparties' liabilities. A sharp increase in value of the claim increases the liabilities of the counterparty and increases the risk of default. A claim which, by rising in value, entitles the holder to twice the payoff as before, is therefore not worth twice as much. Similarly, increasing the principal (promised value) by a factor of two increases the total value of the claim by less than a factor of two. This means that pricing is *non-linear*. Such non-linearity in the face of credit risk was first addressed by Johnson and Stulz (1987) in the context of vulnerable options where the recovery rate was affected by the value of the claim. In the more formal setting of Duffie, Schroder, and Skiadas (1996), the price of the claim is the solution of a quasi-linear PDE.

We usually assume that the value of one claim is negligible compared to the total value of the firm's liabilities. Hence, non-linearity is not an issue. The model, however, could easily be extended to incorporate non-linear pricing.[5]

---

[4] This is consistent with Merton (1974) and Pitts and Selby (1983), who showed that risk premium of risky debt was an increasing function of the quasi-debt-to-firm value ratio in the original firm value model.

[5] In Section 2.1, we established the equivalence of the existence of a price system and absence of arbitrage. A price system was defined as a linear map from a

5.1.3.2 Recovery Rate. In the previous section the recovery rate was defined such that it could be a function of a state variable. We specify the state variable to be the value of the firm's assets, $V(t)$. The recovery rate is specified as

$$\delta(t) = \min\left(\frac{V(t)}{D}, 1\right), \tag{5.13}$$

where D is a constant representing the liabilities of the firm. The definition of $\delta(t)$ implies that all claims participate proportionally in the proceeds from the liquidation of the firm's assets.

The specification of the recovery rate in expression (5.13) is similar to Merton (1974) but allows more than one claim. In Merton's case, a bond is characterized as a default-free bond minus a put option on the firm value struck at the bond's face value. In this case, the payoff of the bond with promised payoff of D is

$$\Lambda = D - \max(D - V, 0). \tag{5.14}$$

This is equal to $D\,(1 - \max(1 - V/D, 0))$. It can easily be verified that this equals $D \min(V/D, 1)$. If there is more than one claim outstanding and if all claims rank equally, a claim with promised payoff D_i is worth $D_i \min(V/D, 1)$ with $D = \sum D_i$ where D is the sum of all outstanding claims.

It is assumed that $V(t)$ evolves according to

$$\frac{dV}{V} = r_t\, dt + \sigma_V d\tilde{W}(t) \tag{5.15}$$

under the risk-neutral measure. Although V may not be a traded asset, a derivative of V is — the stock of the firm. It was shown by Merton (1974) that valuation of derivatives of V is independent of investors' risk preferences in this case. It is therefore reasonable to assume that there exists an equivalent martingale measure.

The recovery made on a claim does not depend on its face value but on the no-default market value of the claim. This definition of recovery rate is in the tradition of Black and Scholes (1973) and Merton (1974). Although in practice there are many claims with cross-default provisions that ensure that all debt becomes due immediately at face value if a default occurs, the recovery rate as specified may nonetheless be a reasonable model of reality because rarely is a recovery made immediately after default.[6] In fact, it seems reasonable to assume that, after a default event, the evolution of the firm value continues such that it might even be possible, if unlikely, for the firm

random variable to a real non-negative number. This result remains valid. Non-linear pricing in the context of credit risk means non-linearity with respect to the promised payoff.

[6] Cf. Helwege (1999) for an empirical investigation of the duration of bankruptcy proceedings and restructurings

value to rise again above the level of liabilities by the time the claim expires. In this case, the default event would not result in a financial loss for the creditors.

5.1.3.3 Bankruptcy Costs. It is also an empirical observation that recovery rates tend to be significantly below unity even for senior and senior secured debt.[7] This observation suggests that once the firm defaults and bankruptcy or restructuring occurs, either the firm value has already fallen far below the amount of outstanding liabilities or the event of default and bankruptcy is associated with such high cost as to eradicate more than half of the firm value on average.

Empirical evidence seems to lend support to the latter conjecture. Altman (1984) finds high costs associated with default. More recently, Asquith, Gertner, and Scharfstein (1994) and Gilson (1997) come to similar conclusions. Default and bankruptcy costs, however, vary depending on circumstances and reasons for default. Jensen (1991) finds that default costs have generally increased in the past few decades due to regulatory changes. Shleifer and Vishny (1992) examine bankruptcy cost within an equilibrium model and argue that the cost of financial distress rises when industry performance decreases. Andrade and Kaplan (1998) attempt to separate default events into default caused by pure financial distress and default caused by economic distress. In their study of highly leveraged firms that defaulted on their obligations, they conclude that bankruptcy costs are much lower if caused by pure financial distress than if caused by economic problems of the firm.

We do not attempt to explain bankruptcy costs within our model. However, because of the empirical evidence of low recovery rates and costs of financial distress, it may be justified to implement bankruptcy costs in the model nonetheless. We propose two ways to model such an implementation.

First, bankruptcy cost can be modeled by implementing a jump upon default. In this case, care has to be taken not to allow arbitrage. We ensure absence of arbitrage by modifying the drift of the process of V such that V is still a martingale under the risk-neutral measure even with the possibility of a jump in the process.

Second, the recovery rate might be modeled directly. In this case V_t/D represents the estimated recovery rate if the firm defaulted immediately. The liquidation value approach can also be used to model the recovery rate for different classes of debt. This could be done by choosing different values of V_t/D for classes with different priority. This does not necessarily result in an inconsistent model since the absolute priority rule is often violated.[8] By modeling the recovery rate directly, we take a step in the direction taken by Jarrow and Turnbull (1995) and related approaches, where the recovery rate is an exogenous constant. In our case, however, it is stochastic, allowing the modeling of correlation effects.

[7] See Altman and Kishore (1996) for a comprehensive study of recovery rates.

[8] Cf. Eberhart, Moore, and Roenfeldt (1990) or Weiss (1990).

5.2 Implementations

In this section we discuss some implementations of models within the framework proposed in Section 5.1. We use multivariate binomial trees and lattices to implement a model in the credit risk framework.

Boyle (1988), who develops a five-jump model to value options with a payoff dependent on two state variables, first uses multidimensional lattice models in option pricing. Boyle, Evnine, and Gibbs (1989), Rubinstein (1991), and Rubinstein (1994) further develop and simplify the multi-dimensional lattice approach. Amin and Bodurtha (1995) and Ho, Stapleton, and Subrahmanyam (1993) use multivariate lattice models for option valuation with stochastic interest rates. As will be seen below, our approach deviates slightly from these approaches.

We first present a bivariate lattice structure for vulnerable equity options. Interest rates are assumed constant. In this model, moments are matched exactly by a numerical procedure. Second, a model is presented where moments are matched asymptotically, as proposed originally by Cox, Ross, and Rubinstein (1979) and for the bivariate case by Rubinstein (1994) and others. This asymptotic model is then extended to allow stochastic interest rates.

5.2.1 Lattice with Deterministic Interest Rates

In this section, we present the multidimensional lattice model as a discrete-time equivalent to Proposition 4.2.1. An advantage of using this lattice model over the explicit formula of Proposition 4.2.1 is the ability to price American options.

From expression (4.3) we have the process definitions for the underlying variable of an option and the value of the counterparty's assets,

$$\begin{aligned} S_T &= S_t \exp\left((r - \frac{1}{2}\sigma_S^2)(T-t) + \sigma_S\big(\tilde{W}_S(T) - \tilde{W}_S(t)\big)\right), \\ V_T &= V_t \exp\left((r - \frac{1}{2}\sigma_V^2)(T-t) + \sigma_V\big(\tilde{W}_V(T) - \tilde{W}_V(t)\big)\right). \end{aligned} \tag{5.16}$$

$\tilde{W}(t)$ denotes Brownian motion under the martingale measure $\mathbf{Q}$. The covariation is $\langle \tilde{W}_S(t), \tilde{W}_V(t)\rangle = \rho t$, as in Chapter 4.

To model the joint evolution of S and V we build a multivariate lattice. Because we are working within a deterministic interest rate and variance framework, the lattice recombines naturally without any transformations.

To illustrate the structure of the lattice, let the pair of initial values be (S, V). Four different states are possible after the first period: (S_u, V_u), (S_u, V_d), (S_d, V_u), and (S_d, V_d). S_u and V_u denote the values of the underlying asset and the firm value, respectively, after an up move in the lattice, S_d and V_d the values after a down move. u and d are the magnitudes of the jumps expressed as factors of process values. There are nine states after the second period: (S_{uu}, V_{uu}), (S_{uu}, V_{ud}), (S_{uu}, V_{dd}), (S_{ud}, V_{ud}), (S_{ud}, V_{uu}), (S_{ud}, V_{dd}), (S_{dd}, V_{uu}), (S_{dd}, V_{ud}), (S_{dd}, V_{dd}), etc. The tree is recombining with $S_{ud} = S_{du}$ and $V_{ud} = V_{du}$. Each node in the tree has four branches.

The lattice structure is illustrated by Figure 5.1. The process starts in the center node. Assume that state variable 1 is S, state variable 2 is V. Then the move to the upper left node, (S_u, V_u) is indicated by u_1u_2. The move to the upper left node, (S_u, V_d) is denoted by u_1d_2, and so on from each node. The recombining property manifests itself by the fact that sequential moves u_1u_2 and d_1d_2, or u_1d_2 and d_1u_2, etc., lead back to the originating node in the illustration.

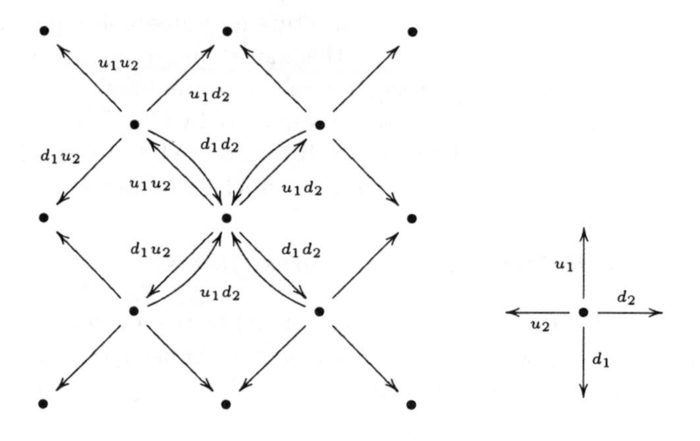

Fig. 5.1. Bivariate lattice

In the lattice, p_i denotes the probability that the state variable i takes an up-jump in the next period. Since four states can be reached from each node in the lattice, there are four probabilities $p_S p_V$, $(1-p_S)p_V$, $p_S(1-p_V)$, and $(1-p_S)(1-p_V)$. For notational simplicity, these probabilities are referred to as p_1, p_2, p_3, p_4, respectively. Thus, the risk-neutral probabilities p_1, p_2, p_3, p_4 are associated with moves (u_S, u_V), (u_S, d_V), (d_S, u_V), (d_S, d_V), respectively.

From the properties of a probability space, a first consistency constraint is derived.

$$p_1 + p_2 + p_3 + p_4 = 1 \text{ and } p_i \geq 0, \quad \forall i \in \{1, 2, 3, 4\}. \tag{5.17}$$

Correlation is introduced by using two sources of uncertainty such that $\epsilon_3 = \rho\epsilon_1 + \sqrt{(1-\rho^2)}\epsilon_2$, where ϵ denotes a standard normal variable. ϵ_3 and ϵ_1 are joint standard normally distributed with correlation ρ. Refer to Section A.4 for a more formal introduction to correlation. Since we approximate one standard normal variable with a binomial variable represented by an up and down move, the second variable Z is approximated by four move sizes as it is composed of two independent standard normal variables, X and Y. Therefore, a total of six different move variables is needed: u_S, d_S, $u_V[u_S]$, $u_V[d_S]$, $d_V[u_S]$, $d_V[d_S]$. $u_V[u_S]$ denotes the size of the up move of V if S also moves up in the same period. For notational simplicity, we define $u_{V+} \equiv u_V[u_S]$, $u_{V-} \equiv u_V[d_S]$, $d_{V+} \equiv d_V[u_S]$, $d_{V-} \equiv d_V[d_S]$.

To approximate the processes in (5.16), the binomial random walks need to be defined such that their moments match those of the continuous processes. From the properties of the lognormal distribution we can derive the following conditions on expectation, variance, and covariance.

$$\begin{aligned} E_{\mathbf{Q}}(X_{t+h}|\mathcal{F}_t) &= X_t e^{rh}, \\ Var(X_{t+h}|\mathcal{F}_t) &= X_t^2 e^{2rh}(e^{\sigma_X^2 h} - 1), \\ Cov(S_{t+h}, V_{t+h}|\mathcal{F}_t) &= V_t S_t e^{2rh}(e^{\rho\sigma_V\sigma_S h} - 1), \end{aligned} \tag{5.18}$$

for processes $X \in (V, S)$. h is the length of the time step in the discrete process. Applied to the lattice parameters, (5.18) gives

$$\begin{aligned} (p_1 + p_2)u_S + (p_3 + p_4)d_S &= e^{rh}, \\ p_1 u_{V+} + p_2 u_{V-} + p_3 d_{V+} + p_4 d_{V-} &= e^{rh}, \\ (p_1 + p_2)u_S^2 + (p_3 + p_4)d_S^2 &= e^{(\sigma_S^2 + 2r)h}, \\ p_1 u_{V+}^2 + p_2 u_{V-}^2 + p_3 d_{V+}^2 + p_4 d_{V-}^2) &= e^{(\sigma_V^2 + 2r)h}, \\ p_1 u_S u_{V+} + p_2 u_S d_{V+} + p_3 d_S u_{V-} + p_4 d_S d_{V-} &= e^{(\rho\sigma_S\sigma_V + 2r)h}. \end{aligned} \tag{5.19}$$

Because of the degrees of freedom of this equation system we need to impose additional, arbitrary constraints. The canonical choice is to set

$$p \equiv p_1 = p_2 = p_3 = p_4 = \frac{1}{4}. \tag{5.20}$$

We might also have used a constraint similar to $u = 1/d$ used by Cox, Ross, and Rubinstein (1979) in their original univariate model. Our choice is often more convenient because negative probabilities are automatically excluded in this way. Had we chosen a restriction on the move sizes as proposed by Cox, Ross, and Rubinstein (1979), then the drift would have been obtained by the appropriate choice of the risk-neutral probabilities, while in our case the lattice realizes the drift by the up and down move sizes. Depending on the constraints chosen, the number of up and down sizes or the number of probabilities changes. Regardless of whether the drift is implemented via move sizes or martingale probabilities, the discounted process is a martingale under the equivalent measure and therefore the system is arbitrage-free.

Given (5.20), expression (5.19) simplifies to

$$\begin{aligned} 2p(u_S + d_S) &= e^{rh}, \\ p(u_{V+} + u_{V-} + d_{V+} + d_{V-}) &= e^{rh}, \\ 2p(u_S^2 + d_S^2) &= e^{(\sigma_S^2+2r)h}, \\ p(u_{V+}^2 + u_{V-}^2 + d_{V+}^2 + d_{V-}^2) &= e^{(\sigma_V^2+2r)h}, \\ p(u_S u_{V+} + u_S d_{V+} + d_S u_{V-} + d_S d_{V-}) &= e^{(\rho\sigma_S\sigma_V+2r)h}. \end{aligned} \tag{5.21}$$

Because there are six variables but only five equations, another constraining equation is required. Since our pricing model is Markovian, the tree can be designed to recombine, i.e.,

$$u_{i,t}d_{i,t+1} = d_{i,t}u_{i,t+1}, \quad \forall i \in (S, V),$$

where $t < T$ denotes a time point in the lattice. This equation is obviously satisfied by $u_{S,t}d_{S,t+1} = d_{S,t}u_{S,t+1}$. However, since there are four move sizes for the V process, for the recombining property to hold, we need to impose explicitly

$$u_{V+}d_{V-} = d_{V+}u_{V-}. \tag{5.22}$$

Equation (5.22) serves as the last restriction imposed on the move sizes.

Armed with equations (5.21) and (5.22), it is possible to solve for u_S, d_S, u_{V+}, u_{V-}, d_{V+}, d_{V-}. The equation system can be solved symbolically by ignoring terms of order smaller than h. In this case the moments of the simulated distribution converge to the moments of the lognormal distribution asymptotically. In this case the model suggested by Rubinstein (1994) is obtained. However, we may obtain a better approximation of (5.16) by computing exact values. Since multivariate lattices are generally of smaller depths than univariate ones for computational reasons, the gain from using exact over asymptotic solutions may be significant.

We compute the solutions to (5.21) numerically. The quadratic form of some of the equations suggests that solutions will not be unique. We need to restrict the solutions further. An acceptable solution satisfies

$$u_S > d_S \quad \text{and} \quad u_{V+} > d_{V+}. \tag{5.23}$$

With the move sizes thus obtained, the lattice can be constructed. Option prices can then be obtained by the standard recursive valuation technique first outlined in Cox, Ross, and Rubinstein (1979).

As with European options in the univariate case, there is a simple representation of the price of a bivariate European option. Since European options cannot be exercised prior to maturity of the contract, it is unnecessary to traverse all intermediate nodes. It is sufficient to have payoff values at maturity to derive the price of the option. Hence, we arrive at the following proposition.

Proposition 5.2.1. *The time t price of a derivative security with payoff function $\Lambda(S(T), V(T))$ at maturity date T is given by*

$$\Lambda(t) = e^{-rnh} \sum_j^n \sum_k^n \binom{n}{k} \binom{n}{j} p_1^a p_2^b p_3^c p_4^d \Lambda(S_T, V_T), \tag{5.24}$$

where

$$S_T = S(t) u^j d^{n-j},$$
$$V_T = V(t) u_{V+}^a u_{V-}^b d_{v+}^c d_{V-}^d,$$

and

$$a = \max(k + j - n, 0), \qquad b = k - \max(k + j - n, 0),$$
$$c = j - \max(k + j - n, 0), \qquad d = n - j - k + \max(k + j - n, 0).$$

n is the number of time steps of length h in the lattice and is calculated $n = (T - t)/h$. p_S and p_V are the risk-neutral probabilities of an up-move in each period for the state variables S and V, respectively.

Because of the initial assumption in equation (5.20), the probability condition in equation (5.17), and $a + b + c + d = n$, equation (5.24) simplifies to

$$\Lambda(t) = e^{-rnh} \sum_j^n \sum_k^n \binom{n}{k} \binom{n}{j} \left(\frac{1}{4}\right)^n \Lambda(S_T, V_T), \tag{5.25}$$

where S_T and V_T are defined as before.

American options are different from European options in that their prices cannot be computed by simply analyzing terminal process values. As for standard default-free options, it is necessary to traverse the lattice from the leaves to the root and to check for exercise at each node. In fact, in a credit-risky environment, early exercise may be optimal at nodes where it would not be if the contract were free of counterparty default risk.

5.2.2 The Bankruptcy Process

The previous section did not include a bankruptcy process and was therefore a discrete-time replication of Proposition 4.2.1. In this section, we introduce a bankruptcy process as proposed in (5.1). The process is driven by a default intensity $\lambda(t)$, which is a function of the firm value as described in Section 5.1.3. Figure 5.2 illustrates the bankruptcy process for a stochastic recovery rate. The process starts at the upper node denoted by 0. It is assumed that the firm is not bankrupt yet. The process can then jump to four states, (V_u, n), (V_d, d), (V_u, b), (V_d, b). b indicates that the firm is in default, n means no default. Alternatively, we write 0 for no-default because the bankruptcy process

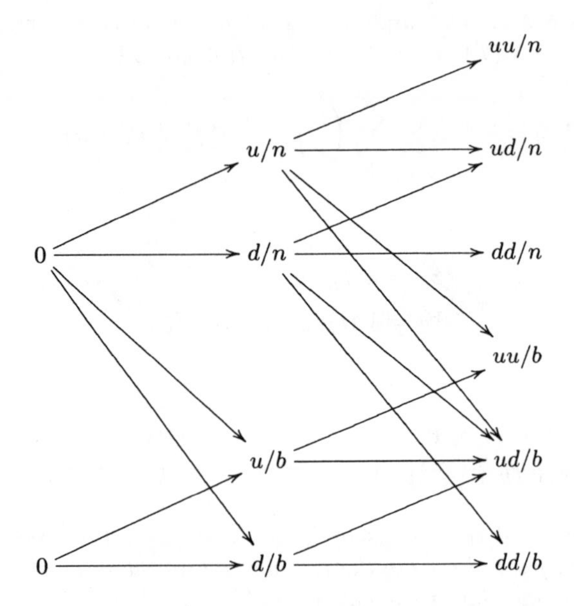

Fig. 5.2. Bankruptcy process with stochastic recovery rate

$\mathbf{1}_{\{\mathcal{T} \leq t\}}$, as defined in expression (5.1), is equal to 0 before default and 1 after default. Once default has occurred, the firm stays in default. From that point on, there are therefore only two branches to go from each node.

When valuing vulnerable derivatives, we have an additional state variable — the underlying asset of the derivative instrument. Therefore, the dimensionality of the lattice increases and the processes of Figure 5.1 and Figure 5.2 combine into Figure 5.3.

States in the lattice are now defined by the triple $(X, V, 1)_t$, which represents the values of the processes for X, V, and the default indicator function $\mathbf{1}_{\{\mathcal{T} \leq t\}}$, all evaluated at t. The initial state at the root of the lattice is given by $(X_0, V_0, 0)$. X and V can move either up or down, denoted by X_u, X_d, and V_u, V_d, respectively. Note that subscripts u and d only indicate the direction of a move, not its magnitude. The value of the default indicator function can change from 0 to 1 or remain unchanged.

After the first time step, a total of eight distinct states is attainable, namely: $(X_u, V_u, 0)$, $(X_u, V_d, 0)$, $(X_d, V_u, 0)$, $(X_d, V_d, 0)$ as no-default states and $(X_u, V_u, 1)$, $(X_u, V_d, 1)$, $(X_d, V_u, 1)$, $(X_d, V_d, 1)$ as default states. After the next time step, there is a total of eighteen states, etc. Because we require the tree to recombine, values X_{ud} and X_{du} and V_{ud} and V_{du} are indistinguishable.

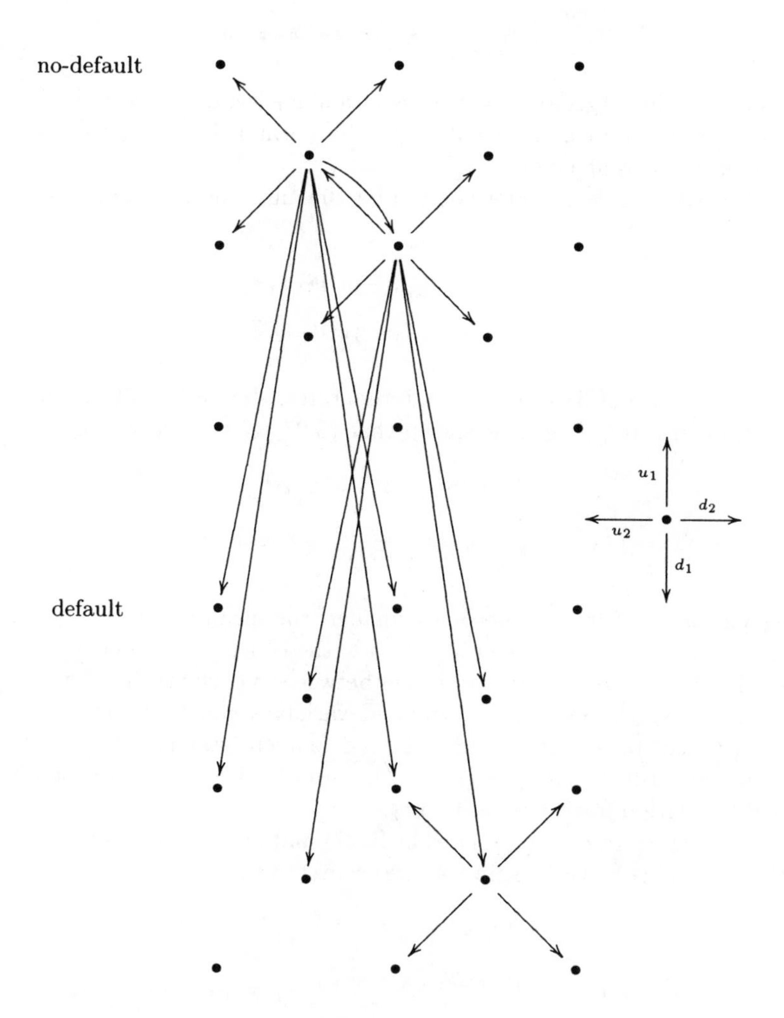

Fig. 5.3. Multidimensional lattice with bankruptcy process

5.2.3 An Extended Lattice Model

In this section the lattice model is simplified and extended to stochastic interest rates. Instead of the exactly matched moments, an asymptotic approximation is used. First the lattice is developed for vulnerable equity options with deterministic interest rates, then it is extended to vulnerable interest-rate options.

The process for the underlying security is given by expression (5.16). We can write the discretized process as

$$X_{t+h} = X_t e^{(r-\frac{1}{2}\sigma^2)h+\sigma\epsilon\sqrt{h}}, \tag{5.26}$$

where h is the length of time between t and $t+1$ and ϵ the standard binomial distribution as an approximation to the normal distribution. The discrete process for V is analogous.

By setting $\epsilon = (1, -1)$ we obtain for the multiplicative move sizes u_X and d_X,

$$\begin{aligned} u_X &= e^{(r-\frac{1}{2}\sigma^2)h+\sigma\sqrt{h}}, \\ d_X &= e^{(r-\frac{1}{2}\sigma^2)h-\sigma\sqrt{h}}. \end{aligned} \tag{5.27}$$

In the case of deterministic interest rates, the probability of up or down moves is set to $\frac{1}{2}$. It can be shown that (5.27) gives expectations

$$\begin{aligned} \lim_{h\to 0} E(X_{t+h}|\mathcal{F}_t) &= X_t e^{rh}, \\ \lim_{h\to 0} E(X_{t+h}^2|\mathcal{F}_t) &= X_t^2 e^{(2r+\sigma^2)h}, \end{aligned} \tag{5.28}$$

which ensure that, as h becomes smaller, the mean and the variance of the lattice process converge to the values of the continuous process.

To obtain the desired correlation between underlying and firm value process, we use the fact that correlated variables can be manufactured from independent ones. Let ϵ_1 and ϵ_2 be independent standard binomials. It can easily be verified that $\epsilon_3 = \rho\epsilon_1 + \sqrt{1-\rho^2}\epsilon_2$ is also a standard binomial and has correlation ρ with respect to ϵ_1.

Setting again $\epsilon_1 = (1, -1)$ as in (5.27) and now additionally $\epsilon_2 = (1, -1)$, the firm value move sizes in the lattice are given as

$$\begin{aligned} u_{V|u_X} &= e^{(r-\frac{1}{2}\sigma^2)h+\sigma\sqrt{h}(\rho+\sqrt{1-\rho^2})}, & \epsilon_1 &= +1, & \epsilon_2 &= +1, \\ d_{V|u_X} &= e^{(r-\frac{1}{2}\sigma^2)h+\sigma\sqrt{h}(\rho-\sqrt{1-\rho^2})}, & \epsilon_1 &= +1, & \epsilon_2 &= -1, \\ u_{V|d_X} &= e^{(r-\frac{1}{2}\sigma^2)h-\sigma\sqrt{h}(\rho-\sqrt{1-\rho^2})}, & \epsilon_1 &= -1, & \epsilon_2 &= +1, \\ d_{V|d_X} &= e^{(r-\frac{1}{2}\sigma^2)h-\sigma\sqrt{h}(\rho+\sqrt{1-\rho^2})}, & \epsilon_1 &= -1, & \epsilon_2 &= -1. \end{aligned} \tag{5.29}$$

The probability for one of the moves in (5.29) to occur is set to $\frac{1}{4}$. It can be shown that, in addition to the equivalent of (5.28) for V, the following condition is satisfied.

$$\lim_{h\to 0} E(X_{t+h}V_{t+h}|\mathcal{F}_t) = X_t V_t e^{(2r+\rho\sigma_X\sigma_V)h}. \tag{5.30}$$

Conditions (5.28) and (5.30) imply that the lattice processes asymptotically approximate the continuous processes.

5.2.3.1 Stochastic Interest Rates. For interest rate claims, interest rates instead of a stock price are modeled as state variable X. In this implementation, we assume that the short rate is Markovian such that a recombining lattice can be constructed.[9] It is well known that a class of Gaussian models satisfies the conditions for recombination. Examples of Gaussian one-factor models include Merton (1973) and Vasicek (1977) and their respective time-inhomogeneous versions proposed by Ho and Lee (1986) and Hull and White (1990). The following exposition is based on the model by Ho and Lee (1986). However, similar recombining lattices can be constructed for any of said Gaussian one-factor models. The method described below also readily extends to models of the Black, Derman, and Toy (1990) type.

Specifically, the short rate is given by

$$r_t = r_0 + \int_0^t \alpha(s)ds + \sigma \tilde{W}_t,$$

where α is defined as a predictable process and σ is constant. In discrete time,

$$r_{t+h} = r_t + \alpha_t h + \sigma \epsilon \sqrt{h}. \tag{5.31}$$

Let the probability for the short rate to move up or down again be $\frac{1}{2}$. Setting $\epsilon = (1, -1)$ we obtain for the move sizes u_X and d_X,

$$\begin{aligned} r_u &= r_0 + \alpha_t h + \sigma \sqrt{h}, \\ r_d &= r_0 + \alpha_t h - \sigma \sqrt{h}. \end{aligned} \tag{5.32}$$

Here, u and d are absolute move sizes. The standard deviation in one period is $\sigma\sqrt{h}$. In other words,

$$r_u = r_d + 2\sigma\sqrt{h}. \tag{5.33}$$

Thus, the volatility parameter immediately determines the distance of the two attainable rates from each other.

The risk-neutral default intensity is given by λ. The default probability in one lattice interval is λh. In the limit,

$$\lim_{h \to 0} (1 - \lambda h)^{t/h} = e^{-\lambda t}.$$

This is the survival probability over time length t for constant λ. Thus, in the limit, default time is exponentially distributed. For time-varying λ, we have

$$\lim_{h \to 0} (1 - \lambda(t)h)^{t/h} = e^{-\int_0^t \lambda(s)\, ds}.$$

Hence, the survival probability converges asymptotically to the specification in expression (5.2). Option prices can now be obtained by the standard method of traversing the lattice.

---

[9] See Carverhill (1994) or Jeffrey (1995) for necessary and sufficient conditions on the volatility function with respect to the Markov property.

5.2.3.2 Recombining Lattice versus Binary Tree. It is not difficult to see that the lattice as defined above recombines such that the number of nodes increases linearly with the number of tree steps. This naturally recombining property of the lattice is lost for more complicated volatility functions. A non-recombining binomial lattice is a binary tree for which the number of nodes grows exponentially with the number of time steps. Pricing options in such a tree is a problem that can only be solved in non-polynomial time. It is therefore desirable that the tree recombine.

Provided that the volatility function implies a Markovian evolution of the short rate, recombining lattices can be constructed even for non-constant volatility functions. Nelson and Ramaswamy (1990) propose a transformation to the process that removes its heteroscedasticity. Tian (1992) simplifies this approach by ensuring non-negative probabilities without the need for multiple jumps in one step. There are other approaches that allow time-varying volatility. For example, Amin (1991) uses variable-sized tree steps to ensure recombination.

In the multivariate context, there is an additional problem. Although the interest rate lattice is recombining for our specification of the short rate or can be made to recombine for other specifications as mentioned above, the interest rate process also affects the evolution of the firm value V with the result that the multivariate lattice does not recombine naturally if interest rates are random. There are several ways of dealing with this problem. First, a simple approximation can used to make the lattice recombine. The approximation interpolates between each two nodes obtained by different paths and sets the value of the node to the approximate value obtained by interpolation. Precision can be increased by storing more than one value.[10] This simple interpolation technique was used to compute the numerical examples in Section 5.6.

Alternatively, we can transform the tree to eliminate any path-dependence. This involves changing risk-neutral probabilities[11] as in the univariate case. Examples of such methods are Amin (1991), Amin and Bodurtha (1995), and Ho, Stapleton, and Subrahmanyam (1995), in increasing order of generality. In addition to changing volatility, Ho, Stapleton, and Subrahmanyam (1995) allow for changing correlation between assets.

[10] Such methods are described in Hull and White (1993a) and Dharan (1997) for path-dependent options and path-dependent interest-rate trees, respectively.

[11] The equivalence of modifying probabilities or move sizes to match expected value, variance, and correlation in the multivariate case, of continuous processes was explained in Section 5.2.1.

5.3 Prices of Vulnerable Options

The price of a vulnerable European option can be expressed in terms of Green's function.[12] Let $p_u(X)$ the payoff function of a claim at time u. If default probability and firm value are independent, the price of this claim at time t allows the simple representation

$$p_t = \Omega_u + \Big(\prod_{k=t+h}^{u} (1-\lambda_k)\Big)(\Sigma_u - \Omega_u), \tag{5.34}$$

where

$$\begin{aligned} \Sigma_u &= \sum_i \sum_j G_u(i,j,0) p_u(i,j,0), \\ \Omega_u &= \sum_i \sum_j G_u(i,j,1) p_u(i,j,1), \end{aligned} \tag{5.35}$$

where $G_u(i,j,0)$ is the discretized Green's function evaluated at time u and in states $(i,j,0)$. 0 denotes no-default, 1 default. The product $\prod$ is taken over index increments of size h. The values from evaluating the Green's function for a given state are also called Arrow-Debreu prices since they represent the price at time zero of a claim paying unity in that state and nothing in all other states (elementary state claim.) In other words, this is the discounted probability of that state under the martingale measure. The prices for the elementary state claims can be obtained by forward induction given the risk-neutral probabilities of up moves for each factor.

The valuation expression in (5.34) has a rather intuitive interpretation. Ω can be interpreted as the price of the claim if default is certain to occur. Σ, on the other hand, is the price of the claim if default is certain not to occur. The price of the claim is the price if default is certain plus the difference between the default-free price and the sure-default price multiplied by the cumulated survival probability.

Expression (5.35) is the discrete-time equivalent representation to expression (5.8). $(\prod_{k=t+h}^{u}(1-\lambda_k))$ is the cumulated survival probability denoted by $S(t,u)$ in the continuous-time case. $\Sigma_u - \Omega_u$ corresponds to $\Lambda(t)(1 - \mathbf{E_Q}[\delta(u)|\mathcal{F}_t])$.

With constant interest rates, the value of the Green's function would simplify to

$$G_t = e^{-rt}\left(\frac{1}{4}\right)^n \sum_i^n \sum_j^n \binom{n}{i}\binom{n}{j} p_t(i,j), \tag{5.36}$$

[12] Refer to Jamshidian (1991b) for an introduction to forward induction and the use of Green's function in this context. Cf. Bronštein and Semendjaev (1995), Chapter 9 for the role of Green's function in the context of PDEs.

where

$$n = \frac{u - t}{h}$$

is the number of time steps. In a stochastic interest rate framework, the lattice is traversed by forward induction to obtain the Green's functions.

Prices for American options are computed in the standard fashion by traversing the lattice by backward induction. At each node it is checked if the **Q**-expected and discounted value is greater than the profit from immediate exercise. If not, the option is exercised. Vulnerable options may be exercised at times when default-free options would not. Vulnerable options are never exercised later than default-free options but may be exercised sooner.

The pricing model presented is not restricted to defaultable options. Symmetrical claims can also be modeled in the given structure. A forward contract, for example, can be assembled by a call and a put option. Each of the two options is exposed to default risk from a different counterparty. This problem was studied in Section 4.6. The principle of valuation is the same, whether the vulnerable options that are used to model the forward contract are themselves modeled by a firm value or a hybrid credit risk model.

5.4 Recovering Observed Term Structures

Ideally, the initial term structures implied by the multivariate lattice fit observed market term structures. This result can be obtained by fitting both the risk-free term structure and the risky term structure separately.

5.4.1 Recovering the Risk-Free Term Structure

The risk-free term structure can be recovered in a standard way. The drift of the process is time-variable and chosen such that the zero-coupon bond prices implied by the model coincide with observed market prices. To obtain the interest rates for each node, we use a standard technique based on forward induction, as proposed by Jamshidian (1991b). First, the Green's functions at the end of the first time step are calculated. This is easily done as the applicable interest rate for the first period is known. The price of a bond which matures after the second time step is given by the sum of the Green's functions in all states at the end of the second time step. For the bond, the Green's functions at time t are equal to those of time $t - h$ multiplied by the discount rate. Thus, the bond price at time t is

$$B_0(t) = \sum_k G_{t-h}(k) e^{-(r_{t-h}(0) + 2k\sigma\sqrt{h})h}. \tag{5.37}$$

The interest rates $r(k)$ in different states k can be expressed in terms of one rate at that time because of the distance between rates which was established in expression (5.33). This equation can now be solved for $r(0)$. This procedure is repeated for each time step in the lattice.

Alternatively, the short rate can be modeled within the Heath, Jarrow, and Morton (1992) framework. In that case no fitting is necessary because the initial term structure is an input to the model.

5.4.2 Recovering the Defaultable Term Structure

While fitting the risk-free term structure is straightforward, fitting the credit-risky term structure poses some problems. The first problem arises if the recovery rate, and not the actual firm value, is modeled. In this case, an initial value has to be estimated for the recovery rate. The size, however, of the credit spread is determined by two factors: the expected rate of recovery in the case of default and the probability of such a default. It is not generally possible to derive an estimation for the recovery rate and for default intensities from observed market prices of risky bonds. Hence, one variable has to be estimated, for example from historical data and then the other one can be derived from market prices. This approach was taken for some of the numerical examples in Section 5.6, where the recovery rate was estimated and the deterministic default intensities were derived from bond prices.

If prices of credit spread options or similar credit derivative instruments exist, then it is possible to derive both recovery rate and implied default intensity from market prices. It is important that the credit derivative be a true credit derivative,[13] not an instrument which can be priced with simple bond credit models.

Provided an estimate for the recovery rate exists, the hybrid credit risk model can be calibrated such that the implied credit spread is equal to an observed exogenous credit-risky term structure. This is achieved by an iterative procedure which derives the risk-neutral default intensities for each period from market bond prices.

Expression (5.34) applies to any claim, in particular also to credit-risky zero-coupon bonds. Replacing p_t with the observed price of the bond maturing at u and rearranging terms, we obtain

$$\lambda_u = \frac{B_0(u) - \Omega_u}{(\prod_{k=t+h}^{u-h}(1-\lambda_k))(\Omega_u - \Sigma_u)} + 1, \tag{5.38}$$

where Σ_u and Ω_u are defined as in (5.34).

The procedure to calculate λ_u for all relevant u starts at the first time step in the lattice. Once λ_{t+h} has been calculated, the formula can be applied to λ_{t+2h}, and so on, until λ_u is known for the last time step. Of course, this procedure requires zero-coupon bond prices for each maturity. They need to be recovered from debt market data. If no market data is available, the credit

[13] In the terminology of Chapter 6, this is a credit instrument of the second class.

term structure has to be estimated or a different approach has to be used, such as the specific functional definition of default risk.

The recursive procedure just described can only be employed if default intensities are assumed to be deterministic. This assumption restricts the model in some important ways. Default intensities can no longer be made dependent on state variables such as the interest rate or the firm value. Dependencies and correlation effects consequently cannot be modeled by means of default probabilities. Correlations, however, can still be incorporated by the stochastic behavior of the recovery rate, though.

On the other hand, keeping the functional dependency between default intensities and state variables makes fitting the implied risky term structure more difficult. In this case, parameters have to be estimated such as to achieve a good fit without sacrificing other desirable characteristics. The problem is similar to estimating parameters of risk-free term structures. In that case, if not carefully chosen, the parameters may imply an unrealistic term structure of volatilities, for example.

5.5 Default-Free Options on Risky Bonds

Although not a special case, default-free options on credit-risky bonds can be handled in the hybrid credit risk model in a straightforward fashion. The key to the solution is the fact that a credit-risky bond is a special case of a vulnerable option. If bond and option maturity dates coincide, a bond is a vulnerable option that promises to pay unity at maturity. While a normal vulnerable option is a bivariate option, the bond is a univariate option because the promised payoff is deterministic. A risky bond can be viewed as a vulnerable option whether credit risk is modeled in a firm value framework as in Chapter 4 or as a bankruptcy process model as in Jarrow and Turnbull (1995), or as a hybrid model as in this chapter. If a bankruptcy process determines credit risk, then the underlying variable of the option is a jump process instead of a diffusion. The principle, however, is the same.

From expressions (5.4) and (5.5), the price of the bond maturing at u is given by

$$\begin{aligned} \Lambda^d(t) &= B(t)\mathbf{E}_{\mathbf{Q}}[B^{-1}(u)g(u)\mathbf{1}_{\{u<\tau\}}|\mathcal{F}_t] \\ &\quad + B(t)\mathbf{E}_{\mathbf{Q}}[B^{-1}(u)\,g(u)\delta(u)\mathbf{1}_{\{\tau\leq u\}}|\mathcal{F}_t] \\ &= \Lambda(t) - B(t)\mathbf{E}_{\mathbf{Q}}[B^{-1}(u)\,g(u)\,N(u)\,(1-\delta(u))|\mathcal{F}_t], \end{aligned}$$

with $g(u) = 1$. At the same time, this expression can be interpreted as an option on the default-free bond with promised payoff $g(u)$.

Realistically, however, the maturity date of the option is before the maturity date of the bond. If the option expires at $T < u$, we have

$$\Lambda(t) = B(t)\mathbf{E}_{\mathbf{Q}}[B^{-1}(T)\,g(P^d(T,u))|\mathcal{F}_t],$$

with

$$P^d(T,u) = P(T,u) - B(T)\mathbf{E}_{\mathbf{Q}}[B^{-1}(u)N(u)(1-\delta(u))|\mathcal{F}_T].$$

$P(T,u)$ is the time T price of the riskless bond, $P^d(T,u)$ of the risky bond. For example, in the case of a call option, we have

$$g(P^d(T,u)) = \max\left(P^d(T,u) - K, 0\right).$$

Because the payoff of the option is itself determined by an option value, a default-free option on the risky bond can be treated as a compound option. The underlying variable of this compound option is not only the option embodying credit risk but also the default-free bond. Hence, the option on risky bonds is an option on a combination of assets, namely the risk-free bond and credit risk. Because a risky bond is a special case of a vulnerable option, the default-free option on the risky bond can also be viewed as an option on a vulnerable claim with a deterministic underlying variable.

The compound option can be valued within the lattice approach suggested in Section 5.2.3. The lattice was constructed to value vulnerable options. An option expiring at T on a vulnerable option expiring at u can be handled in the following way: the lattice is constructed by forward induction until time u. When traversing backwards through the tree determining the prices of the underlying option, the payoff of the compound option can be determined in all time T nodes. The price of the compound option is then obtained in the usual way by backward induction.

5.5.1 Put-Call Parity

While the put-call parity does not generally hold for vulnerable options, it is valid for default-free options on risky bonds. Let $P^d(t,T)$, $P^d(t,u)$, $P(t,T)$, and $P(t,u)$ be bond prices with $t < T < u$. For options on riskless bonds, the put-call parity is calculated as

$$p(x) + P(t,u) = c(x) + xP(t,T), \tag{5.39}$$

where p and c denote put and call option prices, respectively, and x is the strike price of the option. If this parity relationship holds, a portfolio consisting of the riskless long-term bond and a put option is equivalent to a portfolio consisting of a riskless short-term bond and a call option.

If the underlying bond is subject to credit risk, (5.39) changes to

$$p'(x) + P^d(t,u) = c'(x) + xP(t,T). \tag{5.40}$$

The prime indicates a price based on the credit model. While the long-term bond is now the risky bond, the short-term bond is still riskless. The short-term bond is used for discounting purposes only. If it were not riskless, the equality would no longer hold because $xP(t,T)$ might not be worth x at time T. The parity relationship can be verified by standard methods.

5.6 Numerical Examples

This section provides a number of numerical sample prices. In the first subsection, the implementation of Section 5.2.1 is used with a bankruptcy process according to Section 5.2.2. The second subsection includes stochastic interest rates.

5.6.1 Deterministic Interest Rates

Tables 5.1 to 5.3 show percentage reductions of vulnerable options compared to their default-free counterparts. The prices are based on the model developed in Section 5.2.1 and include a bankruptcy process. With respect to the bankruptcy process, the additional assumption is made that default is certain at maturity of the option if the value of the firm is below the value of the liabilities. In other words, the default intensity rises to infinity at maturity if $V < D$. With this assumption, prices for European options approximate the prices obtained when using the explicit formulae of Proposition 4.2.1. The advantage of the lattice method in this case is the ability to price vulnerable American options.

The American options marked with an asterisk denote prices computed without bankruptcy process. Whenever the option is exercised prior to expiration, the payoff is multiplied by the recovery rate, i.e.,

$$\Lambda(t) = g(t) \min\left(\frac{V_t}{D}, 1\right),$$

where $g(t) = \max(S(t) - X, 0)$ for call options and $g(t) = \max(X - S(t), 0)$ for put options. This definition of the payoff prior to maturity can be interpreted as default occurring immediately after valuation of the contract. As a consequence, these prices form a lower bound of possible prices for American options. Similarly, assuming that default does not occur until maturity gives an upper bound of prices. This is, however, a trivial upper bound as the option can be exercised immediately prior to expiration. Hence, the upper bound corresponds to the default-free value.

Prices of American options not marked by an asterisk include a simple bankruptcy process. Default intensity is chosen to be $\lambda(t) = D/V(t)$. This parameter choice implies that the risk-neutral probability of a default decreases with increasing V relative to a given D. As the ratio of D/V can be interpreted as the debt-to-assets ratio, an increasing debt-to-assets ratio also increases the probability of default, thus recognizing that the ability of the firm to retire debt decreases with an increased debt-to-assets ratio, as discussed in Section 5.1.3. However this direct specification gives somewhat unrealistic default probabilities and would have to be refined for practical purposes. For example, $\lambda(t) = (D/V(t))^2$ may be a more realistic specification. The simple specification is, however, sufficient for our purposes in that it demonstrates the effect of a bankruptcy process on American option prices.

The only difference between Tables 5.1, 5.2, and 5.3, is the underlying strike price with respect to which the prices were computed. The ratios between strike price and initial value of the underlying are chosen to be 1, 0.8, and 1.2, for Tables 5.1, 5.2, and 5.3, respectively.

The option prices displayed in these tables imply a risky two-year zero-coupon bond price of 88.63 for a face value of 100. This is 2.1% less than the default-free price of 90.48. In terms of yield spreads, this corresponds to a spread of 100 basis points over the default-free bond. Note that the prices for European options when underlying and firm assets are uncorrelated also result in a reduction of 2.1%. This finding is consistent with Proposition 5.1.1 on page 145.

Table 5.1. Price reductions for different correlations

in %	$X/S = 1$			ρ			n
		-1	-0.5	0	0.5	1	
Calls	European	9.8	5.2	2.1	0.4	0.0	50
	American*	9.6	5.0	2.0	0.4	0.0	50
	American	7.7	5.0	2.0	0.4	0.0	12
Puts	European	0.0	0.4	2.1	5.0	8.9	50
	American*	0.0	0.2	1.2	2.4	3.1	50
	American	0.0	0.2	0.9	1.6	1.8	12

Parameters: $X/S = 1$, $K/V = 0.9$, $\sigma_S = 0.25$, $\sigma_V = 0.15$, $r = 0.05$, $t = 2$. The asterisk denotes American options computed without bankruptcy process. n is the number of intervals in the lattice.

Table 5.2. Price reductions for different correlations

in %	$X/S = 0.8$			ρ			n
		-1	-0.5	0	0.5	1	
Calls	European	7.2	4.3	2.1	0.6	0.0	50
	American*	6.7	4.0	1.9	0.6	0.0	50
	American	5.2	3.3	1.7	0.5	0.0	12
Puts	European	0.0	0.2	2.1	6.5	15.0	50
	American*	0.0	0.2	1.6	4.8	11.3	50
	American	0.0	0.1	1.3	3.9	8.4	12

Parameters: $X/S = 0.8$, $K/V = 0.9$, $\sigma_S = 0.25$, $\sigma_V = 0.15$, $r = 0.05$, $t = 2$. The asterisk denotes American options computed without bankruptcy process. n is the number of intervals in the lattice.

As expected, the impact of correlation on prices is significant. While call options contain virtually no credit risk if V and S are perfectly correlated, there is considerable credit risk involved if the processes are negatively correlated. For put options, the opposite relationship holds. For uncorrelated

Table 5.3. Price reductions for different correlations

in %	$X/S = 1.2$			ρ			n
		-1	-0.5	0	0.5	1	
Calls	European	13.5	6.2	2.1	0.3	0.0	50
	American*	13.5	6.1	2.0	0.3	0.0	50
	American	10.9	5.1	1.8	0.2	0.0	12
Puts	European	0.0	0.6	2.1	4.0	6.1	50
	American*	0.0	0.2	0.6	0.7	0.0	50
	American	0.0	0.2	0.4	0.5	0.1	12

Parameters: $X/S = 1.2$, $K/V = 0.9$, $\sigma_S = 0.25$, $\sigma_V = 0.15$, $r = 0.05$, $t = 2$. The asterisk denotes American options computed without bankruptcy process. n is the number of intervals in the lattice.

processes, the reduction in the price of the option always corresponds to the price reduction of the risky bond regardless of the strike price of the underlying and whether call or put options are considered.

An example of options where underlying and firm value are perfectly correlated are call warrants issued by a firm. Such warrants give the holder the right to receive new shares of the firm. The underlying is therefore the equity value of the firm. Equity and asset value can be assumed to be perfectly correlated if liabilities are deterministic, as they are in this model. Our model confirms the intuition that warrants can be treated as default-free since they can only be in-the-money if the firm is not bankrupt. Quite a different situation prevails if the firm chooses to sell put warrants. In this case underlying and firm value are perfectly negatively correlated, resulting in a option price spread far more pronounced than the corresponding bond price spread.

It is interesting that in our examples, in-the-money calls are slightly riskier than than out-of-the-money calls for positive correlation. The opposite is true for negative correlation. For put options, in-the-money claims display lower credit risk than out-of-the-money claims for positive correlation. This is another intuitive result since it is more likely for the option to contain any intrinsic value when default occurs for in-the-money call if correlation is positive. For call options, credit risk is greatest for high strikes with perfectly negative correlation. In this situation it is most unlikely that the option acquires intrinsic value without default occurring. For put options, the same statement applies to perfectly correlated processes.

The fact that it may pay to exercise American options prior to maturity even for call options which are written on assets that do not pay dividends is reflected in the equal or smaller price reductions for American options. The difference in reduction of call option prices between European and lower-bound American options is minimal, although slightly rising with decreasing strike prices. A vulnerable option is a bivariate option consisting of the default-free option and a short option portraying credit risk. Regarding early exercise, the two components contain opposite effects. While the default-free option by itself is not exercised early, the combination with the other option may make early exercise optimal. That is why the price reduction is slightly lower even for lower-bound American options.

If a bankruptcy process is added to the model, the difference between European and American options becomes even greater. This is because the option holder can exercise the option if the discounted expected value under the equivalent measure is less than the payoff from immediate exercise and receive the full default-free payoff if the firm has not defaulted at that time. The fact that the full default-free value can be obtained by exercising before default makes early exercise for attractive relative to the lower-bound American options.

Put options generally display a greater difference between reductions for European and lower-bound American options. As put options are often optimally exercised prior to maturity even if no default risk is present, the amount of credit risk involved is smaller than for a corresponding call option which is optimally held until maturity.

Interestingly, the reduction for American put options with high correlation is zero for the in-the-money puts while for European options the reduction is highest for those correlations. The reason for this phenomenon is the fact that, as the option moves further in-the-money, it becomes optimal to exercise it before the firm can default. However, if it is always optimal to exercise before default can occur, the option is free of credit risk.

The difference between the prices computed without bankruptcy process and those computed with bankruptcy process seems small. Part of the reason is the simple choice of the default intensity function. $\lambda = D/V_t$ implies a rather high default probability even for normal debt-equity ratios. More realistic λ functions would slightly decrease the price reductions for American options. Another reason is our assumption that default is certain at maturity if the firm value is lower than the value of liabilities.

Table 5.4. Price reductions with default cost

in %	$X/S = 1$			ρ			n
		-1	-0.5	0	0.5	1	
Calls	European	13.4	8.4	4.6	1.7	0.1	50
	American	12.8	8.1	4.5	1.7	0.1	12
Puts	European	0.0	1.7	4.6	8.3	13.2	50
	American	0.0	1.6	3.4	5.2	7.7	12

Parameters: $X/S = 1.2$, $K/V = 0.9$, $\sigma_S = 0.25$, $\sigma_V = 0.15$, $r = 0.05$, $t = 2$. n is the number of intervals in the lattice. Default cost is defined to be 25% of asset value at time of default.

Table 5.4 shows reductions in option prices if bankruptcy costs are included in the model. Bankruptcy costs are assumed to reduce the value of V by 25% at the time of default.

Table 5.4 shows that this assumption implies much higher price reductions than in Table 5.1, where the same parameters were used but bankruptcy costs were not included. However, it should be noted that these increased reductions also imply a higher credit spread on the bond price since bankruptcy costs are effectively increasing the volatility of V.

5.6.2 Stochastic Interest Rates

We assume there exists an exogenously given term structure for both default-free debt and credit-risky debt. The spot rate term structures are assumed to be defined by the following expression:

$$\begin{aligned} s_t &= 0.08 - 0.04e^{-0.07t}, \\ d_t &= 0.10 - 0.05e^{-0.08t}. \end{aligned} \tag{5.41}$$

s_t denotes the default-free spot rate of maturity t and d_t is the corresponding risky rate. The lattices are fitted to these two term structures by using the techniques described in section 5.4. The term structure is specified in this functional form to construct an example. Normally, the term structure is derived from market bond prices.

Table 5.5. Term structure data

bond prices		default free	credit risky
spot	2 years	91.35	89.16
spot	4 years	81.95	77.51
forward	2-4	89.71	86.94

This data was derived from the spot rates definition in expression (5.41).

Table 5.6 shows vulnerable option prices implied by the term structure of default spreads. The prices are for a bond option which expires in two years. The underlying default-free zero-coupon bond matures in four years and has a face value of 100. Option values are computed for strike prices of 88, 90, and 92. The default-free prices are marked with f, the vulnerable ones with v. For convenience, absolute interest volatility is assumed constant at 0.01 in a Ho and Lee (1986) model of the short rate. As pointed out in other sections, this is not a prerequisite of the credit model. The lattice can be adapted to account for changing volatility if volatility is a known function of time.

The recovery rate in the form of the firm value is estimated to be 0.5, initially. This variable is stochastic with a volatility of 15%. The prices are computed based on a lattice of forty intervals.

Table 5.6. Prices of vulnerable bond options

			strike		
			88	90	92
Calls	European	f	1.91	0.81	0.23
		v	1.86	0.79	0.23
	American	f	1.91	0.81	0.23
		v	1.86	0.79	0.23
Puts	European	f	0.35	1.07	2.33
		v	0.34	1.05	2.27
	American	f	6.05	8.05	10.05
		v	6.05	8.05	10.05

Prices of two-year options on a four-year zero-coupon bond for a given credit spread structure. v vulnerable. f default-free. Parameters: $V_0/K = 0.5$, $\sigma = 0.01$, $\sigma_V = 0.15$. 40 lattice steps.

Most vulnerable options exhibit a small reduction in price compared to the default-free options, with the exception of the American put options. These options are optimally exercised immediately. Since the counterparty is assumed to be solvent at the time of the pricing, the American puts do not contain any credit risk and therefore have the same price as their default-free counterparts.

The forward price of the bond can be derived from expression (5.41). Table 5.5 shows the relevant bond prices. Bonds of the counterparty trade at roughly 2.4% less than Treasuries for the same maturity of two years. Note that the vulnerable option prices are also valued at a discount of approximately 2.4%. This observation confirms again Proposition 5.1.1. By fitting the credit-risky lattice to the observed term structure, we implicitly assume deterministic default intensities. As a consequence, we also assume that the default process and the short rate process are uncorrelated.

Table 5.7 depicts the default probabilities derived from fitting the lattice to the exogenous term structures. Default probabilities are shown for several estimates of the initial recovery rates. The table shows that the probabilities are fairly stable even for widely differing initial recovery rates. Recall that these are risk-neutral probabilities. The probabilities were computed as cumulated values over six months of time. The values are annualized and given as percentages. This means that the first row shows the yearly probability under the equivalent measure that default occurs in the first six months. The probabilities increase moderately. Obviously, the change in the probabilities is determined by the term structure from which they were derived. Slowly monotonically increasing default probabilities are typical of high-grade bonds, as observed by Fons (1994), Sarig and Warga (1989), and others.

Table 5.8 shows prices of default-free options on a bond with credit risk. The underlying bond is derived from the credit-risky term structure defined in expression (5.41).

Table 5.7. Implied default probabilities

Time	V_0/K 0	0.2	0.5	1
0 - 0.5	1.05	1.06	1.06	1.07
0.5 - 1	1.16	1.17	1.17	1.19
1 - 1.5	1.26	1.27	1.28	1.30
1.5 - 2	1.36	1.37	1.39	1.42

Risk-neutral default probabilities in % for several time ranges in the lifetime of the 2-year claim. Default probabilities are cumulated and annualized in their time range. V_0/K is the estimated recovery rate for time 0.

Prices marked by f are based on the default-free term structure. p denotes pseudo-risky prices. Those prices were calculated in the same fashion as the prices on the riskless bond, but they were derived from the risky term structure. Finally, v denotes prices calculated with the credit risk model.

Default-free prices differ significantly from the prices computed by the other two methods. Clearly, they do not price the option correctly. They assume that in two years from now, all credit risk has vanished and that the option holder has the right to buy or sell a default-free bond. As can be seen in Table 5.5, this assumption results in a forward price rather different from the market forward price. The other two methods give a forward price equal to the market forward price. Option prices, however, need not be equal, as the table shows.

The prices derived from the risky term structure and by the credit model are similar for call option values. American put option values are identical since this option is exercised immediately and therefore does not contain credit risk. European put options, on the other hand, display significantly higher values if computed by the credit risk model. This reflects the chance of a large negative price change of the bond from the occurrence of default. Ignoring the possibility of default therefore results in underpricing European put options.

Note that the put-call parity holds in both cases. Recall that expressions (5.39) and (5.40) differ in the present value calculation of the strike price. When pricing bond options directly from the credit-risky term structure, the implicit assumption is made that the credit-risky interest rate is the only interest rate in the market. Therefore, the credit-risky rate is used for compounding. On the other hand, in the second case, the two portfolios on both sides of the equation are only equivalent if the investor's term investment is guaranteed to be worth x at maturity of the claim. This can only be achieved by investing in the riskless bond. Thus, the riskless rate must be used for compounding. If $P^d(t,u)$, $P^d(t,T)$, $P(t,T)$ are set to their respective values given by Table 5.5, it can easily be verified that computed option

Table 5.8. Prices of options on credit-risky bond

			strike			
			84	86	88	90
Calls	European	f	5.22	3.45	1.88	0.81
		p	2.73	1.37	0.50	0.11
		v	2.72	1.36	0.49	0.10
	American	f	5.22	3.45	1.88	0.81
		p	2.73	1.37	0.50	0.11
		v	2.72	1.36	0.49	0.10
Puts	European	f	0.01	0.07	0.32	1.07
		p	0.11	0.53	1.45	2.84
		v	1.95	2.42	3.38	4.82
	American	f	2.05	4.05	6.05	8.05
		p	6.49	8.49	10.49	12.49
		v	6.49	8.49	10.49	12.49

Prices of 2-year default-free options on 4-year credit-risky zero-coupon bond for a given credit spread structure. f default-free. p pseudo. v credit-risky. Parameters: $V_0/K = 0.5$, $\sigma = 0.01$, $\sigma_V = 0.15$. 20 lattice steps.

prices are consistent with the put-call parity equations as specified in (5.39) and (5.40).

5.7 Computational Cost

The algorithms used to implement the proposed credit risk models tend to be computationally costly. For univariate recombining lattices, the number of nodes grows linearly with the number of steps and the computational cost is of order $O(n^2)$,[14] n being the number of steps in the lattice. For bivariate lattices, the number of terminal nodes is of order $O(n^2)$ and the total computational cost $O(n^3)$. If a bankruptcy process is added, terminal nodes and total cost further increase, although the total cost is less than $O(n^4)$ because once bankruptcy has occurred, the growth of the lattice decreases. This effect can be seen in Figure 5.3 on page 155. For any node representing a state in which bankruptcy has not occurred, there are eight new tree branches whereas only four branches grow out of nodes representing a state of bankruptcy.

These indications of computational cost assume that the lattice is bivariate with a credit risk state variable such as the firm value and an underlying state variable such as the price of a stock or the short rate. Of course, if more complicated instruments are valued, such as multi-factor options or interest rate derivatives within a two-factor interest rate model, the computational complexity of the algorithms increases even further.

[14] $f = O(g)$ means that there is a constant c such that $f \leq cg$.

Moreover, if the processes of the variables are non-Markovian, recombining trees are not possible. A popular example of a non-Markovian model is a general specification of the evolution of the forward rate in according to Heath, Jarrow, and Morton (1992). If the volatility structure is assumed to be stochastic or interest rates are assumed not to be normally distributed, then only a non-Markovian specification is possible.[15]

Table 5.9 shows price reductions of vulnerable bond options computed on the basis of a Heath-Jarrow-Morton evolution of forward rates in a non-recombining binary tree. The evolution of the forward rates follows the description in Section 2.4. To be able to compare the results, initial term structure and volatility function are chosen to match the specification of the Vasicek model in Figure 4.3, although this specification would allow a recombining tree. The results in Table 5.9 are directly comparable to Table 4.4. The maturity of the option is in 2 years, of the underlying bond in 5 years. The credit risk model also corresponds to that used for the results in Table 4.4, i.e., no bankruptcy process is used.

Table 5.9. Price reductions for various ρ_{VP} under HJM

in %		ρ_{VP}				
		- 0.9	-0.5	0	0.5	0.9
K=84.73	Call	12.0	6.4	2.2	0.5	0.1
	Put	0.1	0.6	2.2	4.4	6.3

Percentage price reductions of vulnerable zero-coupon bond options with respect to the corresponding default-free option price. 10 tree steps.

The price reductions displayed in Table 5.9 deviate slightly from the exact value from Table 4.4. The main reason for the lack of precision is the shallow depth of the binary tree used to model firm value and interest rates. While our implementation of the Heath-Jarrow-Morton tree in C++ allowed up to 22 tree steps — which translates to approximately four million terminal nodes — for a standard non-defaultable bond option before all computer memory was used up, only 10 steps were feasible in the credit risk model. Incorporating a bankruptcy process would have further reduced the number of feasible steps.

This example demonstrates the importance of recombination when implementing credit risk models in lattice structures. Even then the credit risk models can be computationally demanding.

[15] Ritchken and Sankarasubramanian (1995) and Anlong Li and Sankarasubramanian (1995) propose a method to capture path-dependency in a single sufficient statistic such that using recombining lattices is possible again. However, this method is only applicable to a special class of volatility functions.

5.8 Summary

In this chapter we propose an extended credit risk framework consisting of two components: the recovery rate and the bankruptcy process. A bankruptcy process is a one-jump process determining the time of default. The bankruptcy process itself is governed by a default intensity. The credit risk is called hybrid because it combines features of firm value and intensity-based models. The bankruptcy process is used in intensity models while the firm value is the basis of firm value models. The advantage of this hybrid model relative to the firm value model is the existence of a bankruptcy process. A bankruptcy process allows for explicit modeling of bankruptcy time and thus provides a better model for the valuation of American-type securities or securities with cash-flows at more than one time.

Unlike other intensity models, our hybrid framework is fully general in that both intensity and recovery rate can depend on the firm value. It is therefore a firm value model with a bankruptcy process determining the time of default while the firm value determines the recovery made on the claim.

We propose lattice structures to implement the credit risk model consisting of the firm value process and the bankruptcy process. In addition, the process of the underlying security of the option also has to be modeled. This results in an 8-jump lattice. The underlying variable of the derivative can be an equity or an interest rate process.

The lattice can be calibrated such that the implied riskless and risky term structures match actual observed term structures. This calibration is achieved by adjusting default intensities such that, in each time step, the implied risky bond price equals its observed market price. Such a calibration imposes restrictions on the process of default intensities. Specifically, we require default intensities to be deterministic when calibrating the model. Alternatively, a stochastic intensity structure could be assumed. In this case, model parameters need to be estimated such that the model bond prices match market prices.

This chapter also addresses the valuation of default-free options on credit-risky bonds. Such options can be valued within the same framework as vulnerable options, with a straightforward extension. Credit risk can be viewed as a type of option, as done in the firm value model. In an intensity or hybrid model, this model of credit risk can be maintained, although the option is modeled directly by a default process instead of a firm value process. Given our lattice to model credit risk and — as underlying variable in the case of bond options — interest rate risk, the option on the risky bond can be priced as a compound option. There is no need for an additional structure; a modification of the backward induction algorithm to compute the compound option prices is sufficient. Numerical examples show that prices of default-free options on risky bonds can be quite different if calculated on the basis of a credit risk model rather than an adhoc approach using only the risky term structure and otherwise pricing the options as if they were written on a riskless bond.

6. Pricing Credit Derivatives

The previous chapters consider the risk that a counterparty to a derivative contract defaults on its contractual obligations. Additionally, in Section 5.6.2, we address the pricing of default-free derivatives on credit-risky bonds. In this chapter, we look at derivative instruments with credit risk as their underlying variable determining the payoff of the instrument. Such instruments are commonly called *credit derivatives.*

Some types of credit derivatives have existed for a long time, albeit under a different name. An example is the large variety of claim insurance contracts, such as loan guarantees and deposit insurance, which have a long tradition in many countries.

The differences between those traditional credit instruments and today's credit derivatives are both institutional and contractual. While it was possible to transfer the credit risk of an eligible loan to a different agency such as a governmental loan and deposit insurance institutions, the transfer of credit risk to other parties was difficult if not impossible; it was unthinkable that credit risk could be freely traded between financial institutions and as such become a commodity similar to interest rate risk.[1]

Furthermore, the contractual specifications can be different. While the traditional instruments tended to insure losses from defaults and bankruptcies only, credit derivatives as they are used today allow more flexible payoff patterns. For example, a credit derivative may protect from an increasing yield spread or a change in the credit rating.

Credit derivatives are currently issued and traded OTC[2], implying that credit derivatives themselves are also subject to counterparty default risk.

---

[1] See Duffee and Zhou (1996) for an analysis of the benefits and drawbacks of credit derivatives markets with respect to the sharing of risk among financial institutions. Their argument follows Hart (1975), who noted that, within an incomplete market setting, new markets do not necessarily increase the utility of participants.

[2] A few derivatives on aggregated credit risk can be traded on exchanges. For example, the Chicago Mercantile Exchange recently announced options and futures trading on a bankruptcy index. These contracts will be struck on a quarterly consumer bankruptcy index which tracks the number of new bankruptcy filings in U.S. courts. Cf. Chicago Mercantile Exchange (1998). Moreover, some forms of credit derivatives can be constructed synthetically from exchange-traded products. For example, by simultaneously taking positions in both Eurodollar and

This risk, as considered in previous chapters, is completely ignored in this chapter, i.e., all contracts are priced as if they were free of any counterparty default risk, with the exception of Section 6.7.

6.1 Credit Derivative Instruments

There has recently been a surge of new derivative products in the area of credit risk called credit derivatives. Although credit derivatives are often presented as exotic products, we show in this chapter that many credit derivatives can be treated as normal derivative instruments with credit risk as the underlying variable. We call instruments with a payoff dependent only on credit risk *pure* credit derivatives. In practice, credit derivatives are often not pure, but are instruments sensitive to both credit risk and interest rate risk.

We distinguish between two types or classes of credit derivatives. One class is the default or bankruptcy-based class. This class encompasses default risk swaps, default options, and general credit insurance products. These derivatives compensate the creditor from losses suffered by a default. The maturity of such instruments usually coincides with the maturity of the debt insured. The second class of credit derivatives are instruments with a payoff determined by changes in the credit quality of the underlying security instead of a default event.

6.1.1 Credit Derivatives of the First Type

Examples of credit derivatives of the first type that are very popular in the market are credit risk swaps, default put options, and total return swaps.[3]

In a *credit risk* or *default swap*, the insurance-taking party pays a regular insurance premium (credit swap rate) to the insuring counterparty until a credit event occurs or the underlying security (reference asset) matures. The counterparty owes a payment contingent on a credit event. Contract specifications define the credit event and specify the insurance premium and the amount due to the insurance-taking counterparty in case of a credit event.

A *default put option* is very similar to a credit risk swap, but combines the recurring premium payments into a lump-sum payment. The payoff of

Treasury futures, a spread derivative on the credit risk involved in Eurodollar deposits (TED spread) can be traded.

[3] Cf. Smithson (1995) or British Bankers' Association (1996). A dealer poll by the British Bankers' Association (1996) in the London market gave estimates of 35% credit default swaps, 17% total return swaps, 15% spread derivatives, 33% others, such as credit-linked notes. These numbers have to be considered as rough estimates as the variance of estimates among dealers was high. The British Bankers' Association reports, however, that market participants generally expected the share of spread derivatives to rise considerably.

the option is defined in the same way as the swap payment in case of a credit event. Because the default swap requires premium payments only until a triggering default event occurs, default option and default swap are not identical instruments.

A number of variations to the contractual specifications of default swaps and options are conceivable. For example, contractual specifications may require the insurance-taker to continue with the premium payments until maturity of the reference asset. In this case, a credit swap and a default option are very similar because there is no insecurity regarding the number of premium payments and therefore the total amount of premia which the default insurance provider receives.

The specification of the default payment in case of a credit event also influences the characteristics of the default swap or the default option. For example, if a credit event entitles the insurance taker to receive the difference of the face value of the reference asset and the post-default market price of it, as common in actual credit derivative contracts, the payoff of the credit derivative also depends on the term structure of default-free interest rates. For a par bond as a reference asset, the insurance taker also receives compensation for the loss caused by the change of default-free interest rates if interest rates rise between the inception of the contract and the credit event. Moreover, if the reference asset is not a par bond but, for example, a zero-coupon bond, the payoff on the credit derivative exceeds the loss on the reference asset for early credit events because the default-free value is less than the face value of the bond before maturity. As pointed out in Section 6.2.1, pure credit derivatives can be constructed by defining the payoff of the instrument in terms of an equivalent default-free instrument, such as a Treasury bond with equal maturity, instead of the face value.

Sometimes, the payoff of a credit swap or default option is fixed in the contract specifications and is independent of the post-default value (recovery rate) of the reference asset. Such instruments are called digital or binary credit derivatives. Another variation of the standard instruments are credit swaps or default put options that do not insure the entire loss from default. For example, a default put option may only have a payoff if the loss from default exceeds a predetermined threshold value which serves as a strike recovery rate and can be interpreted as a deductible on the insurance.

A *total return swap* is a synthetic exchange of assets. One party pays the returns from a risky credit asset to the insuring party and receives in return the returns from a less risky or risk-free asset. If one asset is risk-free, then this structure is again very similar to the default swap.

The payoff of pure credit derivatives of the first type is usually defined such that it is equal to the loss on the reference asset caused by the credit event. Of course, there may be contractual nuances which change the characteristics of the instrument, but in principle, total return swaps, credit risk swaps, and default put options are all credit insurance instruments very similar to traditional debt insurance instruments.

6.1.2 Credit Derivatives of the Second Type

The second type of credit derivatives are not based on default, but on changes in the credit quality of the debtor. Examples are spread derivatives or rating-based derivatives. The payoff of these instruments is not contingent on default but on rating or credit spread changes.

The payoff of a credit spread option or a credit spread forward contract is determined by the difference between a strike credit spread and the actual credit spread in the market when the option is exercised. The raw spread payoff is then used to compute the actual payoff based on the face value of the claim and possibly the time of exercise. Section 6.2.1 describes the calculation of spread option payoff for the case of spread options on zero-coupon bonds.

Calculating spread derivatives' payoffs requires observable credit spreads. Because such spreads are not always available, credit ratings are sometimes used as proxy variables. In this case, a credit spread is assigned to each credit rating, giving the payoff of the instrument for each credit class. The relationship between credit ratings and payoff (i.e., credit spreads) is stated in the contract specifications and can be fixed or depend on current market values at the time of exercise.

Unlike instruments of the first type, these derivatives generally expire before their underlying assets mature. Instruments of the first type can sometimes be viewed as special cases of a credit spread option. For example, a default put is a spread option with a strike spread of zero and an an expiration date that coincides with the maturity of the reference asset.

6.1.3 Other Credit Derivatives

Other credit derivatives include, for example, credit-linked notes and credit derivatives with exotic features such as knock-in or knock-out provisions. Descriptions of some examples of such products can be found in Flesaker, Hughston, Schreiber, and Sprung (1994) or Howard (1995). For applications and a description of the market, see also Hart (1995) and Irving (1996).

Although they are not pure credit derivative instruments, put options on credit-risky bonds are also affected by credit risk and therefore a combination of interest-rate and credit derivative. We discuss the valuation of put options on credit-risky bonds in Section 5.5 starting on page 162.

6.2 Valuation of Credit Derivatives

Pure credit derivatives of the first class such as simple credit swaps and default options are very similar to traditional default insurance products.

Whether the insurance premium becomes due as a lump-sum payment or as a regular premium payment over the lifetime of the contract is irrelevant for valuation purposes as lump-sum payments can easily be converted in regular payments and vice versa given that premium payments are owed until maturity of the reference asset and no counterparty risk exists.

Default insurance products and standard credit derivatives of the first class can therefore be valued with similar models. It turns out that they can be priced with any credit risk model that is capable of pricing the underlying credit-risky bond. The price of the credit derivative or insurance product, in this case, is the difference between the price of the risk-free bond and the risky bond. The reason for this simple relationship is the fact that a standard credit risk swap or default put option insures all the credit risk inherent in the risky bond and therefore commands exactly the difference in price compared to the riskless bond as an insurance premium.

This simple valuation method explains why it is possible to price credit derivatives such as loan and deposit insurance with the original valuation model proposed by Merton (1974) and its adaptation to default insurance in Merton (1977) and Merton (1978). A large number of papers follow Merton (1977) or Merton (1978), or variants thereof, for the pricing of deposit insurance or loan guarantees in various countries. Examples are Acharya and Dreyfus (1989), Fischer and Grünbichler (1991), Ronn and Verma (1986), Sosin (1980). Grünbichler (1990) applies the model of Merton (1974) to the valuation of pension insurance and Schich (1997) to export credit insurance.

The valuation of the second class of credit derivatives is more involved. Whereas the payoff of credit swaps and default options depends only on the recovery made on the reference asset, i.e., on the credit loss, making it sufficient to model the recovery rate and the probability of default with a firm value model such as Merton (1974) or an intensity model such as Jarrow and Turnbull (1995), the payoff of spread derivatives depends on the credit risk inherent in the reference asset. Spread derivatives therefore expire before the maturity of the reference asset. Hence, the payoff of spread derivatives depends on the market's assessment of the underlying credit risk at the time of expiration or exercise, making necessary the modeling of the process of credit quality change over time.

Some contractual features of derivatives of the first class also make the modeling of a credit risk process necessary. For example, a default put option may only have a payoff if the credit loss exceeds a threshold loss. In other words, the strike price of the default option may not be zero. In such a case, the distribution of credit losses has to be known to price the derivative. Similarly, if the credit swap or default option expires before the maturity date of the reference asset, the process of credit quality change also needs to be modeled to determine the price effect of the chance that default occurs after expiration of the derivative. Furthermore, if credit swap rate payments are only owed until a credit event occurs, default time has to be modeled, making models such as Merton (1974) unsuitable to use.

In the following, however, we restrict the exposition to the valuation of second class derivatives, namely credit spread options and forward contracts.

6.2.1 Payoff Functions

This section presents payoff function of a variety of similar spread derivatives. We begin by defining forward spread contracts on zero-coupon bonds and then define credit spread options on zero-coupon bonds.

Although, in practice, spread options on coupon bonds are more common, we use zero-coupon bonds to illustrate the compound approach to credit spread option valuation. Zero-coupon bonds have the advantage of allowing a closed-form solution as a simple compound option in the case of deterministic interest rates and as a straightforward multi-variate lattice in the case of stochastic rates whereas coupon bonds require more involved lattice structures.

We will make use of some definitions from Section 2.3.3. A bond price can be expressed in terms of forward rates such that

$$P(t,T) = e^{-\int_t^T f(t,s)ds}.$$

The corresponding continuously compounded yield is given by

$$y_{t,T} = -\frac{\ln P(t,T)}{T-t}.$$

Of course, the same definitions apply for credit-risky bonds. We then use the following notation.

$$P^d(t,T) = e^{-\int_t^T f^d(t,s)ds}.$$

The corresponding continuously compounded yield is given by

$$y_{t,T}^d = -\frac{\ln P^d(t,T)}{T-t}.$$

Superscript d indicates that the underlying instrument is subject to default risk.

6.2.1.1 Credit Forward Contracts. Consider a forward contract on a bond. Such a contract can essentially be viewed as a forward rate agreement. The payoff of a forward contract at its maturity date T is given by

$$\begin{aligned} \Lambda(T) &= P(T,U) - P^*(t,T,U) \\ &= e^{-y_{T,U}(U-T)} - e^{-y^*_{t,T,U}(U-T)}, \end{aligned} \tag{6.1}$$

for one unit currency of notional value. $y^*_{t,T,U}$ is the reference interest rate determined at t for a bond from T to U. $P^*(t,T,U)$ is the corresponding reference bond. $y_{T,U}$ is the spot yield at time T for a zero bond with maturity at U. $P(T,U)$ is the time T spot price of the corresponding zero-coupon bond.

Alternatively, the payoff could be defined relative to the reference bond price, i.e.,

$$\begin{aligned}\Lambda(T) &= \frac{P(T,U)}{P^*(T,U)} - 1 \\ &= e^{-y_{T,U}(U-T)+y^*_{t,T,U}(U-T)} - 1.\end{aligned}$$

If the notional value of the forward contract is set to $B^*(T,U)$, it can be seen that the payoff is again $\Lambda(T) = B(T,U) - B^*(T,U)$.

A credit forward contract is very similar. However, the payoff of the credit forward is independent from the level of the yield, or equivalently, the price of bond, but depends only on the difference between the yields of two bonds, or equivalently, the difference between two bond prices. The payoff of a credit forward is as such not directly dependent on the term structure although there might be a dependency through correlation effects between interest rates and credit risk.

The payoff of a credit forward is

$$\begin{aligned}\Lambda(T) &= (P(T,U) - P^d(T,U)) - (P^*(t,T,U) - P^{d*}(t,T,U)) \\ &= e^{-y_{T,U}} - e^{-y^d_{T,U}} - e^{-y^*_{t,T,U}} + e^{-y^{d*}(t,T,U)}.\end{aligned}$$

Define $\psi(t,T) = P(t,T) - P^d(t,T)$. ψ denotes the absolute credit risk effect on the bond. It can be seen that Λ simplifies to

$$\Lambda(T) = \psi(T,U) - \psi^*(t,T,U). \tag{6.2}$$

The payoff of the credit forward is therefore only dependent on the change in the absolute bond price reduction caused by credit risk.

Alternatively, we might work with relative values. In this case, the payoff of the credit forward is

$$\begin{aligned}\Lambda(T) &= \exp\left(((y_{T,U} - y^d_{T,U}) - (y^*_{t,T,U} - y^{d*}_{t,T,U}))(U-T)\right) \\ &= \frac{P^d(T,U)P^*(t,T,U)}{P(T,U)P^{d*}(t,T,U)} - 1.\end{aligned} \tag{6.3}$$

Define $\phi(t,T) = P^d(t,T)P(t,T)^{-1}$. ϕ measure the relationship between risky and risk-free bond prices in relative terms. If ϕ is substituted into (6.3), we obtain

$$\Lambda(T) = \frac{\phi(T,U)}{\phi^*(t,T,U)} - 1.$$

Similar to (6.1), by multiplying the notional value with

$$\phi^*(t,T,U) = P^{d*}(t,T,U)P^*(t,T,U)^{-1},$$

the payoff becomes

$$\begin{aligned} \Lambda(T) &= \frac{P^d(T,U)}{P(T,U)} - \frac{P^{d*}(t,T,U)}{P^*(t,T,U)} \\ &= \phi(T,U) - \phi^*(t,T,U). \end{aligned}$$

This payoff is the difference between the relative bond credit spread at time T and the relative reference bond credit spread agreed upon at time t. To obtain as payoff the absolute difference between the credit risky bond and the reference credit-risky bond price, we can multiply by the risk-free bond value at time T, $P(T,U)$. The result is

$$\Lambda(T) = P^d(T,U) - \phi^*(t,T,U)P(T,U).$$

This is the payoff chosen by Das (1995). Because the spot bond price is part of the strike price, the strike price is stochastic when interest rates are stochastic. As shown above, by using other payoff specifications, the stochastic strike level can be circumvented although it does not generally pose a problem for valuation purposes.

The reference yield or reference price is usually chosen such that the contract has no value at inception. For a standard forward contract, this implies

$$P^*(t,T,U) = \frac{P(t,U)}{P(t,T)} = \exp\left(-y_{t,U}(U-t) + y_{t,T}(T-t)\right).$$

For a credit forward contract, we also need the forward price of the risky bond. It is given by

$$P^{d*}(T,U) = \frac{P^d(t,U)}{P(t,T)} = \exp\left(-y^d_{t,U}(U-t) + y^d_{t,T}(T-t)\right).$$

The forward yield is therefore

$$y^d_{t,T,U} = \frac{y^d_{t,U}(U-t) - y_{t,T}(T-t)}{U-T}.$$

6.2.1.2 Credit Spread Options. The payoffs of credit spread options follow trivially from the payoffs of the forward contracts. For example, the payoff of an credit spread option in absolute values is derived from (6.2) and becomes

$$\begin{aligned} \Lambda(T) &= (\psi(T,U) - \psi^*(t,T,U))^+ \\ &= \left((P(T,U) - P^d(T,U)) - (P^*(t,T,U) - P^{d*}(t,T,U))\right)^+. \end{aligned} \tag{6.4}$$

The other option payoffs are obtained similarly from their corresponding forward payoffs.

6.3 The Compound Pricing Approach

We propose a compound approach to pricing credit derivatives of the second class and illustrate it within a firm value model of credit risk.

Credit risk is the risk that the debtor cannot fulfill its obligations because the amount of liabilities due surpasses the amount of funds available. It is in this context irrelevant if the firm defaults because of insufficient liquidity or simply because the total value of the firms assets is less then the total amount of liabilities. Since assets and liabilities are usually stochastic variables, one way of modeling credit risk is by modeling the stochastic behavior of assets and liabilities. The price of credit risk is a function of those variables, $f(V, D, t)$, and the payoff is always of the form

$$\Lambda(V, D, T) = (V - D)^+ .$$

Hence, credit risk can be expressed as an option, as discussed previously. It does not matter how the stochastic variables determining the payoff of the option are modeled. Some approaches do not require the modeling of asset values but model default directly by a Poisson default process. Nonetheless, the basic structure is always that of an option pricing model whether the underlying is an asset process or a default process. In the case of an intensity model, the credit derivative is a derivative on the default process instead of on the firm value option.

Hence, because credit risk, in its general form, can be modeled as an option, credit derivatives can be viewed and modeled as compound derivatives.

6.3.1 Firm Value Model

It is assumed that a defaultable bond pays

$$P^d(U, U) = P(U, U)\left(1 - \left(1 - \frac{V_U}{D}\right)^+\right).$$

This is our standard firm value model assumption from previous chapters. Refer, for example, to (5.14) on page 147. V_t denotes the firm value, D the bankruptcy threshold at maturity of the claim.

The time T value of this defaultable bond is

$$P^d(T, U) = P(T, U) - p_T(V_T),$$

where $p_T(V_T)$ denotes the time T value of a put option with payoff $p_U(V_U) = P(U, U)\left(1 - \frac{V_U}{D}\right)^+$. This result was established in (4.10). $P(U, U)$ is subsequently normalized, i.e., set to 1.

Assume that the payoff of the spread option to be priced is given by (6.4). We write

$$\Lambda(T) = (\psi(T,U) - \psi^*(t,T,U))^+ , \tag{6.5}$$

where ψ^* denotes the strike spread. Because

$$\psi(T,U) = P(T,U) - P^d(T,U)$$

is the spread between defaultable and risk-free bonds at time T, we can write

$$\psi(T,U) = p_T(V_T).$$

It is evident that the spread option is an option on a put option.

If we make the same assumption as to the firm value process under the risk-neutral measure as in Section 4.2, i.e.,

$$V_U = V_t \exp\left((r - \frac{1}{2}\sigma_V^2)(U-t) + \sigma_V(\tilde{W}_U - \tilde{W}_t) \right),$$

where r is assumed to be a constant interest rate, then the following closed-form pricing formulae can be derived.

Proposition 6.3.1. *A European credit spread option that pays out*

$$\Lambda(T) = (p_T(V_T) - \psi^*)^+$$

at maturity T is called a call *spread option. Its price is given by*

$$\begin{aligned}\Lambda_t = &-\frac{V_t}{D}N_2(-d_1, d_2, \rho)\\ &+ e^{-r(U-t)}N_2(-d_1 + \sigma_V\sqrt{T-t}, -d_2 + \sigma_V\sqrt{U-t}, \rho)\\ &- e^{-r(T-t)}\psi^* N(-d_1 + \sigma_V\sqrt{T-t}),\end{aligned}$$

with

$$\begin{aligned}d_1 &= \frac{\ln\frac{V_t}{V^*} + (r + \frac{1}{2}\sigma_V^2)(T-t)}{\sigma_V\sqrt{T-t}},\\ d_2 &= \frac{\ln\frac{V_t}{D} + (r + \frac{1}{2}\sigma_V^2)(U-t)}{\sigma_V\sqrt{U-t}},\end{aligned}$$

where $\rho = \sqrt{\frac{T-t}{U-t}}$ and V^* is the solution to

$$-\frac{V^*}{D}N(d_3) + e^{-r(U-T)}N(d_3 - \sigma_V(U-T)) = \psi^*, \tag{6.6}$$

with

$$d_3 = \frac{\ln\frac{V^*}{D} + (r + \frac{1}{2}\sigma_V^2)(U-T)}{\sigma_V\sqrt{U-T}}.$$

T denotes the expiration date of the compound options, U of the underlying option.

Proof. The formula for compound options was first derived by Geske (1977) and Geske and Johnson (1984). Below we give a different proof for the general compound option problem. The proof of the proposition readily follows from the general proof.

We start with a standard call option on a call option. Let the underlying option expire at time T_1 and the compound option at T_2 such that $T_1 > T_2$. The value of the underlying call option at time T_2 can be written

$$C_{T_2} = e^{-r(T_1-T_2)}\mathbf{E_Q}[(V_{T_1} - K_1)\mathbf{1}_{\{V_{T_1}>K_1\}}|\mathcal{F}_{T_2}].$$

This is our standard representation for the price of a contingent claim. The compound option can be expressed similarly.

$$CC_t = e^{-r(T_2-t)}\mathbf{E_Q}[(C_{T_2} - K_2)\mathbf{1}_{\{C_{T_2}>K_2\}}|\mathcal{F}_t]. \tag{6.7}$$

Substituting for C_{T_2} in (6.7) gives

$$CC_t = e^{-r(T_2-t)}\mathbf{E_Q}[e^{-r(T_1-T_2)}\mathbf{E_Q}[(V_{T_1} - K_1)\mathbf{1}_{\{V_{T_1}>K_1\}}|\mathcal{F}_{T_2}] \\ - K_2\mathbf{1}_{\{C_{T_2}>K_2\}}|\mathcal{F}_t].$$

By the linearity of expectations and the law of iterated expectations, this is equal to

$$\begin{aligned} CC_t &= e^{-r(T_1-t)}\mathbf{E_Q}[V_{T_1}\mathbf{1}_{\{V_{T_1}>K_1\}}\mathbf{1}_{\{C_{T_2}>K_2\}}|\mathcal{F}_t] \\ &- e^{-r(T_1-t)}\mathbf{E_Q}[K_1\mathbf{1}_{\{V_{T_1}>K_1\}}\mathbf{1}_{\{C_{T_2}>K_2\}}|\mathcal{F}_t] \\ &- e^{-r(T_2-t)}\mathbf{E_Q}[K_2\mathbf{1}_{\{C_{T_2}-K_2\}}|\mathcal{F}_t]. \end{aligned}$$

It is convenient to specify the indicator function $\mathbf{1}_{\{C_{T_2}-K_2\}}$ in a different way. Determine V^* such that

$$C_{T_2} = e^{-r(T_1-T_2)}\mathbf{E_Q}[(V^* - K_1)\mathbf{1}_{\{V^*>K_1\}}|\mathcal{F}_{T_2}] = K_2. \tag{6.8}$$

C_{T_2} is a function of a single random variable, V_{T_2}. Because the function is strictly monotonic, i.e., increasing V_{T_2} increases C_{T_2}, the following equality holds:

$$\mathbf{1}_{\{C_{T_2}>K_2\}} = \mathbf{1}_{\{V_{T_2}>V^*\}}.$$

Consequently, we can write $CC_t = E_1 - E_2 - E_3$ with

$$\begin{aligned} E_1 &= e^{-r(T_1-t)}\mathbf{E_Q}[V_{T_1}\mathbf{1}_{\{V_{T_2}>V^*\}}\mathbf{1}_{\{V_{T_1}>K_1\}}|\mathcal{F}_t], \\ E_2 &= e^{-r(T_1-t)}\mathbf{E_Q}[K_1\mathbf{1}_{\{V_{T_2}>V^*\}}\mathbf{1}_{\{V_{T_1}>K_1\}}|\mathcal{F}_t], \\ E_3 &= e^{-r(T_2-t)}\mathbf{E_Q}[K_2\mathbf{1}_{\{V_{T_2}>V^*\}}|\mathcal{F}_t]. \end{aligned}$$

E_1, E_2, and E_3 are then evaluated. Substituting for V_{T_1} gives

$$E_1 = \mathbf{E_Q}[V_t e^{-\frac{1}{2}\sigma_V^2(T_1-t)+\sigma_V(\tilde{W}_{T_1}-\tilde{W}_t)}\mathbf{1}_{\{V_{T_2}>V^*\}}\mathbf{1}_{\{V_{T_1}>K_1\}}|\mathcal{F}_t].$$

A new measure is defined as

$$\zeta_T = \frac{d\dot{\mathbf{Q}}}{d\mathbf{Q}} = \exp\left(\sigma_V W_T - \frac{1}{2}\sigma_V^2 T\right).$$

We therefore have

$$E_1 = \zeta_t^{-1}\mathbf{E}_{\mathbf{Q}}[V_t \zeta_{T_1} \mathbf{1}_{\{V_{T_2} > V^*\}} \mathbf{1}_{\{V_{T_1} > K_1\}} | \mathcal{F}_t],$$

which is equal to

$$E_1 = e^{-r(T_1 - t)}\mathbf{E}_{\dot{\mathbf{Q}}}[V_t \mathbf{1}_{\{V_{T_2} > V^*\}} \mathbf{1}_{\{V_{T_1} > K_1\}} | \mathcal{F}_t].$$

It follows, as before, that

$$E_1 = V_t N_2(a_1, a_2, \rho)$$

because $\dot{\eta}_V(t, T_1) = \sigma_V(\dot{W}_{T_1} - \dot{W}_t)$ and $\dot{\eta}_V(t, T_2) = \sigma_V(\dot{W}_{T_2} - \dot{W}_t)$ are joint normally distributed.
Noticing that

$$\dot{z}_i = \frac{\dot{W}_{T_i} - \dot{W}_t}{\sqrt{T_i - t}}$$

has law $N(0,1)$ and $\dot{z}_1$ and $\dot{z}_2$ have joint law $N(0, 1, \sqrt{(T_2 - t)/(T_1 - t)})$, we have

$$\rho = \sqrt{(T_2 - t)/(T_1 - t)}\,.$$

The indicator functions determine a_1 and a_2. We evaluate the indicator functions under measure $\dot{\mathbf{Q}}$ and obtain

$$\begin{aligned}
\mathbf{E}_{\dot{\mathbf{Q}}}[\mathbf{1}_{\{V_{T_2} > V^*\}}] &= \dot{\mathbf{Q}}\,(V_{T_2} > V^*) \\
&= \dot{\mathbf{Q}}\left(\dot{z}_1 < \frac{\ln V_t - \ln V^* + (r + \frac{1}{2}\sigma_V^2(T_2 - t))}{\sigma_V\sqrt{T_2 - t}}\right), \\
\mathbf{E}_{\dot{\mathbf{Q}}}[\mathbf{1}_{\{V_{T_1} > K_1\}}] &= \dot{\mathbf{Q}}\,(V_{T_1} > K_1) \\
&= \dot{\mathbf{Q}}\left(\dot{z}_2 < \frac{\ln V_t - \ln K_1 + (r + \frac{1}{2}\sigma_V^2)(T_1 - t)}{\sigma_V\sqrt{T_1 - t}}\right).
\end{aligned}$$

Evaluating E_2, we immediately find that

$$E_2 = e^{-r(T_1 - t)} K_1 N(b_1, b_2, \rho).$$

The indicator functions can be evaluated as for E_1, but without change of measure. We then obtain $b_1 = a_1 - \sigma_V\sqrt{T_2 - t}$ and $b_2 = a_2 - \sigma_V\sqrt{T_1 - t}$.

Evaluation of E_3 is straightforward because it is a univariate problem and no measure changes are necessary. We obtain $E_3 = e^{-r(T_2 - t)} K_2 N(c)$ with $c = b_1$.

The price for a call option on a call option is therefore

$$CC_t = V_t N_2(a_1, a_2, \rho) - e^{-r(T_1-t)} K_1 N(b_1, b_2, \rho) - e^{-r(T_2-t)} K_2 N(b_1),$$

where

$$a_1 = \frac{\ln V_t - \ln V^* + (r + \frac{1}{2}\sigma_V^2(T_2 - t))}{\sigma_V \sqrt{T_2 - t}},$$
$$a_2 = \frac{\ln V_t - \ln K_1 + (r + \frac{1}{2}\sigma_V^2)(T_1 - t)}{\sigma_V \sqrt{T_1 - t}},$$
$$b_1 = a_1 - \sigma_V \sqrt{T_2 - t},$$
$$b_2 = a_2 - \sigma_V \sqrt{T_1 - t}.$$

The proofs for a call option on a put option, a put option on a call option, or a put option on a put option, are analogous. The price of a call option on a put option is given by $CC = -E_1 + E_2 - E_3$.

This completes the general proof for compound options. Proposition 6.3.1 can be seen to be a special case of the proof for call options on put options. In particular, $p_T(V_T)$ is the put option equivalent to the call option C_{T_1} and V_t is set to V_t/D , K_2 to 1. The notation also deviates somewhat with $K_2 = \psi^*$, $T_1 = U$, $T_2 = T$, $a_1 = d_1$, $a_2 = d_2$. Expression (6.8) represents a standard option of the Black-Scholes type. Hence, it is equal to (6.6) when evaluated. This completes the proof of Proposition 6.3.1.

Proposition 6.3.2. *A European credit spread option that pays out*

$$\Lambda(T) = (\psi^* - p_T(V_T))^+$$

at maturity T is called a put *spread option. Its price is given by*

$$\Lambda_t = \frac{V_t}{D} N_2(d_1, d_2, -\rho)$$
$$- e^{-r(U-t)} N_2(d_1 - \sigma_V \sqrt{T-t}, -d_2 + \sigma_V \sqrt{U-t}, -\rho)$$
$$+ e^{-r(T-t)} \psi^* N(d_1 - \sigma_V \sqrt{T-t}).$$

Proof. The proof is analogous to the one of Proposition 6.3.1.

6.3.2 Stochastic Interest Rates

Assuming deterministic interest rates in a model designed to facilitate the pricing of credit spread options may be hard to justify. It is therefore tempting to modify Propositions 6.3.1 and 6.3.2 such as to allow Gaussian interest rates. Although a formula for compound options under stochastic interest rates has been proposed by Geman, Karoui, and Rochet (1995), their extension to Gaussian rates is not without problems. A critical assumption in the proof to Proposition 6.3.1 was the possibility to identify a unique V^* such that

$$p_T(V^*) - \psi^* = 0.$$

In other words, V^* was the value of V_T for which the compound option was exactly at-the-money at maturity. If interest rates are stochastic, the price of the underlying option varies with the bond price, or, more precisely, the correlation between prices of bonds and underlying security. Thus, V^* may no longer be unique. Instead, there may be a set of pairs $(V^*, P^*(T,U))$ for which an extended version of expression (6.6) is satisfied. As Frey and Sommer (1998) have shown, this problem can be overcome if the compound option is written not on the underlying option directly, but on the time U forward price of the underlying option. However, as such a definition of the underlying option is not well-suited to model the credit risk problem, we have to resort to numerical methods for a compound option price under stochastic interest rates.[4]

One method to compute compound option prices with stochastic interest rates are lattice structures. It turns out that compound options are easily priced in a lattice constructed to value the underlying option. Given the lattice for the underlying option, the compound option can be computed while traversing back through the lattice. From the maturity date of the underlying option, the lattice is traversed back in the usual way until the maturity date of the compound option. At this date, in each node the payoff can be calculated with the value of the underlying option in that node. By the usual backward induction, the price of the compound option can then be obtained. We can therefore use a lattice of the kind proposed in Section 5.2.3 to value the compound option. In fact, with respect to pricing, spread derivatives are very similar to default-free options on risky bonds, as discussed in Section 5.5. While a default-free option on a risky bond is a compound option on both a risk-free bond and credit risk, a credit spread derivative is an option on credit risk only.

6.3.3 Intensity and Hybrid Credit Risk Models

When credit risk is not modeled by a firm value process, but by an exogenous bankruptcy process such as in Jarrow and Turnbull (1995), the compound approach as described in the last section still applies. In that case credit risk is not modeled by a firm value process, but by a bankruptcy jump process. While in the former case a credit spread option is an option of the firm value, in the latter case credit risk is modeled directly by the jump process. In this sense, the jump process embodies the option representing credit risk. Consequently, the spread option can be viewed as an option on the bankruptcy jump process. Credit model implementations with jump processes in the form of lattices can be extended in the same way as for firm value processes.

[4] For an simplification of the evaluation of nested normal distributions as they arise in the valuation of multiple compound options, refer to Selby and Hodges (1987).

6.4 Numerical Examples

This section contains some numerical examples of credit spread options. The prices that assume deterministic interest rates were computed with the explicit formula given in Proposition 6.3.1. The prices with Gaussian interest rates were computed numerically. The algorithm for credit spread options with Gaussian interest rates uses a recombining lattice as described in Chapter 5. Prices for default swaps or default options were computed implicitly, as special cases of spread options.

6.4.1 Deterministic Interest Rates

Table 6.1 shows prices of credit spread options for three different maturities and various strike prices. The prices are based on the assumption of a constant continuously compounded interest rate of 5%. The prices of the table were calculated by the compound option formula of Proposition 6.3.1. The strike prices are given in terms of yield spreads over the riskless yield. The forward yield spread is also given in the table.

Table 6.1. Prices of credit spread options on yield spread

%						bp			
	forw		0	50	100	150	200	300	400
1y	bp	Call	3.24	1.90	1.08	0.62	0.36	0.12	0.04
	106	Put	0	0.20	0.90	1.92	3.11	5.69	8.32
2y	bp	Call	3.24	2.37	1.80	1.40	1.09	0.69	0.44
	141	Put	0	0.29	0.87	1.59	2.39	4.16	6.01
3y	bp	Call	3.24	2.74	2.39	2.11	1.88	1.50	1.22
	212	Put	0	0.28	0.70	1.18	1.69	2.80	3.97

Prices of 1-year and 2-year credit spread options in percent of face value for a 5-year bond with face value of 100. Strike spreads given in yield basis points. Parameters: $\sigma_V = 0.2$, $r = 0.05$, $D/V = 0.8$.

The credit-risky term structure implied by the parameters $\sigma_V = 20\%$ and $DV^{-1} = 0.8$ is shown in Figure 6.1. These parameters imply that the risky 5-year bond currently trades at 74.64% of principal. A risk-free bond of equal maturity is worth 77.88% of principal value.

Note that the forward yield spreads given in Table 6.1 do not match the term structure given in this figure. The term structures given in this figure are spot and instantaneous forward term structures, whereas the forward yield spreads of Table 6.1 refer to the strike yield spread of a specific risky 5-year bond.

Figure 6.2 shows the forward yield spread for the risky 5-year bond underlying the option prices of Table 6.1. It can be seen that the forward yield spread tends to infinity as the time approaches the maturity date of the bond,

Fig. 6.1. Implied credit-risky term structure

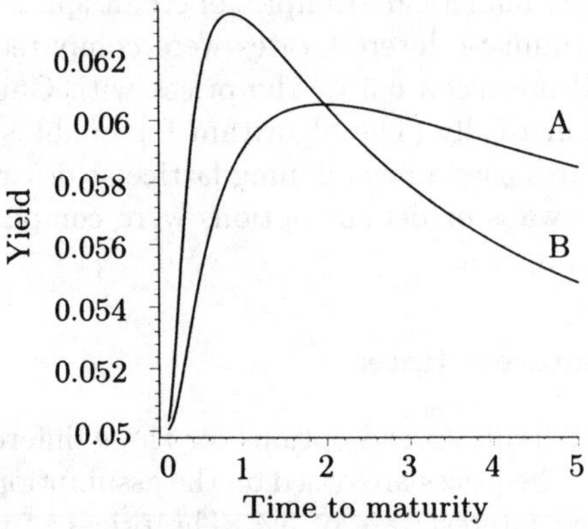

Implied zero bond rates and instantaneous forward rates. Parameters: $r = 0.05$, $D/V = 0.8$, $\sigma_V = 0.2$. A: Spot rate. B: Instantaneous forward rate.

which is in five years. The reason for this effect is that the forward bond price spread increases in absolute terms. This means that the forward yield spread has to increase even more because of the shorter time. The forward risky bond price is given by

$$P^d(t,T,U) = \frac{P^d(t,U)}{P(t,T)},$$

where $t \leq T \leq U$. $P(t,T)$ is the price of a riskless bond. The forward yield spread can be calculated as

$$y^d(t,T,U) - y(t,T,U) = -\frac{\ln P^d(t,T,U)}{U-T} - y(t,T,U),$$

whereas the forward absolute bond spread is

$$P(t,T,U) - P^d(t,T,U).$$

Because the forward absolute bond spread for a time near maturity of the does not converge to zero while $U - T$ converges to zero, the forward yield spread tends to infinity. Although a close comparison is impossible because of the scale of the graph, it can be seen that the forward yield spreads in Table 6.1 seem to be consistent with Figure 6.2.

Table 6.2 displays corresponding forward absolute bond price spreads for the forward yield strike spreads used in Table 6.1. Table 6.3 shows similar spread options on the same 5-year bond, but this time the strike is given as an absolute bond price spread instead of a yield spread.

Fig. 6.2. Forward yield spread

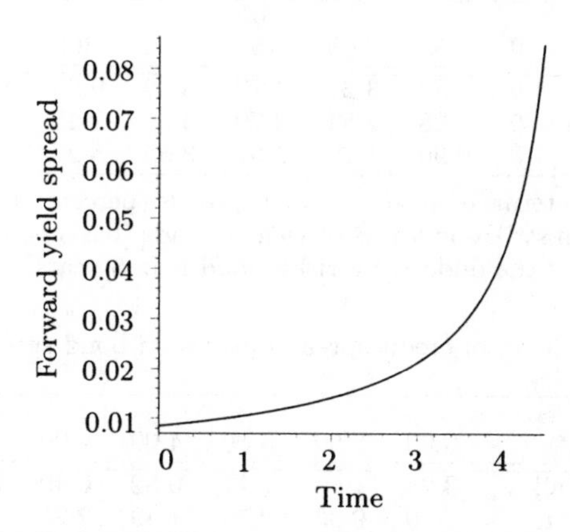

Forward yield spread of a 5-year bond. Parameters: $r = 0.05$, $D/V = 0.8$, $\sigma_V = 0.2$.

Fig. 6.3. Absolute forward bond spread

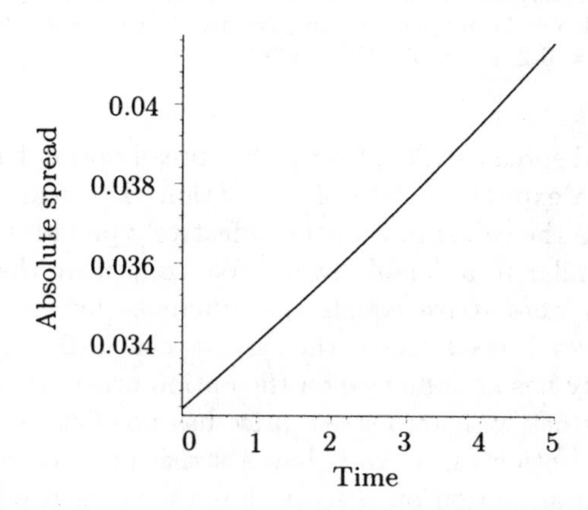

Absolute forward bond spread for a 5-year bond with parameters: $r = 0.05$, $D/V = 0.8$, $\sigma_V = 0.2$.

Table 6.2. Yield spread to bond spread conversion

		bp						
		0	50	100	150	200	300	400
	1 year	0	1.62	3.21	4.77	6.30	9.26	12.11
Strike	2 year	0	1.28	2.54	3.79	5.01	7.41	9.73
	3 year	0	0.90	1.79	2.67	3.55	5.27	6.70

Strike prices in terms of absolute bond spreads (percent of face value) for given strikes in terms of yield spreads (basis points.) The maturity of the underlying risky bond is in 5 years.

Table 6.3. Prices of credit spread options on bond spread

			%					
	forw		0.00	2.00	3.00	4.00	6.00	8.00
1y	%	Call	3.24	1.66	1.17	0.82	0.40	0.19
	3.40	Put	0	0.33	0.79	1.39	2.87	4.56
2y	%	Call	3.24	2.02	1.64	1.34	0.90	0.61
	3.58	Put	0	0.60	1.12	1.72	3.10	4.62
3y	%	Call	3.24	2.32	2.02	1.77	1.37	1.07
	3.76	Put	0	0.81	1.37	1.97	3.30	4.72

Prices of credit spread options with maturities of one to three years given in percent of face value for a 5-year bond. Strike spreads are also to be interpreted in percent of face value. Parameters: $\sigma_V = 0.2$, $r = 0.05$, $D/V = 0.8$.

For a yield or bond spread strike of zero, the spread option has the same price regardless of the expiration date of the option. The reason is that a zero strike price means the call spread option effectively provides full default insurance coverage similar to a default swap.[5] We conjecture that therefore the option price for a zero strike equals the difference between risky and riskless spot bond prices. Indeed, this is the case, as can easily be confirmed. Consequently, maturity has no influence on the option price just as maturity of a call option on a stock with zero strike price has no effect on the option price either. If a call option on a stock has a strike price of zero, we are effectively not holding an option on a stock, but the stock itself. This fact also applies to credit derivatives. A strike of zero means we are holding the underlying which, in the case of credit derivatives, is credit risk. In turn, credit risk is expressed by the difference between riskless and risky bond price. It is obvious that a put option with a strike of zero has no value.

[5] Thus, default insurance in the form of a default swap or default option can be viewed as a special case of a spread option.

6.4.2 Stochastic Interest Rates

Table 6.4 shows credit spread option prices for a 2-year credit spread option on a 5-year risky bond. Prices are given for three different strike spreads and for various correlations. These prices are based on the stochastic evolution of the short rate according to the model by Vasicek (1977). The compound pricing model used to price these yield spread options was implemented as a bivariate lattice modeling firm value and short rate according to the description of bivariate lattices in Chapter 5.

Fig. 6.4. Implied risky term structures for different correlations

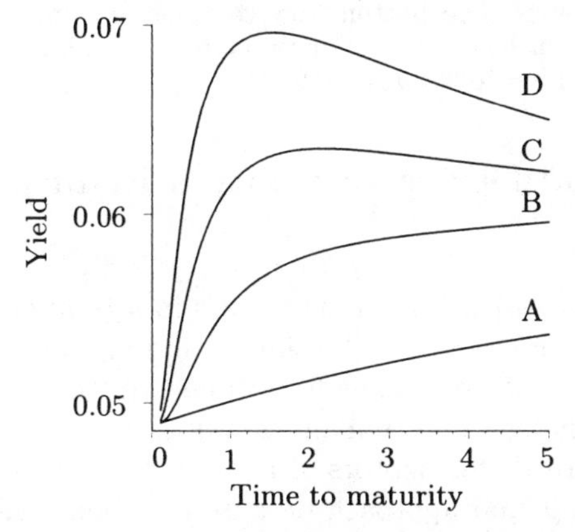

Implied zero bond yields for various correlations between bond price and firm value. A: riskless. B: $\rho = 0.5$. C: $\rho = 0$. D: $\rho = -0.5$. Parameters: $r = \ln 1.05$, $b = \ln 1.08$, $a = 0.1$, $\sigma = 0.02$, $D/V = 0.8$, $\sigma_V = 0.2$.

As seen in previous chapters, the correlation between interest rates and firm value has a significant effect on credit risk and thus on the prices of credit spread options. Negative correlation of the firm value with interest rates, i.e., positive correlation with bond prices, reduces the credit spread. The term structure of spot yields for riskless bonds and risky bonds with different correlations is plotted in Figure 6.4. The credit risk effect of correlation can also be seen in the first row of the table. This row shows the prices of spread options with a strike spread of zero. As was discussed in the previous section, a strike spread of zero implies an insurance for the total amount of credit risk involved in case of a call option. The price of the option is therefore equal to the spot price deduction on the risky bond relative to the riskless bond. The forward yield spread, shown in the last row, rises with increasing credit risk.

Table 6.4. Prices of credit spread options with Vasicek interest rates

%			ρ_{VP}				
	strike		-0.9	-0.5	0.0	0.5	0.9
	bp	Call	6.38	5.48	4.32	3.13	2.19
	0	Put	0	0	0	0	0
2y	bp	Call	4.49	3.66	2.63	1.61	0.85
	100	Put	0.51	0.59	0.72	0.89	1.06
	bp	Call	3.27	2.56	1.70	0.91	0.37
	200	Put	1.63	1.82	2.13	2.52	2.91
		Forward	272	232	182	131	91

Prices of 2-year credit spread options in percent of face value for a 5-year bond with face value of 100. Correlations are between firm value and bond price. Strike spreads given in yield basis points. The bottom row contains forward yield credit spreads in basis points. Parameters: $r = \ln 1.05$, $b = \ln 1.08$, $a = 0.1$ $\sigma = 0.02$, $\sigma_V = 0.2$, $D/V = 0.8$.

6.5 Pricing Spread Derivatives with a Reduced-Form Model

Pricing credit derivatives with firm value models in a compound option approach is, while elegant in theory, hard to implement in practice. First, there are the implementation problems of any firm value models. Unobservable and hard to estimate parameter values remain as well as problems with explaining empirical credit spreads for debtors of high quality. These problems are exacerbated in the compound approach because any model misspecification in the first stage translates into possibly even larger misspecification in the second stage of the modeling problem. In other words, because credit spread options are viewed as options on credit risk, which itself is an option the firm value, any model inaccuracies of the firm value model compared to empirical behavior of credit spreads result in an inaccurate process for the underlying of the compound option. The compound approach for credit spread derivatives is therefore rather hard to implement satisfactorily.

Although reduced-form models lack the economic foundation of structural models, they may be more suitable for the computation of prices for spread derivatives in practice because of the reduced model complexity and easier adaptation to empirical data. The following approach was proposed by Das and Sundaram (1998). It directly models the forward rates and the credit spreads in a bivariate modeling framework that can be easily implemented in a bivariate tree. The approach uses a discrete-time HJM model as introduced in Section 2.4.2.

The process of the forward rates $f(t,T)$ is assumed to be

$$f(t+h,T) = f(t,T) + a(t,T,f(t,T))h + \sigma(t,T,f(t,T)\epsilon\sqrt{h} \tag{6.9}$$

for all T. ϵ is a binomial random variable taking values $(-1,1)$.

The credit spread is denoted by $s(t,T)$. The credit risky short-rate is therefore given by $r^d(t) = r(t) + s(t,t)$. The credit spread is assumed to follow

$$s(t+h,T) = s(t,T) + \beta(t,T,s(t,T))h + \nu(t,T,s(t,T))\epsilon_2\sqrt{h} \tag{6.10}$$

for all T. The value for the money market account $B(t)$ is given by

$$B(t) = \exp\left(\sum_{i=0}^{t/h-1} r(ih)h\right).$$

By definition of the money market account, the single-period return is predictable such that

$$\begin{aligned}\frac{B(t+h)}{B(t)} &= \exp\left(r(t)h\right) \\ &= \exp\left(f(t,t)h\right)\end{aligned} \tag{6.11}$$

A bond free of credit risk is given by

$$B(t) = \exp\left(-\sum_{i=t/h}^{T/h-1} f(t,ih)h\right). \tag{6.12}$$

Using a risk-neutral valuation argument, under the martingale measure $\mathbf{Q}$, we have

$$\frac{P(t,T)}{B(t)} = \mathbf{E_Q}[\frac{P(t+h,T)}{B(t+h)}]$$

Equivalently, the expected return under the martingale measure equals the return on the riskless money market account, i.e.,

$$\frac{B(t+h)}{B(t)} = \mathbf{E_Q}[\frac{P(t+h,T)}{P(t,T)}] \tag{6.13}$$

Substituting expressions (6.12) and (6.11) into (6.13) we obtain

$$\begin{aligned}\exp\left(f(t,t)h\right) &= \mathbf{E_Q}[\exp\left(-\sum_{i=t/h+1}^{T/h-1} f(t+h,ih)h\right)\exp\left(-\sum_{i=t/h}^{T/h-1} f(t,ih)h\right)] \\ &= \mathbf{E_Q}[\exp\left(\left(-\sum_{i=t/h+1}^{T/h-1}(f(t+h,ih)-f(t,ih))h\right)+f(t,t)h\right)].\end{aligned} \tag{6.14}$$

Because $f(t,t)$ is known at time t, we have

$$1 = \mathbf{E_Q}[\exp\left(- \sum_{i=t/h+1}^{T/h-1} (f(t+h, ih) - f(t, ih))h \right)].$$

Substituting from expression (6.9) we obtain

$$1 = \mathbf{E_Q}[\exp\left(- \sum_{i=t/h+1}^{T/h-1} \left(a(ih)h + \sigma(ih)\epsilon_1\sqrt{h}\right)h \right)]. \tag{6.15}$$

The drift term of the forward rate process under measure $\mathbf{Q}$ can therefore be recursively computed as

$$\sum_{i=t/h+1}^{T/h-1} a(ih) = \frac{1}{h^2} \ln \mathbf{E_Q}[\exp\left(- \sum_{i=t/h+1}^{T/h-1} \sigma(ih)\epsilon_1\sqrt{h}h \right)]. \tag{6.16}$$

So far, the analysis represents an alternative to Section 2.4.2 to obtain the arbitrage-free drift term in the discrete version of the HJM methodology. A similar approach can be used to find the arbitrage-free drift term for the credit spread process.

The price of a credit-risky bond $P^d(t,T)$ is given by

$$P^d(t,T) = \exp\left(- \sum_{i=t/h}^{T/h-1} (f(t, ih) + s(t, ih))h \right). \tag{6.17}$$

According to Duffie and Singleton (1999), an adjustment of the discounting short rate accounts for the hazard process. Then the credit-risk adjusted money market account is given by

$$B^d(t) = \exp\left(- \sum_{i=0}^{t/h-1} (f(ih, ih) + s(ih, ih))h \right). \tag{6.18}$$

The martingale condition also applies to credit-risky assets. Thus, the credit-risky bond is a martingale under measure $\mathbf{Q}$ and the adjusted numéraire money market account. We obtain

$$\frac{P^d(t,T)}{B^d(t)} = \mathbf{E_Q}[\frac{P^d(t+h,T)}{B^d(t+h)}].$$

This can also be written

$$1 = \mathbf{E_Q}[\frac{P^d(t+h,T)}{P^d(t,T)} \frac{B^d(t)}{B^d(t+h)}]. \tag{6.19}$$

From (6.17) and (6.18) it follows that

$$\frac{P^d(t+h,T)}{P^d(t,T)} = \exp\Bigg(- \sum_{i=t/h+1}^{T/h-1} \big(f(t+h,ih) + s(t+h,ih)\big)h$$

$$+ \sum_{i=t/h}^{T/h-1} \big(f(t,ih) + s(t,ih)\big)h \Bigg)$$

$$= \exp\Bigg(- \sum_{i=t/h+1}^{T/h-1} \big(f(t+h,ih) + s(t+h,ih)$$

$$- f(t,ih) - s(t,ih)\big)h \Bigg) \exp\left(f(t,t)h + s(t,t)h\right)$$

and

$$\frac{B^d(t)}{B^d(t+h,T)} = \exp\left(-f(t,t)h - s(t,t)h\right)$$

Substituting into expression (6.19) yields

$$1 = \mathbf{E_Q}\Bigg[\exp\Bigg(- \sum_{i=t/h+1}^{T/h-1} \big(f(t+h,ih) + s(t+h,ih) - f(t,ih) - s(t,ih)\big)h \Bigg)\Bigg]$$

$$= \mathbf{E_Q}\Bigg[\exp\Bigg(- \sum_{i=t/h+1}^{T/h-1} \big(a(ih)h + \beta(ih)h + \sigma(ih)\epsilon_1\sqrt{h} + \nu(ih)\epsilon_2\sqrt{h}\big)h \Bigg)\Bigg]$$

Solving for the drift term gives

$$\sum_{i=t/h+1}^{T/h-1} \beta(ih) = \frac{1}{h^2}\ln\Bigg(\mathbf{E_Q}\Bigg[\exp\Bigg(- \sum_{i=t/h+1}^{T/h-1} \big(\sigma(ih)\epsilon_1\sqrt{h} + \nu(ih)\epsilon_2\sqrt{h}\big)h \Bigg)\Bigg]$$

$$- \sum_{i=t/h+1}^{T/h-1} a(ih)\Bigg)$$

Because the credit-risky assets are expressed in terms of risk-free forward rates and a credit spread, the drift of the credit spread process depends on the drift of the risk-free forward rates. Consequently, the drift of the respective risk-free forward rate has to be computed first, as shown in expression (6.16), before the drift of the credit spread process can be computed.

$[\epsilon_1, \epsilon_2]$ is a bivariate binomial random variable. Correlation between the forward rate process and the credit spread process can be introduced in several ways. For example, we can assume that ϵ_1 and ϵ_2 are independent and define $\epsilon_3 = \rho\epsilon_1 + \sqrt{1-\rho^2}\epsilon_2$. Then ϵ_1 and ϵ_3 are correlated with a correlation coefficient of ρ. Thus, ϵ_2 in the interest rate and credit spread process equations above would be replaced by ϵ_3 and the bivariate variable could be expressed as

$$[\epsilon_1, \epsilon_3] = \{(1, \rho + \sqrt{1-\rho^2}), (1, \rho - \sqrt{1-\rho^2}), \\ (-1, -\rho + \sqrt{1-\rho^2}), (-1, -\rho - \sqrt{1-\rho^2})\}$$

with the risk-neutral probability of each outcome being fixed at 0.25 for constant correlation.

Alternatively, the independent ϵ_1 and ϵ_2 can be used without transformation and correlation be introduced by the probabilities in the tree. In this case, the bivariate variable is given by

$$[\epsilon_1, \epsilon_2] = \{(1,1), (1,-1), (-1,1), (-1,-1)\}$$

with respective probabilites $\{\frac{1+\rho}{4}, \frac{1-\rho}{4}, \frac{1-\rho}{4}, \frac{1+\rho}{4}\}$ for constant correlation. In either case, correlation could be made a function of interest rates, credit spreads, or time.

6.6 Credit Derivatives as Exchange Options

Credit risk can be viewed as an exchange option. When a debtor refuses or is unable to fulfill an obligation, creditors can take over the debtor firm's assets. In other words, the equity holders have a short position in a riskless bond and an option to deliver the firm's assets in lieu of the bond value.

This view of credit risk leads to a simple extension of the model by Merton (1974). While Merton assumed that bond holders have a short position in a put option on the value of the firm, this approach assumes that bond holders are short an option to exchange the obligation for the firm's assets.

Unlike the original firm-value approach, the exchange-option approach takes into account the stochastic nature of firms' liabilities and can incorporate correlation between the value of the obligation and the value of the firm's assets.

A credit derivative that insures the buyer against a potential credit loss from debtor default effectively replaces the exchange option that the creditor have given to the debtor. In other words, for a fee, a third party assumes the obligations (i.e., loss on the bond position) that may arise from the creditor's short position in the exchange option. Therefore, the price of the credit derivative equals the value of the exchange option.

6.6.1 Process Specifications

Let a filtered probability space $(\Omega, \mathcal{F}, (\mathcal{F}_t)_{0 \leq t \leq T}, \mathbf{P})$ be given and assume that there is a probability measure $\mathbf{Q}$ equivalent to $\mathbf{P}$ such that asset prices deflated by the money market account $B(t)$ are martingales.

The evolution of the instantaneous forward rates under $\mathbf{Q}$ is assumed to be

$$f(t,T) = f(0,T) + \int_0^t \sigma(s,T) \cdot \sigma'(s,T)\, ds + \int_0^t \sigma(s,T) \cdot d\tilde{W}(t), \quad (6.20)$$

where $\tilde{W}_t$ is a Brownian motion in $\mathbb{R}^d$ defined on $(\Omega, \mathcal{F}, \mathbf{Q})$. $\sigma'(t,T) = \int_t^T \sigma(t,s)\, ds$ where $\sigma(t,s)$ is a bounded $\mathbb{R}^d$-valued function. This is an arbitrage-free forward rate process as shown by Heath, Jarrow, and Morton (1992). Expression (6.20) implies

$$P(t,T) = P(0,T) \exp\left(\int_0^t r(s) ds - \frac{1}{2} \int_0^t \|\sigma'(s,T)\|^2\, ds - \int_0^t \sigma'(s,T) \cdot d\tilde{W}(s) \right) \quad (6.21)$$

for the process of the prices of zero-coupon bonds with maturity $T \in [0, \mathcal{T}]$ under the risk-neutral measure $\mathbf{Q}$. $r(t)$ is the short rate. $\cdot$ denotes the inner product and $\|.\|$ is the Euclidian norm. We assume throughout that that the volatility function $\sigma(t,T)$ is deterministic, thus implying Gaussian interest rates and lognormal bond prices.

Similarly, the evolution of a stock price under measure $\mathbf{Q}$ is given by the process

$$S(t) = S(0) \exp\left(\int_0^t r(s) ds - \frac{1}{2} \int_0^t \|\sigma(s)\|^2\, ds + \int_0^t \sigma(s) \cdot d\tilde{W}(s) \right)$$

For the analysis of exchange options, it is convenient, though not necessary, to start the analysis from the process specifications under numéraire $P(t,T)$ and the forward neutral measure $\mathbf{F}$. The time-T forward price of a traded asset $S_i(t)$ is defined as

$$S_i^F(t,T) = \frac{S_i(t)}{P(t,T)}$$

Under the measure $\mathbf{F}$ and numéraire $P(t,T)$, the forward price process of a traded asset $S_i^F(T,T)$ is

$$S_i^F(T,T) = S_i^F(t,T) \exp\left(\int_t^T \beta_{S_i}(s,T) \cdot dW^{\mathbf{F}}(s) - \frac{1}{2} \int_t^T \|\beta_{S_i}(s,T)\|^2\, ds \right)$$

or in SDE notation

$$\frac{dS_i^F(t,T)}{S_i^F(t,T)} = \beta_{S_i}(t,T) \cdot dW^{\mathbf{F}}(t).$$

For a bond as given in (6.21), but maturing at time $\tau > T$, the forward price volatility vector is given by

$$\beta(t,T,\tau) = b(t,\tau) - b(t,T)) = -\int_T^{\tau} \sigma(t,u)\,du$$

with $b(t,T) = -\int_t^T \sigma(t,u)\,du$. For a stock, we specify the forward price volatility vector as

$$\beta_{S_i}(t,T) = \sigma_{S_i}(t) - b(t,T) \tag{6.22}$$

6.6.2 Price of an Exchange Option

The following proposition and corollaries quantify the credit risk price by determining the price of the option to exchange one asset for another.

Proposition 6.6.1. *The price $X(t)$ of a claim to receive stock S_1 in exchange for stock S_2 at time T is given by*

$$X(t) = S_1(t)N(d) - S_2(t)N(d - \chi(t,T))$$

with

$$d = \frac{\ln \frac{S_1(t)}{S_2(t)} + \frac{1}{2}\chi^2(t,T)}{\chi(t,T)}$$

and

$$\chi^2(t,T) = \int_t^T \|\sigma_1(s) - \sigma_2(s)\|^2\,ds$$

$N(d)$ denotes the standard normal distribution function.

Proof. The price of the exchange claim can be written as

$$X_t = P(t,T)\mathbf{E}_{\mathbf{F}}[(S_1(T,T) - S_2(T,T))\mathbf{1}_{\{S_1^F(T,T)\geq S_2^F(T,T)\}}|\mathcal{F}_t].$$

$\mathbf{F}$ denotes the forward neutral measure.

This expression can be split into two separate terms such that

$$X_t = E_1 - E_2,$$

where

$$E_1 = P(t,T)\,\mathbf{E}_{\mathbf{F}}[S_1^F(T,T)\,\mathbf{1}_{\{S_1^F(T,T)>S_2^F(T,T)\}}|\mathcal{F}_t],$$
$$E_2 = P(t,T)\,\mathbf{E}_{\mathbf{F}}[S_2^F(T,T)\,\mathbf{1}_{\{S_1^F(T,T)>S_2^F(T,T)\}}|\mathcal{F}_t].$$

Under the forward neutral measure, the dynamics of the forward asset prices S_i^F are given by

$$\frac{dS_i^F(t,T)}{S_i^F(t,T)} = \beta_{S_i}(t,T)\cdot dW_t^{\mathbf{F}}.$$

$\beta_{S_i}(t,T)$ is defined in (6.22). Additionally, we define

$$\eta_i(t,T) = \int_t^T \beta_{S_i}(s,T) \cdot dW^{\mathbf{F}}(s)$$

and

$$\nu_i^2(t,T) = \int_t^T \|\beta_i(s,T)\|^2 \, ds.$$

$S_i^F(T,T)$ can therefore be written

$$S_i^F(T,T) = S_i^F(t,T) \exp\left(\eta_{S_i}(t,T) - \frac{1}{2}\nu_{S_i}^2(t,T)\right), \quad \forall i \in \{1,2\}. \quad (6.23)$$

We also have $S_i^F(T,T) = S(T)$ and $S_i^F(t,T) = S_i(t)P(t,T)^{-1}$, by the definition of the forward price.

E_1 and E_2 can be evaluated similarly. We use an index of i for expressions that apply to the evaluation of both E_1 and E_2. We evaluate the expressions

$$E_i = P(t,T)\, \mathbf{E}_{\mathbf{F}}[S_i^F(T,T)\, \mathbf{1}_{\{S_1^F(T,T) > S_2^F(T,T)\}} | \mathfrak{F}_t].$$

for $i \in \{1,2\}$. By substituting for $S_i^F(T,T)$ we obtain

$$\begin{aligned} E_i &= \mathbf{E}_{\mathbf{F}}[\frac{S_i(t)}{P(t,T)} \exp\left(\int_t^T \beta_{S_i}(s,T) \cdot dW^{\mathbf{F}}(s) - \frac{1}{2}\int_t^T \|\beta_{S_i}(s,T)\|^2 \, ds\right) \\ &\qquad \mathbf{1}_{\{S_1^F(T,T) > S_2^F(T,T)\}} | \mathfrak{F}_t] P(t,T) \\ &= S_i(t)\mathbf{E}_{\mathbf{F}}[\exp\left(\eta_{S_i}(t,T) - \frac{1}{2}\nu_{S_i}^2(t,T)\right) \mathbf{1}_{\{S_1^F(T,T) > S_2^F(T,T)\}} | \mathfrak{F}_t]. \quad (6.24) \end{aligned}$$

We introduce a new measure $\mathbf{F}^{\mathbf{i}}$ defined by

$$\zeta_T = \frac{d\mathbf{F}^{\mathbf{i}}}{d\mathbf{F}} = \exp\left(\int_0^T \beta_{S_i}(s,T) \cdot dW^{\mathbf{F}}(s) - \frac{1}{2}\int_0^T \|\beta_{S_i}(s,T)\|^2 \, ds\right), \quad (6.25)$$

with $\zeta_t = \mathbf{E}_{\mathbf{F}}[\zeta_T | \mathfrak{F}_t]$, $\forall t \in [0,T]$.

Expression (6.24) can be seen to be equal to

$$E_i = \zeta_t^{-1}\, \mathbf{E}_{\mathbf{F}}[S_i(T)\zeta_T \mathbf{1}_{\{S_1^F(T,T) > S_2^F(T,T)\}} | \mathfrak{F}_t].$$

By the Bayes rule, this is equal to

$$E_i = S_i(T)\, \mathbf{E}_{\mathbf{F}^{\mathbf{i}}}[\mathbf{1}_{\{S_1^F(T,T) > S_2^F(T,T)\}} | \mathfrak{F}_t].$$

$S_i^F(T,T)$ are functions of $\eta_{S_i}(t,T)$. By the definition of Brownian motion, $\eta_{S_i}(t,T)$ has $\mathbf{F}$-law $N(0, \nu_{S_i}^2(t,T))$. Furthermore, by the definition of conditional expectation, we have

$$
\begin{aligned}
E_i &= S_i(t) \int_t^T \exp\left(\eta_{S_i} - \frac{1}{2}\nu_{S_i}^2\right) \mathbf{1}_{\{S_1^F(T,T) > S_2^F(T,T)\}} \, d\eta_{S_1} \\
&= S_i(t) \int_t^T \mathbf{1}_{\{S_1^F(T,T) > S_2^F(T,T)\}} \, d\dot{\eta}_{S_i} \\
&= S_i(t) N(a),
\end{aligned}
$$

where $\dot{\eta}_i(t,T) = \int_t^T \beta_{S_i}(s,T) \cdot dW^{\mathbf{F^i}}(s)$ and a is determined by the evaluation of the indicator functions.

By Girsanov's theorem, the change of measure by (6.25) implies that

$$
W^{\mathbf{F^i}}(t) = W^{\mathbf{F}}(t) - \int_0^t \beta_{S_i}(s,T)\, ds
$$

is a standard Brownian motion under $\mathbf{F^i}$.

To determine E_1, the indicator is evaluated under measure $\mathbf{F^1}$, giving

$$
\begin{aligned}
&\mathbf{F^1}(S_1^F(T,T) > S_2^F(T,T)) \\
&\quad = \mathbf{F^1}\Big(S_1^F(t,T) \exp\big(\eta_{S_1}(t,T) - \frac{1}{2}\nu_{S_1}(t,T)^2\big) \\
&\qquad\qquad > S_2^F(t,T) \exp\big(\eta_{S_2}(t,T) - \frac{1}{2}\nu_{S_2}(t,T)^2\big)\Big) \\
&\quad = \mathbf{F^1}\Big(F_1 \exp\Big(\int_t^T \beta_{S_1}(s,T) \cdot (dW^{\mathbf{F^1}}(s) + \beta_{S_1}(s,T) ds) - \frac{1}{2}\nu_{S_1}^2(t,T)\Big) \\
&\qquad\qquad > F_2 \exp\Big(\int_t^T \beta_{S_2}(s,T) \cdot (dW^{\mathbf{F^1}}(s) + \beta_{S_1}(s,T) ds) - \frac{1}{2}\nu_{S_2}^2(t,T)\Big)\Big) \\
&\quad = \mathbf{F^1}\Big(\ln S_1(t) + \eta_{S_1}(t,T) + \frac{1}{2}\nu_{S_1}^2(t,T) \\
&\qquad\qquad > \ln S_2(t) + \eta_{S_2}(t,T) + \int_t^T \beta_{S_1}(s,T) \cdot \beta_{S_2}(s,T)\, ds - \frac{1}{2}\nu_{S_2}^2(t,T)\Big) \\
&\quad = \mathbf{F^1}\Big(\int_t^T (\beta_{S_1}(s,T) - \beta_{S_2}(s,T))\, dW^{\mathbf{F^1}}(s) \\
&\qquad\qquad < \ln\frac{S_1(t)}{S_2(t)} + \frac{1}{2}\nu_{S_1}^2(t,T) - \frac{1}{2}\nu_{S_2}^2(t,T) \\
&\qquad\qquad\qquad - \int_t^T \beta_{S_1}(s,T) \cdot \beta_{S_2}(s,T)\, ds\Big) \\
&\quad = \mathbf{F^1}\left(\zeta < \frac{\ln\frac{S_1(t)}{S_2(t)} + \frac{1}{2}\chi^2(t,T)}{\chi(t,T)} \right)
\end{aligned}
$$

with $F_i(t,T) = \frac{S_i(t)}{P(t,T)}$ and $\chi^2(t,T) = \int_t^T \|\beta_{S_1}(s,T) - \beta_{S_2}(s,T)\|^2\, ds = \nu_{S_1}^2(t,T) + \nu_{S_2}^2(t,T) - \phi_{S_1S_2}(t,T)$. Therefore,

$$E_1 = S_1(t)N(d),$$

where

$$d = \frac{\ln \frac{S_1(t)}{S_2(t)} + \frac{1}{2}\chi^2(t,T)}{\chi(t,T)}.$$

By the definition of the forward volatility, $\beta_{S_i}(t,T)$, as given in expression (6.22), we have $\chi^2(t,T) = \int_t^T \|\beta_{S_1}(s,T) - \beta_{S_2}(s,T)\|^2 \, ds = \int_t^T \|\sigma_1(s,T) - \sigma_2(s,T)\|^2 \, ds$.

To determine E_2, the indicator function is evaluated under measure $\mathbf{F}^2$ instead of $\mathbf{F}^1$. It follows that

$$\begin{aligned}
&\dot{\mathbf{F}}(S_1^F(T,T) > S_2^F(T,T)) \\
&= \dot{\mathbf{F}}\Bigg(\frac{S_1(t)}{P(t,T)} \exp\Bigg(\int_t^T \beta_{S_1}(s,T) \cdot (dW^{\dot{\mathbf{F}}}(s) + \beta_{S_1}(s,T)ds) \\
&\qquad\qquad - \frac{1}{2}\nu_{S_1}^2(t,T)\Bigg) \\
&> \frac{S_2(t)}{P(t,T)} \exp\Bigg(\int_t^T \beta_{S_2}(s,T) \cdot (dW^{\dot{\mathbf{F}}}(s) + \beta_{S_1}(s,T)ds) \\
&\qquad\qquad - \frac{1}{2}\nu_{S_2}^2(t,T)\Bigg)\Bigg) \\
&= \dot{\mathbf{F}}\left(\zeta < \frac{\ln \frac{S_1(t)}{S_2(t)} + \frac{1}{2}\chi^2(t,T)}{\chi(t,T)} \right)
\end{aligned}$$

Thus,

$$E_2 = S_2(t)N(d - \chi(t,T)).$$

This completes the proof of Proposition 6.6.1.

In a credit risk context, an option to exchange one stock for another may not be the appropriate model to price a credit derivative that insures against the loss in case of the default of an debtor because the original obligation is a bond. In a Gaussian framework, however, the exchange option model can easily be extended to options that give the right to exchange a bond into a stock, a stock into a bond, or a bond into a bond. The corresponding pricing formulae are given below. They are straightforward variations of the formula presented above for stock exchange options.

Corollary 6.6.1. *The price $X(t)$ of a claim $X(T) = (S(T) - P(T,\tau))^+$ to receive stock $S(T)$ in exchange for bond $P(T,\tau)$ at time $T < \tau$ can be expressed as*

$$X(t) = S(t)N(d) - P(t,\tau)N(d - \chi(t,T))$$

with

$$d = \frac{\ln S^F(t,\tau) + \frac{1}{2}\chi^2(t,T)}{\chi(t,T)}$$

and

$$\chi^2(t,T) = \int_t^T \|\sigma_1(s) - b(s,\tau)\|^2\, ds = \int_t^T \|\beta_S(t,\tau)\|^2\, ds$$

$S^F(t,\tau)$ is the forward price of S with respect to time τ. $\beta_S(t,\tau)$ is the corresponding forward volatility of S.

Proof. Follows immediately from the proof of Proposition 6.6.1.

Corollary 6.6.2. *The price $X(t)$ of a claim $X(T) = (P(T,\tau_1) - P(T,\tau_2))^+$ to receive bond $P(T,\tau_1)$ in exchange for bond $P(T,\tau_2)$ at time T can be expressed as*

$$X(t) = P(t,\tau_1)N(d) - P(t,\tau_2)N(d - \chi(t,T))$$

with

$$d = \frac{\ln P^F(t,\tau_2,\tau_1) + \frac{1}{2}\chi^2(t,T)}{\chi(t,T)}$$

and

$$\nu^2(t,T) = \int_t^T \|b(s,\tau_1) - b(s,\tau_2)\|^2\, ds = \int_t^T \|\beta(s,\tau_2,\tau_1)\|^2\, ds$$

with $\beta(t,\tau_2,\tau_1) = -\int_{\tau_2}^{\tau_1} \sigma(t,u)\, du$. $T < \tau_1$ and $T < \tau_2$. $P^F(t,\tau_2,\tau_1)$ is the forward price of $P(t,\tau_1)$ with respect to time τ_2. $\beta_S(t,\tau_2,\tau_1)$ is the corresponding forward volatility.

Proof. Follows immediately from the proof of Proposition 6.6.1.

Corollary 6.6.3. *The price $X(t)$ of a claim $X(T) = (P(T,\tau) - S(T))^+$ to receive bond $P(T,\tau_1)$ in exchange for stock $S(T)$ at time T can be expressed as*

$$X(t) = P(t,\tau)N(d) - S(t)N(d - \chi(t,T))$$

with

$$d = \frac{\ln \frac{P(t,\tau)}{S(t)} + \frac{1}{2}\chi^2(t,T)}{\chi(t,T)}$$

and

$$\chi^2(t,T) = \int_t^T \|b(s,\tau) - \sigma_2(s)\|^2\, ds$$

Proof. Follows immediately from the proof of Proposition 6.6.1.

6.7 Credit Derivatives with Counterparty Default Risk

Credit derivatives are used to take a position in or hedge a credit exposure. Sometimes, however, the credit derivative itself is also subject to credit risk. While credit derivatives counterparties tend to be of high credit quality, this may not necessarily always be the case. For example, during the Asian crisis in 1998, several credit derivatives counterparties deteriorated substantially in credit quality — almost in parallel with the firms for which they provided credit protection — and some actually defaulted on their obligations from credit derivatives contracts.

This section provides an approach for integrating counterparty risk into the valuation of credit derivatives. The approach uses the same firm-value approach to model counterparty default risk as Chapter 4. Furthermore, the credit derivative is modeled as an exchange option as in Section 6.6. Note that now two credit risk issues have to be dealt with. First, there is the promised payoff function, which is modeled by the payoff from the exchange option. Second, there is the derivative counterparty. The derivative counterparty and the credit counterparty that determines the promised derivative payoff are distinct entities.

6.7.1 Price of an Exchange Option with Counterparty Default Risk

The following proposition gives a closed-form solution for the price of an exchange option that is subjet to counterparty default risk. To model the credit risk of the counterparty a firm value model in accordance with Merton (1974) is used. The credit risk model for the counterparty is the same as in Chapter 4. In this section, however, the liabilities of the counterparty firm are assumed to be constant although a similar extension to stochastic liabilities as in Chapter 4 is possible.

The bond and stock price processes are specified in Section 6.6.1. The value of the derivative counterparty's assets are assumed to have the usual process, i.e.,

$$V(t) = V(0)\exp\left(\int_0^t r(s)ds - \frac{1}{2}\int_0^t \|\sigma_V(s)\|^2\, ds + \int_0^t \sigma_V(s)\,\cdot d\tilde{W}(s)\right)$$

under the risk-neutral probability measure $\mathbf{Q}$.

Proposition 6.7.1. *The price X_t of vulnerable call option with promised payoff $X_T = (S_1(T) - S_2(T))^+$ and actual payoff $X_T = \theta(T)(S_1(T) - S_2(T))^+$ with $\theta(T) = \frac{V_T}{D}$ in case of bankruptcy is given by*

$$
\begin{aligned}
X_t = & S_1(t)N_2(a_1, a_2, \rho_1(t,T) - \rho_2(t,T)) - S_2(t)N_2(b_1, b_2, \rho_1(t,T) - \rho_2(t,T)) \\
& + \frac{S_1(t)V_t}{DP(t,T)} e^{\phi_{S_1V}(t,T)} N_2(c_1, c_2, \rho_2(t,T) - \rho_1(t,T)) \\
& - \frac{S_2(t)V_t}{DP(t,T)} N_2(d_1, d_2, \rho_2(t,T) - \rho_1(t,T)),
\end{aligned}
$$

with parameters

$$
\begin{aligned}
a_1 &= \frac{\ln \frac{S_1(t)}{S_2(t)} + \frac{1}{2}\chi^2(t,T)}{\chi(t,T)}, \\
a_2 &= \frac{\ln V_t - \ln D - \ln P(t,T) + \phi_{S_1V}(t,T) - \frac{1}{2}\nu_V^2(t,T)}{\nu_V(t,T)}, \\
b_1 &= \frac{\ln \frac{S_1(t)}{S_2(t)} - \frac{1}{2}\chi^2(t,T)}{\chi(t,T)}, \\
b_2 &= \frac{\ln V_t - \ln D - \ln P(t,T) + \phi_{S_2V}(t,T) - \frac{1}{2}\nu_V^2(t,T)}{\nu_V(t,T)}, \\
c_1 &= \frac{\ln \frac{S_1(t)}{S_2(t)} + \frac{1}{2}\chi^2(t,T) + \phi_{S_1V}(t,T) - \phi_{S_2V}(t,T)}{\chi(t,T)}, \\
c_2 &= -\frac{\ln V_t - \ln D - \ln P(t,T) + \frac{1}{2}\nu_V^2(t,T) + \phi_{S_1V}(t,T)}{\nu_V(t,T)}, \\
d_1 &= \frac{\ln \frac{S_1(t)}{S_2(t)} - \frac{1}{2}\chi^2(t,T) + \phi_{S_1V}(t,T) - \phi_{S_2V}(t,T)}{\chi(t,T)}, \\
d_2 &= -\frac{\ln V_t - \ln D - \ln P(t,T) + \frac{1}{2}\nu_V^2(t,T) + \phi_{S_2V}(t,T)}{\nu_V(t,T)},
\end{aligned} \tag{6.26}
$$

and

$$
\begin{aligned}
\nu_i^2(t,T) &= \int_t^T \|\beta_i(s,T)\|^2 \, ds, \qquad \forall i \in \{S_1, S_2, V\}, \\
\phi_{ij}(t,T) &= \int_t^T \beta_i(s,T) \cdot \beta_j(s,T) \, ds, \qquad \forall i, \forall j \in \{S_1, S_2, V\}, \\
\rho_i(t,T) &= \int_t^T \frac{\beta_{S_i}(s,T) \cdot \beta_V(s,T)}{\|\beta_{S_i}(s,T)\| \, \|\beta_V(s,T)\|} ds, \quad \forall i \in \{1, 2\}, \\
\chi^2(t,T) &= \int_t^T \|\beta_{S_1}(s,T) - \beta_{S_2}(s,T)\|^2 \, ds
\end{aligned} \tag{6.27}
$$

β_i, $\forall i \in \{S_1, S_2, V\}$, *is a bounded and deterministic function.*

Remark 6.7.1. In a credit derivative context, $S_1(t)$ and $S_2(t)$ are interpreted as the variables that determine the payoff of the credit derivative (such as the bond obligation and the asset value of the firm that issued the underlying credit) whereas $V(t)$ denotes the asset value of the counterparty firm of the derivative contract. Obviously, the interpretation and use of the pricing formula above is not restricted to credit derivatives. It can be used to value any exchange option that is subject to counterparty risk given that the stochastic processes used for $S_1(t)$ and $S_2(t)$ represent reasonable models for the actual processes of the underlying assets.

Proof. The price of the vulnerable exchange claim can be written as

$$X_t = P(t,T)\mathbf{E_F}[(S_1(T,T) - S_2(T,T))^+ \\ (\mathbf{1}_{\{V_F(T,T)\geq D\}} + \theta_F(T,T)\mathbf{1}_{\{V_F(T,T)<D\}})|\mathfrak{F}_t],$$

with $\theta_F(T,T) = \frac{V_F(T,T)}{D}$. $\mathbf{F}$ denotes the forward neutral measure.

This expression can be split into four separate terms such that

$$X_t = E_1 - E_2 + E_3 - E_4,$$

where

$$E_1 = P(t,T)\,\mathbf{E_F}[S_1^F(T,T)\,\mathbf{1}_{\{S_1^F(T,T)>S_2^F(T,T)\}}\mathbf{1}_{\{V_F(T,T)\geq D\}}|\mathfrak{F}_t],$$
$$E_2 = P(t,T)\,\mathbf{E_F}[S_2^F(T,T)\,\mathbf{1}_{\{S_1^F(T,T)>S_2^F(T,T)\}}\mathbf{1}_{\{V_F(T,T)\geq D\}}|\mathfrak{F}_t],$$
$$E_3 = P(t,T)\,\mathbf{E_F}[S_1^F(T,T)\theta_F(T,T)\,\mathbf{1}_{\{S_1^F(T,T)>S_2^F(T,T)\}}\mathbf{1}_{\{V_F(T,T)<D\}}|\mathfrak{F}_t],$$
$$E_4 = P(t,T)\,\mathbf{E_F}[S_2^F(T,T)\theta_F(T,T)\,\mathbf{1}_{\{S_1^F(T,T)>S_2^F(T,T)\}}\mathbf{1}_{\{V_F(T,T)<D\}}|\mathfrak{F}_t].$$

Under the forward neutral measure, the dynamics of the forward asset prices S_i^F and V_F are given by

$$\frac{dS_i^F(t,T)}{S_i^F(t,T)} = \beta_{S_i}(t,T)\cdot dW_t^{\mathbf{F}}, \qquad \frac{dV_F(t,T)}{V_F(t,T)} = \beta_V(t,T)\cdot dW_t^{\mathbf{F}}.$$

$\beta_{S_i}(t,T)$ and $\beta_V(t,T)$ are defined in (6.22). Additionally, we define

$$\eta_i(t,T) = \int_t^T \beta_i(s,T)\cdot dW^{\mathbf{F}}(s),$$

for $i \in \{S_1, S_2, V\}$. $S_F(T,T)$ and $V_F(T,T)$ can therefore be written

$$i_F(T,T) = i_F(t,T)\exp\left(\eta_i(t,T) - \frac{1}{2}\nu_i^2(t,T)\right). \tag{6.28}$$

We also have $i_F(T,T) = i_T$ and $i_F(t,T) = i_t P(t,T)^{-1}$, by the definition of the forward price.

Evaluation of E_1. We evaluate the expression

$$E_1 = P(t,T)\,\mathbf{E_F}[S_1^F(T,T)\,\mathbf{1}_{\{S_1^F(T,T)>S_2^F(T,T)\}}\mathbf{1}_{\{V_F(T,T)\geq D\}}|\mathcal{F}_t].$$

By substituting for $S_1^F(T,T)$ we obtain

$$\begin{aligned} E_1 &= \mathbf{E_F}[\frac{S_1(t)}{P(t,T)}\exp\left(\int_t^T \beta_{S_1}(s,T)\cdot dW^{\mathbf{F}}(s) - \frac{1}{2}\int_t^T \|\beta_{S_1}(s,T)\|^2\,ds\right) \\ &\qquad \mathbf{1}_{\{S_1^F(T,T)>S_2^F(T,T)\}}\mathbf{1}_{\{V_F(T,T)\geq D\}}|\mathcal{F}_t]\,P(t,T) \\ &= S_1(t)\mathbf{E_F}[\exp\left(\eta_{S_1}(t,T) - \frac{1}{2}\nu_{S_1}^2(t,T)\right) \\ &\qquad \mathbf{1}_{\{S_1^F(T,T)>S_2^F(T,T)\}}\mathbf{1}_{\{V_F(T,T)\geq D\}}|\mathcal{F}_t]. \end{aligned} \tag{6.29}$$

We introduce a new measure $\dot{\mathbf{F}}$ defined by

$$\zeta_T = \frac{d\dot{\mathbf{F}}}{d\mathbf{F}} = \exp\left(\int_0^T \beta_{S_1}(s,T)\cdot dW^{\mathbf{F}}(s) - \frac{1}{2}\int_0^T \|\beta_{S_1}(s,T)\|^2\,ds\right), \tag{6.30}$$

with $\zeta_t = \mathbf{E_F}[\zeta_T|\mathcal{F}_t]$, $\forall t \in [0,T]$.
Expression (6.29) can be seen to be equal to

$$E_1 = \zeta_t^{-1}\,\mathbf{E_F}[S_1(t)\zeta_T\mathbf{1}_{\{S_1^F(T,T)>S_2^F(T,T)\}}\mathbf{1}_{\{V_F(T,T)\geq D\}}|\mathcal{F}_t].$$

By the Bayes rule, this is equal to

$$E_1 = S_1(t)\,\mathbf{E}_{\dot{\mathbf{F}}}[\mathbf{1}_{\{S_1^F(T,T)>S_2^F(T,T)\}}\mathbf{1}_{\{V_F(T,T)\geq D\}}|\mathcal{F}_t].$$

$S_i^F(T,T)$ and $V_F(T,T)$ are functions of $\eta_{S_i}(t,T)$ and $\eta_V(t,T)$, respectively. By the definition of Brownian motion, $\eta_i(t,T)$, $\forall i \in \{S_1, S_2, V\}$, has $\mathbf{F}$-law $N(0,\nu_i^2(t,T))$ with joint normal law $N_2(0,0,\nu_{S_i}^2(t,T),\nu_V^2(t,T),\rho_i(t,T))$. By the definition of conditional expectation and the linearity of the correlation coefficient, we have

$$\begin{aligned} E_1 &= S_1(t)\int_t^T\int_t^T \exp\left(\eta_{S_1} - \frac{1}{2}\nu_{S_1}^2\right) \\ &\qquad \mathbf{1}_{\{S_1^F(T,T)-S_2^F(T,T)>0\}}\mathbf{1}_{\{V_F(T,T)\geq D\}}\,d\eta_V\,d(\eta_{S_1}-\eta_{S_2}) \\ &= S_1(t)\int_t^T\int_t^T \mathbf{1}_{\{S_1^F(T,T)-S_2^F(T,T)>0\}}\mathbf{1}_{\{V_F(T,T)\geq D\}}\,d\dot{\eta}_V\,d(\dot{\eta}_{S_1}-\dot{\eta}_{S_2}) \\ &= S_1(t)\int_t^T\int_t^T \mathbf{1}_{\{S_1^F(T,T)>S_2^F(T,T)\}}\mathbf{1}_{\{V_F(T,T)\geq D\}}\,d\dot{\eta}_V\,d(\dot{\eta}_{S_1}-\dot{\eta}_{S_2}) \\ &= S_1(t)N_2(a_1,a_2,\rho_1(t,T)-\rho_2(t,T)), \end{aligned}$$

where $\dot{\eta}_i(t,T) = \int_t^T \beta_i(s,T) \cdot dW^{\dot{\mathbf{F}}}(s)$, $\forall i \in \{S_1, S_2, V\}$, and a_1 and a_2 are determined by the evaluation of the indicator functions.

By Girsanov's theorem, the change of measure by (6.30) implies that

$$W^{\dot{\mathbf{F}}}(t) = W^{\mathbf{F}}(t) - \int_0^t \beta_{S_1}(s,T)\,ds$$

is a standard Brownian motion under $\dot{\mathbf{F}}$. The indicator functions can therefore be evaluated as

$$\begin{aligned}
&\dot{\mathbf{F}}(S_1^F(T,T) > S_2^F(T,T)) \\
&= \dot{\mathbf{F}}\Big(S_1^F(t,T)\exp\left(\eta_{S_1}(t,T) - \frac{1}{2}\nu_{S_1}^2(t,T)\right) \\
&\qquad > S_2^F(t,T)\exp\left(\eta_{S_2}(t,T) - \frac{1}{2}\nu_{S_2}^2(t,T)\right)\Big) \\
&= \dot{\mathbf{F}}\Bigg(\frac{S_1(t)}{P(t,T)} \exp\bigg(\int_t^T \beta_{S_1}(s,T) \cdot (dW^{\dot{\mathbf{F}}}(s) + \beta_{S_1}(s,T)ds) \\
&\qquad\qquad - \frac{1}{2}\nu_{S_1}^2(t,T)\bigg) \\
&\qquad > \frac{S_2(t)}{P(t,T)} \exp\bigg(\int_t^T \beta_{S_2}(s,T) \cdot (dW^{\dot{\mathbf{F}}}(s) + \beta_{S_1}(s,T)ds) \\
&\qquad\qquad - \frac{1}{2}\nu_{S_2}^2(t,T)\bigg)\Bigg) \\
&= \dot{\mathbf{F}}\Big(\ln S_1(t) + \eta_{S_1}(t,T) + \frac{1}{2}\nu_{S_1}^2(t,T) \qquad\qquad (6.31) \\
&\qquad > \ln S_2(t) + \eta_{S_2}(t,T) + \int_t^T \beta_{S_1}(s,T) \cdot \beta_{S_2}(s,T)\,ds \\
&\qquad\qquad - \frac{1}{2}\nu_{S_2}^2(t,T)\Big) \\
&= \dot{\mathbf{F}}\Big(\int_t^T (\beta_{S_1}(s,T) - \beta_{S_2}(s,T))\,dW^{\dot{\mathbf{F}}}(s) \\
&\qquad < \ln\frac{S_1(t)}{S_2(t)} + \frac{1}{2}\nu_{S_1}^2(t,T) + \frac{1}{2}\nu_{S_2}^2(t,T) \\
&\qquad\qquad - \int_t^T \beta_{S_1}(s,T) \cdot \beta_{S_2}(s,T)\,ds\Big) \\
&= \dot{\mathbf{F}}\left(\zeta < \frac{\ln\frac{S_1(t)}{S_2(t)} + \frac{1}{2}\chi^2(t,T)}{\chi(t,T)} \right)
\end{aligned}$$

with $\chi^2(t,T) = \int_t^T \|\beta_{S_1}(s,T) - \beta_{S_2}(s,T)\|^2\,ds = \nu_{S_1}^2(t,T) + \nu_{S_2}^2(t,T) - \phi_{S_1S_2}(t,T)$.

$$
\begin{aligned}
&\dot{\mathbf{F}}(V_F(t,T) \geq D) \\
&= \dot{\mathbf{F}}\left(V_F(t,T) \exp\left(\eta_V(t,T) - \frac{1}{2}\nu_V^2(t,T) \right) \geq D \right) \\
&= \dot{\mathbf{F}}\Bigg(\frac{V_t}{P(t,T)} \exp\Bigg(\int_t^T \beta_V(s,T) \cdot (dW^{\dot{\mathbf{F}}}(s) + \beta_{S_1}(s,T) ds) \\
&\qquad\qquad\qquad\qquad - \frac{1}{2}\nu_V^2(t,T) \Bigg) \geq \Bigg) \\
&= \dot{\mathbf{F}}\Bigg(\frac{V_t}{P(t,T)} \exp\Bigg(\eta_V(t,T) + \int_t^T \beta_{S_1}(s,T) \cdot \beta_V(s,T)\, ds \\
&\qquad\qquad\qquad\qquad - \frac{1}{2}\nu_V^2(t,T) \Bigg) \geq D \Bigg) \\
&= \dot{\mathbf{F}}\left(\eta_V(t,T) \geq \ln \frac{DP(t,T)}{V_t} - \phi_{S_1 V}(t,T) + \frac{1}{2}\nu_V^2(t,T) \right) \\
&= \dot{\mathbf{F}}\left(\eta_V(t,T) \leq \ln V_t - \ln D - \ln P(t,T) + \phi_{S_1 V}(t,T) - \frac{1}{2}\nu_V^2(t,T) \right).
\end{aligned} \tag{6.32}
$$

Therefore,

$$
E_1 = S_1(t) N_2(a_1, a_2, \rho_1(t,T) - \rho_2(t,T)),
$$

where

$$
\begin{aligned}
a_1 &= \frac{\ln \frac{S_1(t)}{S_2(t)} + \frac{1}{2}\chi^2(t,T)}{\chi(t,T)}, \\
a_2 &= \frac{\ln \frac{V_t}{DP(t,T)} + \phi_{S_1 V}(t,T) - \frac{1}{2}\nu_V^2(t,T)}{\nu_V(t,T)}.
\end{aligned}
$$

Evaluation of E_2.

$$
E_2 = P(t,T)\, \mathbf{E}_{\mathbf{F}}[S_2^F(T,T)\, \mathbf{1}_{\{S_1^F(T,T) > S_2^F(T,T)\}} \mathbf{1}_{\{V_F(T,T) \geq D\}} | \mathcal{F}_t]
$$

The evaluation of this expectation is similar to that of E_1 where S_1 is replaced by S_2. In this case, the auxiliary measure $\dot{\mathbf{F}}$ is therefore defined as

$$
\zeta_T = \frac{d\dot{\mathbf{F}}}{d\mathbf{F}} = \exp\left(\int_0^T \beta_{S_2}(s,T) \cdot dW^{\mathbf{F}}(s) - \frac{1}{2}\int_0^T \|\beta_{S_2}(s,T)\|^2\, ds \right). \tag{6.33}
$$

Accordingly, the evaluation of the indicator functions changes to

$$
\begin{aligned}
&\dot{\mathbf{F}}(S_1^F(T,T) > S_2^F(T,T)) \\
&\quad = \dot{\mathbf{F}}\Bigg(\frac{S_1(t)}{P(t,T)} \exp\left(\int_t^T \beta_{S_1}(s,T) \cdot (dW^{\dot{\mathbf{F}}}(s) + \beta_{S_2}(s,T)ds) - \frac{1}{2}\nu_1^2(t,T) \right) \\
&\qquad > \frac{S_2(t)}{P(t,T)} \exp\left(\int_t^T \beta_{S_2}(s,T) \cdot (dW^{\dot{\mathbf{F}}}(s) + \beta_{S_2}(s,T)ds) - \frac{1}{2}\nu_2^2(t,T) \right) \Bigg) \\
&\quad = \dot{\mathbf{F}}\left(\zeta < \frac{\ln \frac{S_1(t)}{S_2(t)} - \frac{1}{2}\chi^2(t,T)}{\chi(t,T)} \right)
\end{aligned} \tag{6.34}
$$

The second indicator function is also evaluated under measure $\dot{\mathbf{F}}$, giving

$$
\begin{aligned}
&\dot{\mathbf{F}}(V_F(t,T) \geq D) \\
&\quad = \dot{\mathbf{F}}\left(\zeta \leq \frac{\ln V_t - \ln D - \ln P(t,T) + \phi_{S_2V}(t,T) - \frac{1}{2}\nu_V^2(t,T)}{\nu_V(t,T)} \right).
\end{aligned} \tag{6.35}
$$

It follows that

$$
E_2 = S_2(t) N_2(b_1, b_2, \rho_1(t,T) - \rho_2(t,T)),
$$

where

$$
\begin{aligned}
b_1 &= \frac{\ln \frac{S_1(t)}{S_2(t)} - \frac{1}{2}\chi^2(t,T)}{\chi(t,T)}, \\
b_2 &= \frac{\ln \frac{V_t}{DP(t,T)} + \phi_{S_2V}(t,T) - \frac{1}{2}\nu_V^2(t,T)}{\nu_V(t,T)}.
\end{aligned}
$$

Evaluation of E_3. We evaluate the expression

$$
E_3 = P(t,T)\, \mathbf{E}_{\mathbf{F}}[S_1^F(T,T) V_F(T,T) D^{-1}\, \mathbf{1}_{\{S_1^F(T,T) > S_2^F(T,T)\}} \mathbf{1}_{\{V_F(T,T) < D\}} | \mathcal{F}_t].
$$

Substituting for $S_1^F(T,T)$ and $V_F(T,T)$, we obtain

$$
\begin{aligned}
E_3 &= \mathbf{E}_{\mathbf{F}}\left[\exp\left(\eta_{S_1}(t,T) - \frac{1}{2}\nu_{S_1}^2(t,T) \right) \exp\left(\eta_V(t,T) - \frac{1}{2}\nu_V^2(t,T) \right)\right. \\
&\qquad \left. \mathbf{1}_{\{S_1^F(T,T) > S_2^F(T,T)\}} \mathbf{1}_{\{V_F(T,T) < D\}} | \mathcal{F}_t\right] \frac{S_1(t)V(t)}{DP(t,T)} \\
&= \frac{S_1(t)V(t)}{DP(t,T)} \mathbf{E}_{\mathbf{F}}\left[\exp\left(\int_t^T (\beta_{S_1}(s,T) + \beta_V(s,T)) \cdot dW^{\mathbf{F}}(s) \right.\right. \\
&\qquad \left. - \frac{1}{2} \int_t^T \|\beta_{S_1}(s,T)\|^2 + \|\beta_V(s,T)\|^2\, ds \right) \\
&\qquad \left. \mathbf{1}_{\{S_1^F(T,T) > S_2^F(T,T)\}} \mathbf{1}_{\{V_F(T,T) < D\}} | \mathcal{F}_t \right].
\end{aligned} \tag{6.36}
$$

We introduce a new measure $\dot{\mathbf{F}}$ defined by

$$\frac{d\dot{\mathbf{F}}}{d\mathbf{F}} = \exp\left(\int_0^T \beta_{S_1}(s,T) + \beta_V(s,T) \cdot dW^{\mathbf{F}}(s) - \frac{1}{2}\int_0^T \|\beta_{S_1}(s,T) + \beta_V(s,T)\|^2 \, ds\right),$$

with $\dot{\zeta}_t = \mathbf{E_F}[\frac{d\dot{\mathbf{F}}}{d\mathbf{F}}|\mathcal{F}_t] \quad \forall t \in [0,T]$. Since

$$\int_t^T \|\beta_{S_1}(s,T) + \beta_V(s,T)\|^2 \, ds = \int_t^T \|\beta_{S_1}(s,T)\|^2 + \|\beta_V(s,T)\|^2 \, ds + 2\int_t^T \beta_{S_1}(s,T) \cdot \beta_V(s,T) \, ds\,,$$

expression (6.36) can be seen to be equal to

$$\begin{aligned} E_3 &= \frac{S_1(t)V(t)}{DP(t,T)} \dot{\zeta}_t^{-1} \, \mathbf{E_F}[\dot{\zeta}_T \exp(\phi_{S_1V}(t,T)) \\ &\qquad \mathbf{1}_{\{S_1^F(T,T) > S_2^F(T,T)\}} \mathbf{1}_{\{V_F(T,T) < D\}} |\mathcal{F}_t] \\ &= \frac{S_1(t)V(t)}{DP(t,T)} e^{\phi_{S_1V}(t,T)} \dot{\zeta}_t^{-1} \, \mathbf{E_F}[\dot{\zeta}_T \mathbf{1}_{\{S_1^F(T,T) > S_2^F(T,T)\}} \mathbf{1}_{\{V_F(T,T) < D\}} |\mathcal{F}_t], \end{aligned}$$

where $\phi_{S_1V}(t,T) = \int_t^T \beta_{S_1}(s,T) \cdot \beta_V(s,T) \, ds$. By the Bayes rule (cf. Corollary A.6.3), this is equal to

$$E_3 = \frac{S_1(t)V(t)}{DP(t,T)} e^{\phi_{S_1V}(t,T)} \mathbf{E}_{\dot{\mathbf{F}}}[\mathbf{1}_{\{S_1^F(T,T) > S_2^F(T,T)\}} \mathbf{1}_{\{V_F(T,T) < D\}} |\mathcal{F}_t]. \tag{6.37}$$

By the definition of Brownian motion, $\eta_i(t,T)$ has $\dot{\mathbf{F}}$-law $N(0, \nu_i^2(t,T))$ with joint normal law $N_2(0, 0, \nu_{S_i}^2(t,T), \nu_V^2(t,T), \rho_i(t,T))$. We can therefore write

$$\begin{aligned} E_3 &= \frac{S_1(t)V(t)}{DP(t,T)} \int_t^T \int_t^T \exp\left(\eta_{S_1} - \frac{1}{2}\nu_{S_1}^2 + \eta_V - \frac{1}{2}\nu_V^2\right) \\ &\qquad \mathbf{1}_{\{S_1^F(T,T) > S_2^F(T,T)\}} \mathbf{1}_{\{V_F(T,T) < D\}} \, d\eta_V \, d(\eta_{S_1} - \eta_{S_2}) \\ &= \frac{S_1(t)V(t)}{DP(t,T)} e^{\phi_{S_1V}(t,T)} \\ &\qquad \int_t^T \int_t^T \mathbf{1}_{\{S_1^F(T,T) > S_2^F(T,T)\}} \mathbf{1}_{\{V_F(T,T) < D\}} \, d\dot{\eta}_V \, d(\dot{\eta}_{S_1} - \dot{\eta}_{S_2}) \\ &= \frac{S_1(t)V(t)}{DP(t,T)} e^{\phi_{S_1V}(t,T)} N_2(c_1, c_2, \rho_2(t,T) - \rho_1(t,T)). \end{aligned}$$

The indicator functions determine c_1 and c_2. The sign of the correlation coefficient changes according to expression (4.43). By Girsanov's theorem, the change of measure above implies that

$$W^{\dot{\mathbf{F}}}(t) = W^{\mathbf{F}}(t) - \int_0^t (\beta_{S_1}(s,T) + \beta_V(s,T))\, ds$$

is a standard Brownian motion under $\dot{\mathbf{F}}$. Therefore, the indicator functions are evaluated as

$$\begin{aligned}
&\dot{\mathbf{F}}(S_1^F(T,T) > S_2^F(T,T)) \\
&= \dot{\mathbf{F}}\Bigg(S_1^F(t,T) \exp\left(\eta_{S_1}(t,T) - \frac{1}{2}\nu_{S_1}^2(t,T)\right) \\
&\qquad > S_2^F(t,T) \exp\left(\eta_{S_2}(t,T) - \frac{1}{2}\nu_{S_2}^2(t,T)\right)\Bigg) \\
&= \dot{\mathbf{F}}\Bigg(\frac{S_1(t)}{P(t,T)} \exp\Bigg(\int_t^T \beta_{S_1}(s,T) \cdot (dW^{\dot{\mathbf{F}}}(s) + (\beta_{S_1}(s,T) + \beta_V(s,T))ds) \\
&\qquad\qquad - \frac{1}{2}\nu_{S_1}^2(t,T)\Bigg) \\
&\quad > \frac{S_2(t)}{P(t,T)} \exp\Bigg(\int_t^T \beta_{S_2}(s,T) \cdot (dW^{\dot{\mathbf{F}}}(s) + (\beta_{S_1}(s,T) + \beta_V(s,T))ds) \\
&\qquad\qquad - \frac{1}{2}\nu_{S_2}^2(t,T)\Bigg)\Bigg) \\
&= \dot{\mathbf{F}}\Bigg(\ln S_1(t) + \eta_{S_1}(t,T) + \frac{1}{2}\nu_{S_1}^2(t,T) + \phi_{S_1 V}(t,T) \\
&\qquad > \ln S_2(t) + \eta_{S_2}(t,T) + \phi_{S_1 S_2}(t,T) + \phi_{S_2 V}(t,T) - \frac{1}{2}\nu_{S_2}^2(t,T)\Bigg) \\
&= \dot{\mathbf{F}}\Bigg(\int_t^T (\beta_{S_1}(s,T) - \beta_{S_2}(s,T))\, dW^{\dot{\mathbf{F}}}(s) \\
&\qquad < \ln \frac{S_1(t)}{S_2(t)} + \frac{1}{2}\nu_{S_1}^2(t,T) + \frac{1}{2}\nu_{S_2}^2(t,T) \\
&\qquad\qquad - \phi_{S_1 S_2}(t,T) + \phi_{S_1 V} - \phi_{S_2 V}\Bigg) \\
&= \dot{\mathbf{F}}\left(\zeta < \frac{\ln \frac{S_1(t)}{S_2(t)} + \frac{1}{2}\chi^2(t,T) + \phi_{S_1 V} - \phi_{S_2 V}}{\chi(t,T)} \right)
\end{aligned}$$

with $\chi^2(t,T) = \int_t^T \|\beta_{S_1}(s,T) - \beta_{S_2}(s,T)\|^2\, ds$.

$$\begin{aligned}
&\dot{\mathbf{F}}(V_F(t,T) < D)\\
&= \dot{\mathbf{F}}\left(V_F(t,T)\exp\left(\eta_V(t,T) - \frac{1}{2}\nu_V^2(t,T)\right) < D\right)\\
&= \dot{\mathbf{F}}\Bigg(\frac{V_t}{P(t,T)}\exp\Bigg(\int_t^T \beta_V(s,T)\cdot(dW^{\dot{\mathbf{F}}}(s) + (\beta_{S_1}(s,T)+\beta_V(s,T))ds)\\
&\qquad\qquad - \frac{1}{2}\nu_V^2(t,T)\Bigg) < D\Bigg)\\
&= \dot{\mathbf{F}}\Bigg(\frac{V_t}{P(t,T)}\exp\Bigg(\eta_V(t,T) + \int_t^T \beta_{S_1}(s,T)\cdot\beta_V(s,T)\,ds\\
&\qquad\qquad + \frac{1}{2}\nu_V^2(t,T)\Bigg) < D\Bigg)\\
&= \dot{\mathbf{F}}\left(\eta_V(t,T) < \ln\frac{DP(t,T)}{V_t} - \phi_{S_1V}(t,T) - \frac{1}{2}\nu_V^2(t,T)\right)\\
&= \dot{\mathbf{F}}\left(\eta_V(t,T) < -(\ln V_t - \ln D - \ln P(t,T) + \phi_{S_1V}(t,T) + \frac{1}{2}\nu_V^2(t,T))\right).
\end{aligned}$$

Therefore,

$$E_3 = \frac{S_1(t)V_t}{DP(t,T)}N_2(c_1, c_2, \rho_2(t,T) - \rho_1(t,T)),$$

where

$$c_1 = \frac{\ln\frac{S_1(t)}{S_2(t)} + \frac{1}{2}\chi^2(t,T) + \phi_{S_1V}(t,T) - \phi_{S_2V}(t,T)}{\chi(t,T)},$$

$$c_2 = -\frac{\ln V_t - \ln D - \ln P(t,T) + \phi_{S_1V}(t,T) + \frac{1}{2}\nu_V^2(t,T)}{\nu_V(t,T)}.$$

Evaluation of E_4. We evaluate the expression

$$E_4 = P(t,T)\,\mathbf{E_F}[S_2^F(T,T)V_F(T,T)D^{-1}\mathbf{1}_{\{S_1^F(T,T)>S_2^F(T,T)\}}\mathbf{1}_{\{V_F(T,T)<D\}}|\mathcal{F}_t].$$

The evaluation is similar to E_3, however, the new measure introduced is defined by

$$\begin{aligned}
\frac{d\dot{\mathbf{F}}}{d\mathbf{F}} = \exp\Bigg(&\int_0^T (\beta_{S_2}(s,T) + \beta_V(s,T))\cdot dW(s)\\
&- \frac{1}{2}\int_0^T \|\beta_{S_2}(s,T) + \beta_V(s,T)\|^2\,ds\Bigg).
\end{aligned}$$

The evaluation of the indicator function thus changes accordingly and gives

$$E_4 = \frac{S_2(t)V_t}{DP(t,T)} N_2(d_1, d_2, \rho_2(t,T) - \rho_1(t,T)),$$

where

$$d_1 = \frac{\ln \frac{S_1(t)}{S_2(t)} - \frac{1}{2}\chi^2(t,T) + \phi_{S_1V}(t,T) - \phi_{S_2V}(t,T)}{\chi(t,T)},$$

$$d_2 = -\frac{\ln V_t - \ln D - \ln P(t,T) + \phi_{S_2V}(t,T) + \frac{1}{2}\nu_V^2(t,T)}{\nu_V(t,T)}.$$

6.8 Summary

We begin this chapter by describing several credit derivative instruments, namely credit risk swaps, default put options, total return swaps and credit spread options and forward contracts, and identify two types of credit derivatives.

- The first class of credit derivatives consists of default-based instruments. The payoff of these derivatives is triggered by a credit event such as default or bankruptcy. Examples of credit derivatives of this type are default risk swaps, total return swaps, default insurance, and similar instruments.
- The second class of credit derivatives includes instruments with a payoff determined not by realized losses caused by default or bankruptcy, but by the credit quality, or a change thereof, of the reference security. Unlike instruments of the first type, these derivatives usually expire before the maturity date of the underlying reference asset. The most popular representatives of this class of derivatives are credit spread options.

Credit derivatives of the first type can often be valued with standard credit risk models as used to value credit-risky bonds. We therefore focus on derivatives of the second class, which cannot be valued in the same way.

We propose a compound approach for pricing credit derivatives of the second type, such as spread options and forward contracts. Because the underlying of credit spread derivatives is the credit risk of the reference asset, credit spread derivatives are derivatives on credit risk. As demonstrated in previous chapters, credit risk can be modeled in a variety of ways, for example with a firm value or an intensity approach. In this chapter, we describe the compound approach using the firm value to model the credit risk of the reference asset although an intensity model could also be used.

We present a closed-form pricing formula for credit spread options under deterministic interest rates. The formula is the Geske (1977) compound option pricing formula. For stochastic interest rates, we show that the lattice structures introduced in previous chapters can be extended to compound models in a straightforward way. This extension applies to both firm value models and intensity-based or hybrid models.

We give a number of numerical examples for yield spread and bond spread options. If the yield spread strike level of the option is set to zero, the option is equivalent to full credit insurance. We show that, in this case, the price of the option equals the price of credit risk. In the case of stochastic interest rates, we show that the correlation of firm value with interest rates has a considerable effect on credit spreads and thus also on the value of credit spread options, confirming results from previous chapters.

Furthermore, we present a simple reduced-form approach for valuing spread derivatives. This approach models credit spreads directly in a framework that is similar to the discrete version of the Heath-Jarrow-Morton methodology.

Finally, we treat the problem of credit derivatives that are themselves subject to counterparty risk. We use a firm-value approach for modeling the default risk of the derivative counterparty similar to the approach used in Chapter 4. The payoff of the credit derivative is modeled as an exchange option between two stocks, a bond and a stock, or two bonds. Thus, the credit derivative subject to default risk by the derivative counterparty is viewed as a vulnerable exchange option.

7. Conclusion

This book discusses methods and models for the valuation of credit risk. It reviews some of the most common approaches to valuing credit risk and focuses on the application of credit risk valuation to derivative contracts. In particular, it covers four aspects of derivative credit risk, namely

- *Counterparty default risk.* Because derivative instruments are contracts in which the parties agree on future cashflows according to predefined rules, parties which are to receive cash-flows are exposed to credit risk if it is conceivable that the counterparty will not or cannot satisfy its contractual obligations in the future. As a consequence, the fair price of a vulnerable derivative differs from the default-free price.
- *Options on credit-risky bonds.* Credit risk results in lower prices for credit-risky bonds. However, the price distribution does not simply move, it also changes shape because of the low-probability, high-loss property of default risk. Accordingly, options on risky bonds cannot be priced with standard option pricing methods but require a credit risk model.
- *Credit derivatives.* Credit risk may be the underlying variable of derivative contracts. In this case credit risk is not a byproduct of a derivative, but the purpose of the contract itself.
- *Credit derivatives with counterparty default risk.* As typical OTC derivative contracts, credit derivatives themselves are subject to counterparty risk. In this case, two distinct forms of credit risk affect the price of the credit derivative. On the one hand, the promised payoff of the contract is calculated based on a credit risk variable, such as a credit spread or a default loss caused by the default of the party specified in the contract. On the other hand, the counterparty risk of the derivative counterparty can affect the value of the contract.

We propose pricing models for all four distinct credit risk pricing problems identified above although we focus on derivatives with counterparty risk. Options on risky bonds are treated in Section 5.5, credit derivatives and credit derivatives with counterparty risk in Chapter 6.

7.1 Summary

We begin with an introduction to contingent claims pricing in Chapter 2. In that chapter, the standard valuation methodologies for derivative contracts are reviewed. For example, we discuss the pricing of stock options in a Black-Scholes world, exchange options, and fixed income derivatives in a Heath-Jarrow-Morton framework. We continue with a discussion of credit risk models in Chapter 3. In this chapter, we review several of the standard approaches to the valuation of credit risk, such as the basic firm-value model by Merton (1974), first-passage-time approaches such as Black and Cox (1976) and Longstaff and Schwartz (1994), and intensity or reduced-form approaches such as Jarrow and Turnbull (1995) or Jarrow, Lando, and Turnbull (1997). In Chapter 4, we propose a firm value model for pricing vulnerable derivative securities. Credit risk is modeled by the stochastic evolution of the value of the counterparty's assets, as Black and Scholes (1973) and Merton (1974) propose. We assume, however, that the counterparty has more than one outstanding claim, in fact that the size of each claim is negligible compared to the value of the entirety of claims on the firm's assets.

We derive explicit pricing formulae for European options subject to counterparty default risk and for forward contracts with bilateral default risk. Not only do we derive an explicit formula for the case of deterministic interest rates and counterparty liabilities, we also extend the credit risk model to stochastic interest rates and stochastic liabilities of the counterparty. New explicit pricing formulae are derived for the extended model. American options can be priced in a bivariate lattice modeling the evolution of the underlying instrument and the firm value.

The firm value model proposed in Chapter 4 offers a number of important insights with respect to the factors that influence the price effect of counterparty risk of derivatives. For example, the correlation structure between the value of the firm, the underlying asset of the contract, and interest rates is a major determinant of counterparty risk. Such insights are particularly transparent because of the availability of explicit pricing formulae for standard derivatives such as stock and bond options and forward contracts.

The timing of default is difficult to model in firm value approaches. In response to this difficulty, intensity models have been proposed. In Chapter 5, we develop a hybrid model in which the recovery rate in case of default is dependent on the firm value. At the same time, a bankruptcy process models the time of default. The intensity of the process can also depend on the firm value process. This hybrid model is an attempt to alleviate the limitations of firm value models without giving up the obvious advantages of such structural approaches. By using the firm value to model the recovery rate and possibly also the default intensity, a hybrid model allows for more flexibility with respect to dependencies among variables than pure intensity models such as Jarrow and Turnbull (1995).

The general hybrid model does not admit explicit pricing formulae. We therefore implement it using lattice structures. As in standard binomial lattices, derivative securities are priced by backward induction. The lattices, however, tend to be much more complex than in standard models because of the need to model the evolution of several variables, namely the underlying variable, for example the short rate, the firm value, and the bankruptcy process. We give a number of numerical examples for European and American options under deterministic and stochastic interest rates.

Chapter 5 also addresses the problem of valuing options on risky debt. We show that this problem requires only a small extension to the models for vulnerable derivatives. Specifically, the option is valued using a compound approach. The risk-free part of the bond and the credit risk part can be modeled in the lattice structures proposed. An option on a risky bond can then be priced in the same lattice structure as a compound option. Numerical examples show, for some situations, large deviations from the prices obtained by an adhoc approach which uses only the defaultable term structure to model the evolution of the underlying risky bond.

Chapter 6 considers credit derivatives, a recent form of derivative security written on credit risk. We identify two distinct classes of credit derivatives. One class can be priced with standard credit risk models while the other requires extended models. We show that credit derivatives of the second class, such as credit spread options, can be valued within a compound approach. This compound approach is very general and does not make strong assumptions with respect to the underlying credit risk model.

We demonstrate that, under deterministic interest rates, the compound option formula by Geske (1977) can be used to price spread credit derivatives if the underlying credit risk model is a firm value model. Under stochastic interest rates, a lattice approach can be taken to solve the compound problem. Again, the lattice structures proposed in Chapter 5 can be easily extended to price credit spread options. Numerical examples are provided for spread options under deterministic and stochastic interest rates.

We also present a reduced-form approach for the valuation of spread derivatives. The approach models credit spreads directly similar to the direct forward-rate modeling of the Heath, Jarrow, and Morton (1992) methodology.

Finally, we argue that credit risk can often be viewed as an option to exchange one asset for another. We derive prices for exchange options under Gaussian interest rates to obtain prices for options to exchange a bond for a bond, a stock for a bond, or a bond for a stock. As credit derivatives are OTC-issued contracts, they are subject to counterparty default risk. Modeling the credit derivative as an exchange option, we extend the model to allow for counterparty risk in a framework that is based on the methodology presented in Chapter 4.

7.2 Practical Implications

In this monograph we propose several models that include credit risk in the computation of derivative prices. The numerical examples show that credit risk cannot generally be ignored. If credit risk is independent of the underlying variable of the derivative contract, the price reduction on the derivative security is equal to the reduction on a zero coupon bond of the counterparty of equal maturity. The correlation of the underlying asset process with interest rates and with firm value, however, can greatly influence the size of the price reduction. As a consequence, the price effects of credit risk on derivatives are not solely determined by the firm's credit quality, but depend also on the correlation structure. In some circumstances, the relative price reduction of the derivative security can be several times as high as the reduction on the counterparty's bonds.

Although credit risk does not affect all derivatives to the same extent, this work clearly shows that its effects on prices can be substantial. Moreover, the assumption that credit risk affects prices only in extreme or unusual circumstances has to be rejected. In fact, we show that forward contracts, for example, display the highest deviation from their credit risk-free price in presumably very common circumstances. Given the insights provided by the credit risk models, it seems difficult to justify ignoring credit risk when pricing OTC derivatives in practice.

We also demonstrate that prices of options on credit-risky bonds obtained by a credit risk model can be quite different from the prices obtained by an ad-hoc approach which treats the defaultable term structure as a riskless term structure. Again, given the evidence, ignoring the effects of credit risk seems difficult to justify in practice.

With the compound approach, we present a framework within which all currently used credit derivatives can be priced in a consistent way based on a credit risk model. The compound approach is not restricted to a particular credit risk model but can be used with any firm value, intensity, or hybrid model. The adaptation of credit risk models within the compound approach is a consistent and theoretically superior alternative to the adhoc credit derivatives pricing models that have often been used so far.

Which credit risk model to use in practice depends on a variety of factors, such as the type of derivative to be priced, available data for parameter estimation, ease and stability of parameter estimation, and computational resources. In the end, as with interest rate modeling, the problem reduces to assessing the marginal benefit derived from additional model sophistication.

7.3 Future Research

In this monograph we restrict ourselves to theoretical modeling and do not address empirical estimation and calibration issues. However, there are a number of challenging questions that arise when implementing the models we have proposed in the previous chapters.

The main advantage of the hybrid approach over the firm value approach proposed in Chapter 4 is the ability to model the default time explicitly. Particularly vulnerable American derivatives can be modeled in a more realistic way if a bankruptcy process exists. Moreover, a simplified hybrid approach is easier to calibrate to initial risk-free and risky term structures. However, such a calibration can only be achieved easily by imposing restrictions on the correlation structure among default intensity, firm value, and recovery rate. Calibrating a firm value model to observed market data is more difficult, as the entire risky term structure is defined by the process parameters. A similar problem occurs in interest rate modeling when using time-homogeneous models of the short rate.

Although estimating the generally unobservable parameters of a firm value model may not be without problems, a general hybrid model may prove to be even more of a challenge as far as estimation of the parameters is concerned. Computing prices in this model also tends to be computationally expensive, whereas the closed-form solutions of the firm value model obviate expensive computations.

There has been almost no empirical work done in testing credit risk models aside from the Merton model. Much remains to be done before a definite conclusion can be drawn regarding conceivable superiority of highly flexible approaches such as the hybrid model proposed in this book compared to simple intensity or firm value models.

The lack of empirical work emphasizes that credit risk research is still in its infancy and a number of problems have yet to be overcome before advanced credit risk models will become commonplace in everyday practical use.

A. Useful Tools from Martingale Theory

This appendix covers a number of standard results from martingale theory and stochastic calculus which are used in the main body of the text. The concepts are outlined in a very brief fashion. Proofs are often omitted or the reader is referred to the literature. The first section introduces the probabilistic setup of the material used in later sections. See, for example, Williams (1991) for a more detailed introduction to probability. The later sections cover some important results from martingale theory and stochastic calculus. A more extensive treatment of this material can be found in any text on Brownian motion and stochastic calculus. Most of the material presented in this section is adapted from Karatzas and Shreve (1991) and Revuz and Yor (1994).

A.1 Probabilistic Foundations

Definitions. A measurable space $(\Omega, \mathcal{F})$ consists of a sample space Ω and a collection of subsets of Ω called a σ-algebra. A countably additive mapping $\mu : \mathcal{F} \longrightarrow \mathbb{R}^+$ is called a measure on $(\Omega, \mathcal{F})$ and the triple $(\Omega, \mathcal{F}, \mu)$ is called a measure space.

A *$\mathcal{F}$-measurable function* is a mapping h from $(\Omega, \mathcal{F})$ to a state space $(S, \mathcal{S})$ such that for any $\mathcal{A} \in \mathcal{S}, h^{-1}(\mathcal{A}) \in \mathcal{F}$. A Borel σ-algebra is a σ-algebra containing all open sets of a topological space. For example, $\mathcal{B}(\Omega)$ is the Borel σ-algebra of Ω. The mapping h is called a Borel function if it is $\mathcal{B}(\Omega)$-measurable.

A *random variable* is a measurable function. For example, consider a collection of functions h from a topological space Ω with σ-algebra $\mathcal{F}$ to $S = \mathbb{R}^d$ and $\mathcal{S} = \mathcal{B}(\mathbb{R}^d)$. $S = \mathbb{R}^d$ is the Euclidean space and $\mathcal{B}(\mathbb{R}^d)$ is the smallest σ-algebra of Borel sets of the topological space $\mathbb{R}^d$ such that each h is measurable. We say that the σ-algebra $\sigma(h)$ is generated by h. Obviously, for a random variable X, $\sigma(X) \in \mathcal{F}$. Otherwise the mappings would not be measurable.

A *process* is an family of random variables and is written $(X_t)_{0 \le t \le T}$ or simply X_t in shorthand notation. The index t is often interpreted as time defined on the interval $[0, T]$ for $T < \infty$.

For fixed sample points $\omega \in \Omega$, the function $t \to X_t(\omega)$ is called the *sample path* or *trajectory* of the process. However, a process can be viewed as a function of two variables (t, ω) where t is the (time) index and $\omega \in \Omega$, i.e. we have the mapping $(t, \omega) \to X_t(\omega)\ \forall \omega \in \Omega$. Therefore, a process is a joint mapping h from $\mathbb{R}^+ \times \Omega$ onto S. This mapping is *measurable* if we equip it with the corresponding σ-algebrae of Borel sets, i.e., if we have $h : (t, \omega) \to X_t(\omega) : (\mathbb{R}^+ \times \Omega, \mathcal{B}(\mathbb{R}^+) \otimes \mathcal{F} \to (\mathbb{R}^d, \mathcal{B}(\mathbb{R}^d)))$ such that for any $\mathcal{A} \in \mathcal{B}(\mathbb{R}^d)$, $h^{-1}(\mathcal{A}) \in \mathcal{B}(\mathbb{R}^+) \otimes \mathcal{F}$.

A *probability space* is a measure space with $\mu(\Omega) = 1$ and $\mu(\emptyset) = 0$. In this case μ is called a *probability measure*. It associates subsets of Ω (events) with a probability. For example, consider a probability space $(\Omega, \mathcal{F}, \mathbf{P})$. If, for an event $\mathcal{A} \in \mathcal{F}$, $\mathbf{P}(\mathcal{A}) = 1$, then we say that the event occurs *almost surely*, abbreviated as $\mathbf{P}$-a.s. or a.s. An event $\mathcal{A} \in \mathcal{F}$ with $\mathbf{P}(\mathcal{A}) = 0$ is called *almost impossible*. If $\mathcal{A} \notin \mathcal{F}$, then $\mathcal{A}$ is impossible.

A probability measure $\mathbf{Q}$ is said to be *equivalent* to $\mathbf{P}$ if, for any event $\mathcal{A} \in \mathcal{F}$, $\mathbf{P}(\mathcal{A}) = 0 \Leftrightarrow \mathbf{Q}(\mathcal{A}) = 0$. This means that two probability measures $\mathbf{P}$ and $\mathbf{Q}$ are equivalent if and only if for any event which is almost impossible under $\mathbf{P}$, it is also almost impossible under $\mathbf{Q}$. If only $\mathbf{P}(\mathcal{A}) = 0 \Rightarrow \mathbf{Q}(\mathcal{A}) = 0$, then we say that $\mathbf{Q}$ is *absolutely continuous* with respect to $\mathbf{P}$.

A *filtration* $\mathcal{F}_{0 \le t \le T} = \{\mathcal{F}_0, \ldots, \mathcal{F}_T\}$, sometimes written $(\mathcal{F}_t)_{0 \le t \le T}$ or $\mathcal{F}_t$ in shorthand, is a family of sub-σ-algebrae included in $\mathcal{F}$ which is non-decreasing in the sense that $\mathcal{F}_s \subseteq \mathcal{F}_t$ for $s < t$. Often, a filtration can be interpreted as representing the accumulation of information over time. A process generates a filtration. For example, consider the filtration $\mathcal{F}_t = \sigma(X_{s \in [0,t]})$ generated by X_t. This is the smallest σ-algebra with respect to which X_s is measurable for every $s \in [0, t]$. For technical reasons, we usually assume that it also contains all $\mathbf{P}$-null sets, i.e., if $\mathcal{A} \in \mathcal{F}$ and $\mathbf{P}(\mathcal{A}) = 0$, then $\mathcal{A} \in \mathcal{F}_t$, $\forall t \in [0, T]$. In this case it is called the natural filtration of X. Generally, we implicitly assume a filtration to be of the natural kind. A filtration to which the null sets are added is called augmented.

A filtered probability space is a probability space equipped with a filtration and is often written $(\Omega, \mathcal{F}, (\mathcal{F}_t)_{0 \le t \le T}, \mathbf{P})$, $(\Omega, \mathcal{F}, (\mathcal{F}_{t \in [0,T]}), \mathbf{P})$, or $(\Omega, \mathcal{F}, (\mathcal{F}_t)_{0 \le t \le T}, \mathbf{P})$. Consider a filtered probability space. If, for an event $\mathcal{A} \in \mathcal{F}$, $\mathbf{P}(\mathcal{A}) = 1$ for all $t \in [0, T]$, then we say that $\mathcal{A}$ is given *almost everywhere*. In shorthand notation we write $\mathcal{A}$ $\mathbf{P}$-a.e. or a.e.

Let a filtered probability space $(\Omega, \mathcal{F}, (\mathcal{F}_t)_{0 \le t \le T}, \mathbf{P})$ be given. A process $(X_t)_{0 \le t \le T}$ is called *adapted* (to the filtration $\mathcal{F}_t$) if it is $\mathcal{F}_t$-measurable. We also say that the process is $\mathcal{F}_t$-adapted. A process X_t is called predictable if it is $\mathcal{F}_{t^-}$-measurable for continuous processes and $\mathcal{F}_{t-1}$-measurable for discrete processes. A predictable process is $\mathcal{F}_{t^-}$-adapted ($\mathcal{F}_{t-1}$-adapted) and $\mathcal{F}_t$-adapted. For notational simplicity, we sometimes write $\mathcal{F}_s$, $s < t$, for $\mathcal{F}_{t^-}$. The difference between adapted and predictable processes is mostly relevant in the case of discrete processes.

A.2 Process Classes

Definition A.2.1. *Let $\mathcal{L}^\infty$ be the class of all adapted processes. θ is a process. Define the following spaces:*

$$\mathcal{L}^1 = \left\{ \theta \in \mathcal{L}^\infty : \int_0^T |\theta_t| dt < \infty \quad a.s. \right\},$$

$$\mathcal{L}^2 = \left\{ \theta \in \mathcal{L}^\infty : \int_0^T \theta_t^2 dt < \infty \quad a.s. \right\},$$

$$\mathcal{H}^2 = \left\{ \theta \in \mathcal{L}^2 : E\left(\int_0^T \theta_t^2 dt \right) < \infty \right\},$$

$$\mathcal{L}^{1,1} = \left\{ \theta \in \mathcal{L}^\infty : \int_0^{T_1} \int_0^{T_2} |\theta(s,t)| ds dt < \infty \quad a.s. \right\},$$

$$\mathcal{L}^{2,1} = \left\{ \theta \in \mathcal{L}^{1,1} : \int_0^T \theta(s,u)^2 ds < \infty \quad a.s., \right.$$
$$\left. \int_0^{T_1} \left(\int_0^{T_2} |\theta(s,u)| du \right)^2 ds < \infty \quad a.s. \right\}.$$

We write $\theta \in \mathcal{L}^1(\Omega, \mathcal{F}, \mathbf{P})$ or simply $\theta \in \mathcal{L}^1$ if it is clear from context on which probability space the process is defined. In most instances in this text, integrable ($\mathcal{L}^1$) or square-integrable ($\mathcal{L}^2$) processes are used.

A.3 Martingales

Definition A.3.1. *A process M_t adapted to $\mathcal{F}_t$ and satisfying $\mathbf{E}[|M_t|] < \infty$, $\forall t \in [0,T]$ is called a $\mathbf{P}$-submartingale if*

$$\mathbf{E_P}[M_T|\mathcal{F}_t] \geq M_t, \quad \forall t \in [0,T],$$

and a $\mathbf{P}$-supermartingale if

$$\mathbf{E_P}[M_T|\mathcal{F}_t] \leq M_t, \quad \forall t \in [0,T].$$

M_t is a martingale if it is both a submartingale and a supermartingale, i.e., if $\mathbf{E_P}[M_T|\mathcal{F}_t] = M_t$.

Remark A.3.1. Informally, we call the process M_t a local martingale if it does not satisfy the technical condition $E(|M_t|) < \infty$, $\forall t \in [0,T]$.

Theorem A.3.1 (Doob-Meyer). *Let X be a right-continuous submartingale. Under some technical conditions, X has the decomposition*

$$X_t = M_t + A_t, \qquad 0 \leq t < \infty,$$

such that M_t is right-continuous martingale and A_t an increasing process.

Proof. Cf. Karatzas and Shreve (1991), p. 25.

Definition A.3.2. *For a martingale $M_t \in \mathcal{L}^2$, the* quadratic variation *process is $\langle M, M \rangle_t = A_t$, where A_t is the increasing process of the Doob-Meyer decomposition of $X = M^2$. Quadratic variation is also known as* cross-variation.

Remark A.3.2. M is a unique process such that $M^2 - \langle M, M \rangle$ is a martingale. $\langle M, M \rangle$ is often abbreviated $\langle M \rangle = \langle M, M \rangle$.

Definition A.3.3. *Similar to the previous definition, for M and N two different continuous martingales, there exists a unique continuous increasing process $\langle M, N \rangle = \frac{1}{4}(\langle M + N \rangle - \langle M - N \rangle)$ vanishing at zero such that $MN - \langle M, N \rangle$ is a martingale.*

Remark A.3.3. This definition immediately follows from the previous definition. Consider two different martingales M and N. By the previous definition, the process $(M+N)^2 - \langle M + N \rangle$ is a martingale, but so is $(M-N)^2 - \langle M - N \rangle$ and their difference $4MN - \langle M + N \rangle + \langle M - N \rangle$. Therefore, the cross-variation process is $\langle M, N \rangle = \frac{1}{4}(\langle M + N \rangle - \langle M - N \rangle)$.

If M and N are independent martingales, then $\langle M, N \rangle = 0$.

Definition A.3.4. *A continuous* semimartingale *is a continuous adapted process X which has the decomposition $X = X_0 + M + A$ where M is a continuous local martingale and A a non-decreasing, continuous, adapted process of finite variation. X_0 is a $\mathfrak{F}_0$-measurable random variable.*

Theorem A.3.2 (Martingale Representation). *Let M be a continuous local martingale. For any adapted local martingale X, there is a predictable process ϕ such that*

$$X_t = X_0 + \int_0^t \phi_s \, dM_s.$$

Proof. Cf. Revuz and Yor (1994), Chapter IV.

Corollary A.3.1. *Let W_t be a $\mathbb{R}^d$-valued Brownian motion. If X_t is a continuous martingale in $(\mathcal{L}^2)^n$ with $X^0 = 0$, then there exists a unique predictable process ϕ in $(\mathcal{H}^2)^{n \times d}$ such that*

$$X_t = X_0 + \int_0^t \phi_s \cdot dW_s.$$

Proof. Cf. Karatzas and Shreve (1991), Section 3.D.

Remark A.3.4. For X_t is a local martingale, it is sufficient that $\phi_t \in (\mathcal{L}^2)^d$ (cf. Karatzas and Shreve (1991), p. 188).

A.4 Brownian Motion

Definition A.4.1. *A probability space $(\Omega, \mathcal{F}, (\mathcal{F}_t)_{0\le t\le T}, \mathbf{P})$ is given. Let*

$$(W_t)_{t\in[0,T]} = (W_t^1, \dots, W_t^d)$$

be a $\mathbb{R}^d$-valued continuous process adapted to the filtration $\mathcal{F}_t$. W_t is called a Wiener process or a Brownian motion if, $\forall t, s \in [0,T]$ s.t. $s < t$, the process increments $W_t - W_s$ are independent and identically distributed. The distribution is normal with mean zero and covariance matrix $(t-s)V_d$.

Remark A.4.1. We usually define W_t^i and W_t^j, $i \neq j$, to be independent. This implies that a d-dimensional Brownian motion $W = (W^1, \dots, W^d)$ has cross-variation $\langle W^i, W^j\rangle_t = \delta^{ij} t$ where δ is the Kronecker delta. Therefore, $\langle W^i\rangle_t = t$. The covariance matrix is then given by $(t-s)V_d = (t-s)I_d$, where I_d is the $d \times d$ identity matrix.

However, sometimes we define W_t^i and W_t^j, $i \neq j$ to be correlated Brownian motion. In this case, $\langle W^i, W^j\rangle_t = \int_0^t \rho_s^{ij}\, ds$, where ρ_t^{ij} is called the *correlation coefficient* between Brownian motions W_t^i and W_t^j. For constant correlation coefficients, $\langle W^i, W^j\rangle_t = \rho_{ij} t$.

Definition A.4.2. *Given a filtered probability space $(\Omega, \mathcal{F}, (\mathcal{F}_t)_{0\le t\le T}, \mathbf{P})$, fix a Brownian motion $W_t = (W_t^1, \dots, W_t^d)$ and adapted processes $\alpha_t \in (\mathcal{L}^1)^n$ and $\sigma_t^i = (\sigma_t^{i1}, \dots, \sigma_t^{id}) \in (\mathcal{L}^2)^{n\times d}$. The adapted process $X_t = (X_t^1, \dots, X_t^n)$ is called an Itô process or a diffusion and can be represented by the stochastic integral equation*

$$X_t = X_0 + \int_0^t \alpha_s\, ds + \int_0^t \sigma_s\, dW_s, \tag{A.1}$$

where $\alpha : \mathbb{R}^n \times \mathbb{R}^+ \to \mathbb{R}^n$ and $\sigma : \mathbb{R}^{n\times d} \times \mathbb{R}^+ \to \mathbb{R}^{n\times d}$. σ is called volatility matrix.

Sometimes we write (A.1) *with sums, i.e., $\forall i \in \{1, \dots, n\}$,*

$$\begin{aligned} X_t^i &= X_0^i + \int_0^t \alpha_s^i\, ds + \sum_j^d \int_0^t \sigma_s^{ij}\, dW_s^j \\ &= X_0^i + \int_0^t \alpha_s^i\, ds + \int_0^t \sigma_s^i \cdot dW_s. \end{aligned}$$

The cross variation process is given by $\langle X^i, X^j\rangle_t = \int_0^t \sigma_s^i \cdot \sigma_s^j\, ds$, where $\cdot$ denotes the inner product (dot product) of vectors σ^i and σ^j.

The correlation coefficient is defined as

$$\rho_{ij} = \frac{\sigma_i \cdot \sigma_j}{\|\sigma_i\| \|\sigma_j\|}$$

and is also a process. $\|\cdot\|$ denotes the Euclidian norm.

Definition A.4.3. *The representation of Itô processes is slightly different if we use correlated Brownian motion B_t as a building block. In that case we write*

$$X_t = X_0 + \int_0^t \alpha_s \, ds + \int_0^t \upsilon_s \, dB_s,$$

where $\alpha : \mathbb{R}^n \times \mathbb{R}^+ \to \mathbb{R}^n$ and $\upsilon : \mathbb{R}^n \times \mathbb{R}^+ \to \mathbb{R}^n$. The equivalent notation with sums is $X_t = \sum_i^n X^i$ with

$$X_t^i = X_0^i + \int_0^t \alpha_s^i \, ds + \int_0^t \upsilon_s^i \, dB_s^i.$$

This notation relates to the previous definition of Itô processes by the equation

$$\begin{aligned} \upsilon_t^i \, dB_t^i &= \|\sigma_t^i\| \, dB_t^i \\ &= \sigma_t^i \cdot dW_t, \qquad \forall i \in \{1..n\}, \end{aligned}$$

where σ_t^i and W_t are defined as in Definition A.4.2.

Let V_t denote the instantaneous covariance matrix. *Each element of V_t is given by*

$$\upsilon_{ij} = \frac{d\langle X_i, X_j \rangle_t}{dt} = \rho_{ij} \upsilon_i \upsilon_j \, .$$

We also call V_t a covariance matrix if its elements are given by $d\langle X_i, X_j \rangle_t$.

To determine the symmetrical covariance matrix V_t from the triangular volatility matrix σ_t, we compute $\sigma_t \sigma_t^\top = V_t$. Conversely, σ_t can be recovered from V_t by a Cholesky decomposition if V_t is positive definite. Conveniently, a valid covariance matrix is always positive definite.

Example A.4.1. Consider the $\mathbb{R}^n$-valued process $X_t^i = X_0^i + \sigma_i B_t^i$ for $i \in [1, 2]$ with B_t^i a $\mathbb{R}$-valued Brownian motion and constant covariance matrix

$$V = \begin{pmatrix} \sigma_1^2 & \rho\sigma_1\sigma_2 \\ \rho\sigma_1\sigma_2 & \sigma_2^2 \end{pmatrix}.$$

The covariance matrix is given by the fact that $\langle B^1, B^2 \rangle_t / t = \rho$. Applying the Cholesky decomposition, we obtain

$$\sigma = \begin{pmatrix} \sigma_1 & 0 \\ \rho\sigma_2 & \sqrt{1-\rho^2}\sigma_2 \end{pmatrix}.$$

This result can be easily verified by checking $V = \sigma\sigma^\top$. Therefore, $X_t = (X_t^1, X_t^2)$ is equivalent to $Y_t = (Y_t^1, Y_t^2)$ such that

$$\begin{aligned} Y_t^1 &= X_0^1 + \sigma_1 W_t^1 \, , \\ Y_t^2 &= X_0^2 + \rho\sigma_2 W_t^1 + \sqrt{1-\rho^2}\sigma_2 W_t^2, \end{aligned}$$

for a $\mathbb{R}^2$-valued Brownian motion with $\langle W^1, W^2 \rangle = 0$.

Remark A.4.2. Whether correlated or independent Brownian motion is used depends on the application at hand. Since Brownian motion is always explicitly defined as independent or correlated whenever used in the main body of this text, we usually do not use a distinct symbol for correlated Brownian motion.

A.5 Stochastic Integration

Definition A.5.1. *For any $(\theta_t)_{0\le t\le T} = (\theta_t^1, \dots, \theta_t^d)$ in $(\mathcal{H}^2)^d$, we write the stochastic integral with respect to a square-integrable martingale M as*

$$I_t(\theta) = \int_0^t \theta_s \cdot dM_s = \sum_{i=1}^d \int_0^s \theta_s^i dM_s^i.$$

$I_t(\theta)$ is a unique, square-integrable martingale with quadratic variation process $\langle I(\theta)\rangle_t = \int_0^t \|\theta_s\|^2 \, d\langle M\rangle_s$. If $M = W$ a Brownian motion, then $\langle I(\theta)\rangle_t = \int_0^t \|\theta_s\|^2 \, ds$.

Remark A.5.1. It can be shown that the stochastic integral is also defined for the wider class $\theta \in \mathcal{L}^2$ in which case I_t is a local martingale.

If $\langle M\rangle$ is not absolutely continuous, the martingale M has to be progressively measurable. If $\langle M\rangle$ is absolutely continuous, it is sufficient if the integrand is adapted since adaptivity and continuity imply progressive measurability.

For a rigorous construction of the stochastic integral, see Karatzas and Shreve (1991) or Revuz and Yor (1994).

Theorem A.5.1 (Itô). *Let the function $f : \mathbb{R}^+ \times \mathbb{R}^d \to \mathbb{R}$ be of class $\mathcal{C}^{1,2}$. For a continuous d-dimensional semimartingale $X_t = (X_t^1, \dots, X_t^d)$ with decomposition $X_t = X_0 + M_t + A_t$, we have*

$$f(X_t) = f(X_0) + \sum_i \int_0^t \frac{\partial f(X_s)}{\partial x_i} \, dM_s^i + \sum_i \int_0^t \frac{\partial f(X_s)}{\partial x_i} \, dA_s^i + \frac{1}{2} \sum_i \sum_j \int_0^t \frac{\partial^2 f(X_s)}{\partial x_i \partial x_j} \, d\langle X^i, X^j\rangle_s,$$

or in differential notation

$$df(X_t) = \sum_i \frac{\partial f(X_t)}{\partial x_i} \, dX_t^i + \frac{1}{2} \sum_i \sum_j \frac{\partial^2 f(X_t)}{\partial x_i \partial x_j} \, d\langle X^i, X^j\rangle_t.$$

Proof. For example, by an application of a Taylor series. Cf. Karatzas and Shreve (1991), p. 150-153, or Revuz and Yor (1994), p. 139, 145.

Example A.5.1. Consider the special case of X_t a $\mathbb{R}^n$-valued Itô process $X_t^i = X_0^i + \int_0^t \alpha_s^i \, ds + \sum_{j=1}^d \int_0^t \sigma_s^{ij} \, dW_s^j$, for $i = \{1, \ldots, n\}$ with W_t a Brownian motion in $\mathbb{R}^d$. σ^i is an $\mathbb{R}^d$-valued vector. Assuming that σ and α satisfy the same boundedness condition as in Definition A.4.2, then Itô's formula has the following terms:

$$
\begin{aligned}
dA_t^i &= \alpha_t^i \, dt, \\
dM_t^i &= \sigma_t^i \cdot dW_t = \sum_j^d \sigma_t^{ij} \, dW_t^j, \\
d\langle X^i, X^j \rangle_t &= \sigma_t^i \cdot \sigma_t^j \, dt = \sum_k^d \sigma_t^{ik} \sigma_t^{jk} \, dt.
\end{aligned}
$$

This is equivalent to

$$
df_X = f_X dX + \frac{1}{2} \operatorname{trace}(\sigma \sigma^\top f_{XX}) dt,
$$

where σ is in $\mathbb{R}^{n \times d}$, f_X in $\mathbb{R}^n$, and f_{XX} in $\mathbb{R}^{n \times n}$.

Example A.5.2. Consider the special case of X_t a $\mathbb{R}^2$-valued Itô process $X_t^i = X_0^i + \int_0^t \alpha_s^i \, ds + \int_0^t \sigma_s^i \, dW_s^i$, for $i = \{1, 2\}$ with W_t^i a *correlated* Brownian motion in $\mathbb{R}$. Assuming the same boundedness conditions as above, Itô's formula in differential notation is

$$
\begin{aligned}
df(x_1, x_2) = {} & f_{x_1} dx_1 + f_{x_2} dx_2 \\
& + \frac{1}{2} f_{x_1 x_1} d\langle x_1 \rangle + \frac{1}{2} f_{x_2 x_2} d\langle x_2 \rangle + f_{x_1 x_2} d\langle x_1, x_2 \rangle,
\end{aligned}
$$

where

$$
\begin{aligned}
d\langle x_i \rangle &= \sigma_{x_i}^2 \, dt, \qquad \forall i \in \{1, 2\}, \\
d\langle x_1, x_2 \rangle &= \rho_{x_1 x_2} \sigma_{x_1} \sigma_{x_2} \, dt,
\end{aligned}
$$

where $\rho_{x_1 x_2}$ denotes the correlation coefficient between W_t^1 and W_t^2.

Theorem A.5.2 (Integration by parts). *For X_t and Y_t two continuous semimartingales,*

$$
X_t Y_t = X_0 Y_0 + \int_0^t X_s \, dY_s + \int_0^t Y_s \, dX_s + \langle X, Y \rangle_t,
$$

or in differential form,

$$
d(X_t Y_t) = X_t \, dY_t + Y_t \, dX_t + d\langle X, Y \rangle_t.
$$

If $X = Y$,

$$
X_t^2 = X_0^2 + 2 \int_0^t X_s \, dX_s + \langle X \rangle_t.
$$

Proof. Analogous to Itô's formula by setting $f(X_t, Y_t) = X_t Y_t$. Cf. Revuz and Yor (1994), p. 138, for a different proof.

Example A.5.3. A simple example of a special case is that of a standard Brownian motion W_t. Clearly, we have $W_t^2 = 2\int_0^t W_s\, dW_s + t$.

Example A.5.4. Consider the more general special case of two real-valued Itô processes X_t^1 and X_t^2, i.e., $X_t^i = X_0^i + \int_0^t \alpha_s^i\, ds + \int_0^t \sigma_s^i\, dW_s$ for $i \in \{1, 2\}$ and $W_t = (W_t^1, \ldots, W_t^d)$. α and σ are defined as in Definition A.4.2 and satisfy the same boundedness conditions. In this case their product is

$$X_t^1 X_t^2 = X_0^1 X_0^2 + \int_0^t X_s^1\, dX_s^2 + \int_0^t X_s^2\, dX_s^1 + \int_0^t \sigma_s^1 \sigma_s^2\, ds.$$

σ_s^1 and σ_s^2 are vectors in $\mathbb{R}^d$, i.e., $\langle X^1, X^2\rangle_t = \int_0^t \sum_k^d \sigma_s^{1,k} \sigma_s^{2,k}\, ds$. This result is obtained in a straightforward way by recognizing the multiplication rules $dt\, dt = 0$, $dt\, dW_t = 0$, $dW_t^k dW_t^k = dt$, and $dW_t^j dW_t^k = 0$, $j \neq k$. The analogy to the derivation of Itô's formula for Itô processes is evident.

Example A.5.5. Let X_t and Z_t be two semimartingales. Compute the process $F(X, Z) = \frac{X}{Z}$. For $Y = f(X) = \frac{1}{Z}$ we have $F = XY$ and can apply the integration by parts theorem such that $F = \int X\, dY + \int Y\, dX + \langle X, Y\rangle$. Since Y is a function of Z, we apply Itô's formula. Since $f'(X) = -\frac{1}{Z^2}$ and $f''(X) = \frac{2}{Z^3}$, we have

$$dY = -\frac{1}{Z^2}\, dZ + \frac{1}{Z^3}\, d\langle Z\rangle\,, \tag{A.2}$$

by Itô's formula. Substituting for dY, we have

$$F = \int X \left(-\frac{1}{Z^2}\, dZ + \frac{1}{Z^3}\, d\langle Z\rangle\right) + \int \frac{1}{X}\, dX + \langle X, Y\rangle\,.$$

By expression (A.2), the covariation can be expressed as

$$\langle X, Y\rangle = \langle X, \frac{1}{Z}\rangle = \langle X, -\int \frac{1}{Z^2}\, dZ + \int \frac{1}{Z^3}\, d\langle X\rangle\rangle\,.$$

By the multiplication rule, $\langle X, \langle X\rangle\rangle = 0$, and therefore $\langle X, \frac{1}{Z}\rangle = -\frac{1}{Z^2}\langle X, Z\rangle$. Thus,

$$F_t = \frac{X_t}{Z_t} = \int_0^t \frac{1}{Z}\, dX - \int_0^t \frac{X}{Z^2}\, dZ + \int_0^t \frac{X}{Z^3}\, d\langle Z\rangle - \frac{1}{Z^2}\langle X, Z\rangle.$$

Dividing by $\frac{X}{Z}$ and differentiating gives

$$\frac{dF}{F} = \frac{d\frac{X}{Z}}{\frac{X}{Z}} = \frac{dX}{X} - \frac{dZ}{Z} + \frac{d\langle Z\rangle}{Z^2} - \frac{d\langle X, Z\rangle}{XZ}.$$

Theorem A.5.3 (Doléans-Dade). *Let L_t be a continuous local martingale defined on $(\Omega, \mathcal{F}, (\mathcal{F}_t)_{0\leq t\leq T}, \mathbf{P})$. If a process $\mathcal{E}_t(L)$ satisfies the SDE*

$$d\mathcal{E}_t = \mathcal{E}_t \, dL_t,$$

then $\mathcal{E}_t(L)$ is given by the local martingale

$$\mathcal{E}_t(L) = \exp\left(L_t - \frac{1}{2}\langle L\rangle_t\right).$$

$\mathcal{E}_t$ is called the Doléans-Dade exponential or the stochastic exponential.

Corollary A.5.1. *A similar result applies to a driftless Itô process. For a process γ_t in $(\mathcal{L}^2)^d$, define $L_t = \int_0^t \gamma_t \cdot dW_t$. W_t is an $\mathbb{R}^d$-valued Brownian motion. A process $(\mathcal{E}_t)_{0\leq t\leq T}$ satisfying the stochastic differential equation $d\mathcal{E}_t = \mathcal{E}_t\gamma_t \cdot dW_t$ is given by the local martingale*

$$\mathcal{E}_t = \exp\left(\int_0^t \gamma_s \cdot dW_s - \frac{1}{2}\int_0^t \|\gamma_s\|^2 \, ds\right). \tag{A.3}$$

Proof. Consider the transformation $\mathcal{E} = e^X$. Define

$$X = \int_0^t \gamma_s \cdot dW_s - \frac{1}{2}\int_0^t \|\gamma_s\|^2 \, ds.$$

The stochastic differential equation for $\mathcal{E}$ then follows as a straightforward application of Itô's formula. The proof for Theorem A.5.3 is analogous.

Corollary A.5.2 (Novikov). *If $\mathbf{E}[\exp(\frac{1}{2}\langle L\rangle)] < \infty$ then $\mathcal{E}$ in A.5.3 is a martingale.*

Proof. Cf. Revuz and Yor (1994), p. 318.

Remark A.5.2. The same result also applies to $\mathcal{E}$ if defined as in expression (A.3), i.e., if $L_t = \int_0^t \gamma_t \cdot dW_t$. It can easily be seen that the condition in that case is $\mathbf{E}[\exp(\frac{1}{2}\int_0^T \|\gamma_s\|^2 ds)] < \infty$.

Theorem A.5.4 (Fubini). *For a process $h(s,t) \in \mathcal{L}^{2,1}$ such that*

$$\int_0^{t_1}\int_0^{t_2} h(s,t) \, dW_s \, dt$$

is continuous a.e., an interchange of Lebesque and Itô integrals is permissible. Therefore, we have the equalities

$$\int_0^{t_1}\int_0^{t_2} h(s,t)\, dW_s\, dt = \int_0^{t_2}\int_0^{t_1} h(s,t)\, dt\, dW_s$$

$$\int_0^{t_1}\int_0^{t_2\wedge t} h(s,t)\mathbf{1}_{\{s\leq t\}} dW_s\, dt = \int_0^{t_2}\int_s^{t_1} h(s,t)\mathbf{1}_{\{s\leq t\}} dt\, dW_s.$$

Proof. Cf. Karatzas and Shreve (1991), p. 233, Protter (1990), p. 159.

A.6 Change of Measure

Theorem A.6.1. *A probability triple $(\Omega, \mathcal{F}, \mathbf{P})$ is given. Let ζ be the stochastic exponential from A.5.3. If a new probability measure $\mathbf{Q}$ defined on $(\Omega, \mathcal{F})$ is given by*

$$\mathbf{Q} = \int_A \zeta_T \, d\mathbf{P}, \qquad \forall A \in \mathcal{F},$$

then $\mathbf{Q}$ is equivalent to $\mathbf{P}$.

Proof. Cf. Revuz and Yor (1994), p. 325.

Remark A.6.1. If ζ is defined as the stochastic exponential in expression (A.3), then the above statement is valid if and only if γ is in $\mathcal{L}^2$.

Definition A.6.1. $\zeta_T = \frac{d\mathbf{Q}}{d\mathbf{P}}$ *is called the Radon-Nikodým derivative or Radon-Nikodým density of $\mathbf{Q}$ with respect to $\mathbf{P}$.*

Corollary A.6.1. *For arbitrary t, such that $0 \geq t \geq T$, ζ_t is given by*

$$\zeta_t = \mathbf{E_P}[\zeta_T | \mathcal{F}_t].$$

Proof. By the (local) martingale property of ζ (cf. Theorems A.5.3 and A.5.2.).

Corollary A.6.2. *Assume that ζ is a martingale and X an adapted process from $\mathcal{L}^1$ such that*

$$\mathbf{E_Q}[X_t] = \mathbf{E_P}[X_t \zeta_t].$$

Proof. Substituting the definition of $\mathbf{Q}$ into $\int_\Omega X d\mathbf{P}$ gives $\int_\Omega X \zeta d\mathbf{P}$.

Corollary A.6.3 (Bayes). *Similarly, for $0 \leq s \leq t \leq T$, we have*

$$\mathbf{E_Q}[X_t | \mathcal{F}_s] = \zeta_s^{-1} \mathbf{E_P}[X_t \zeta_t | \mathcal{F}_s].$$

Proof. For every $A \in \mathcal{F}_s$,

$$\mathbf{E_Q}[\zeta_s^{-1} \mathbf{E_P}[X_t \zeta_t | \mathcal{F}_s] | \mathcal{F}_s] = \int_A \zeta_s^{-1} \mathbf{E_P}[X_t \zeta_t | \mathcal{F}_s] d\mathbf{Q},$$

by the definition of conditional expectation. By Theorem A.6.1, we obtain

$$\int_A \mathbf{E_P}[X_t \zeta_t | \mathcal{F}_s] \, d\mathbf{P}. \tag{A.4}$$

The law of iterated expectations (cf. Williams (1991), p.88), another application of Theorem A.6.1, and the definition of conditional expectation give the following equalities:

$$\int_A \mathbf{E_P}[X_t \zeta_t | \mathcal{F}_s] \, d\mathbf{P} = \int_A X_t \zeta_t d\mathbf{P} = \int_A X_t d\mathbf{Q} = \mathbf{E_Q}[X_t | \mathcal{F}_s].$$

Theorem A.6.2 (Girsanov). *Given that a probability space is endowed with a filtration, $(\Omega, \mathcal{F}, (\mathcal{F}_t)_{0 \le t \le T}, \mathbf{P})$, assume that the probability measure $\mathbf{Q}$ is absolutely continuous with respect to $\mathbf{P}$. If the Radon-Nikodým derivative $D = \frac{d\mathbf{Q}}{d\mathbf{P}}$ is continuous, then all semimartingales under $\mathbf{P}$ are semimartingales under $\mathbf{Q}$ and, for M a continuous local martingale defined on $(\Omega, \mathcal{F}, (\mathcal{F}_t)_{0 \le t \le T}, \mathbf{P})$,*

$$\tilde{M}_t = M_t - \int_0^t D^{-1} \, d\langle M, D\rangle$$

is a continuous local martingale under $(\Omega, \mathcal{F}, (\mathcal{F}_t)_{0 \le t \le T}, \mathbf{Q})$. For N, another local martingale, we have $\langle \tilde{M}, \tilde{N}\rangle = \langle \tilde{M}, N\rangle = \langle M, N\rangle$.

If $\mathbf{Q}$ is an equivalent probability measure defined by $Q = \int \mathcal{E}(L) \, d\mathbf{P}$ (cf. Theorem A.5.3), then we have

$$\tilde{M} = M - \int D^{-1} \, d\langle M, D\rangle = M - \langle M, L\rangle .$$

Again, $\tilde{M}$ is defined on the same filtration as M.

Corollary A.6.4. *Given a filtered probability space, $(\Omega, \mathcal{F}, (\mathcal{F}_t)_{0 \le t \le T}, \mathbf{P})$, define $\gamma_t = (\gamma_t^1, \ldots, \gamma_T^d)$ to be an adapted process in $(\mathcal{L}^2)^d$ such that*

$$\frac{d\mathbf{Q}}{d\mathbf{P}} = \exp\left(\int_0^T \gamma_s \cdot dW_s - \frac{1}{2} \int_0^T \|\gamma_s\|^2 \, ds \right) \tag{A.5}$$

and $W_t = (W_t^1, \ldots, W_t^d)$ a d-dimensional Brownian motion. If we assume that Novikov's condition holds such that $\zeta_t = \mathbf{E}[\frac{d\mathbf{Q}}{d\mathbf{P}}]$ is a martingale. Then the process

$$\tilde{W}_t = W_t - \int_0^t \gamma_s ds$$

is a d-dimensional Brownian motion defined on $(\Omega, \mathcal{F}, (\mathcal{F}_t)_{0 \le t \le T}, \mathbf{Q})$.

Remark A.6.2. It is easy to see that the corollary is a special case of the theorem. W_t corresponds to M and $L_t = \int_0^t \gamma_s \cdot dW_s$. For $D = \mathcal{E}(L)$, we obtain the SDE $d\mathcal{E} = \mathcal{E}\gamma dW$. We start with $d\langle M, D\rangle = d\langle W, \mathcal{E}\rangle = dW \, d\mathcal{E}$. Then, substituting for $d\mathcal{E}$ and applying the multiplication rules gives $d\langle M, D\rangle = \mathcal{E}\gamma dt$. Simplification and integration give the result of the corollary. Since $D = \mathcal{E}(L)$, we can also use the second part of the theorem, which yields the desired result immediately by the multiplication rules.

We can show that the equivalent probability measure is invariant with respect to time. ζ_t satisfies the integral equation $\zeta_t = 1 + \int_0^t \zeta_s \gamma_s \cdot dW_s$. With Novikov's condition satisfied, we have $\mathbf{E}[\zeta_t] = 1, \ \forall t \in [0, T]$. Since integral and expectation are equivalent, the equivalent probability measure can also be defined as $\mathbf{Q}_t = \mathbf{E}[\mathbf{1}_{\{\mathcal{A}\}} \zeta_t], \forall \mathcal{A} \in \mathcal{F}_t$ and for any $t \in [0, T]$. By the martingale property, $\mathbf{Q}_t(\mathcal{A}) = \mathbf{Q}_s(\mathcal{A})$, for any $s, t \in [0, T]$.

Proof. For a proof of Girsanov's theorem for semimartingales, refer to Revuz and Yor (1994), Chapter 8. For the special case of Itô processes, see Karatzas and Shreve (1991), Section 3.5.B.

References

ABKEN, P. (1993): "Valuation of Default-Risky Interest-Rate Swaps," *Advances in Futures and Options Research*, 6, 93–116.

ABRAMOWITZ, M., AND I. A. STEGUN (1972): *Handbook of Mathematical Functions*. Dover Publications, New York, N.Y.

ACHARYA, S., AND J.-F. DREYFUS (1989): "Optimal Bank Reorganization Policies and the Pricing of Federal Deposit Insurance," *Journal of Finance*, 44(5), 1313–1333.

ALTMAN, E. I. (1984): "A Further Investigation of the Bankruptcy Cost Question," *Journal of Finance*, 39(4), 1067–1089.

ALTMAN, E. I. (1989): "Measuring Corporate Bond Mortality and Performance," *Journal of Finance*, 44(4), 909–921.

ALTMAN, E. I., AND V. M. KISHORE (1996): "Almost Everything You Wanted to Know about Recoveries on Defaulted Bonds," *Financial Analysts Journal*, 52(6), 57–64.

AMIN, K. I. (1991): "On the Computation of Continuous-Time Option Prices Using Discrete Approximations," *Journal of Financial and Quantitative Analysis*, 26(4), 477–495.

AMIN, K. I., AND J. N. BODURTHA (1995): "Discrete-Time Valuation of American Options with Stochastic Interest Rates," *Review of Financial Studies*, 8(1), 193–234.

AMIN, K. I., AND R. A. JARROW (1992): "Pricing Options on Risky Assets in a Stochastic Interest Rate Economy," *Mathematical Finance*, 2(4), 217–237.

AMMANN, M. (1998): "Pricing Derivative Credit Risk," Ph.D. thesis, University of St.Gallen.

AMMANN, M. (1999): *Pricing Derivative Credit Risk*, vol. 470 of *Lecture Notes in Economics and Mathematical Systems*. Springer Verlag, Berlin, Heidelberg, New York.

ANDERSON, R. W., AND S. SUNDARESAN (1996): "Design and Valuation of Debt Contracts," *Review of Financial Studies*, 9(1), 37–68.

ANDERSON, R., AND S. SUNDARESAN (2000): "A Comparative Study of Structural Models of Corporate Bond Yields: An Exploratory Investigation," *Journal of Banking and Finance*, 24, 255–269.

ANDRADE, G., AND S. N. KAPLAN (1998): "How Costly is Financial (Not Economic) Distress? Evidence from Highly Leveraged Transactions That Became Distressed," *Journal of Finance*, 53(5), 1443–1493.

ARTZNER, P., AND F. DELBAEN (1995): "Default Risk Insurance and Incomplete Markets," *Mathematical Finance*, 5(3), 187–195.

ASQUITH, P., R. GERTNER, AND D. SCHARFSTEIN (1994): "Anatomy of Financial Distress: An Examination of Junk-Bond Issuers," *Quarterly Journal of Economics*, 109, 625–658.

BACK, K., AND S. R. PLISKA (1991): "On the Fundamental Theorem of Asset Pricing with an Infinite State Space," *Journal of Mathematical Economics*, 20(1), 1–18.

BHASIN, V. (1996): "On the Credit Risk of OTC Derivative Users," Discussion paper, Board of Governors of the Federal Reserve System.

BLACK, F., AND J. C. COX (1976): "Valuing Corporate Securities: Some Effects of Bond Indenture Provisions," *Journal of Finance*, 31(2), 351–367.

BLACK, F., E. DERMAN, AND W. TOY (1990): "A One-Factor Model of Interest Rates and Its Application to Treasury Bond Options," *Financial Analysts Journal*, 46(1), 33–39.

BLACK, F., AND M. SCHOLES (1973): "The Valuation of Options and Corporate Liabilities," *Journal of Political Economy*, 81(3), 637–654.

BOYLE, P. P. (1988): "A Lattice Framework for Option Pricing with Two State Variables," *Journal of Financial and Quantitative Analysis*, 23(1), 1–12.

BOYLE, P. P., J. EVNINE, AND S. GIBBS (1989): "Numerical Evaluation of Multivariate Contingent Claims," *Review of Financial Studies*, 2(2), 241–250.

BREMAUD, P. (1981): *Point Processes and Queues: Martingale Dynamics.* Springer Verlag, Berlin, Heidelberg, New York.

BRENNAN, M. J., AND E. S. SCHWARTZ (1980): "Analyzing Convertible Bonds," *Journal of Financial and Quantitative Analysis*, 15(4), 907–929.

BRENNER, M. (1990): "Stock Index Options," in *Financial Options*, ed. by S. Figlewski, W. L. Silber, and M. G. Subrahmanyam, chap. 5. Irwin, New York, N.Y.

BRITISH BANKERS' ASSOCIATION (1996): *BBA Credit Derivatives Report 1996.*

BRIYS, E., AND F. DE VARENNE (1997): "Valuing Risky Fixed Rate Debt: An Extension," *Journal of Financial and Quantitative Analysis*, 32(2), 239–248.

BRONŠTEIN, I. N., AND K. A. SEMENDJAEV (1995): *Taschenbuch der Mathematik.* Verlag Harri Deutsch, Thun, Frankfurt am Main, 2nd edn.

BUONCUORE, A., A. NOBILE, AND L. RICCIARDI (1987): "A New Integral Equation for the Evaluation of First-Passage Time Probability Densities," *Advances in Applied Probability*, 19, 784–800.

CARVERHILL, A. (1994): "When is the Short Rate Markovian?," *Mathematical Finance*, 4(4), 305–312.

CHANCE, D. M. (1990): "Default Risk and the Duration of Zero-Coupon Bonds," *Journal of Finance*, 45(1), 265–274.

CHEW, L. (1992): "A Bit of a Jam," *RISK*, 5(8), 86–91.

CHICAGO MERCANTILE EXCHANGE (1998): *Quarterly Bankruptcy Index: Futures and Options.*

CHRISTOPEIT, N., AND M. MUSIELA (1994): "On the Existence and Characterization of Arbitrage-Free Measures in Contingent Claim Valuation," *Stochastic Analysis and Applications*, 12(1), 41–63.

CLAESSENS, S., AND G. PENNACCHI (1996): "Estimating the Likelihood of Mexican Default from the Market Prices of Brady Bonds," *Journal of Financial and Quantitative Analysis*, 31(1), 109–126.

COOPER, I., AND M. MARTIN (1996): "Default Risk and Derivative Securities," *Applied Mathematical Finance*, 3(1), 53–74.

COOPER, I. A., AND A. S. MELLO (1991): "The Default Risk of Swaps," *Journal of Finance*, 46(2), 597–620.

CORNELL, B., AND K. GREEN (1991): "The Investment Performance of Low-Grade Bond Funds," *Journal of Finance*, 46(1), 29–48.

COSSIN, D. (1997): "Credit Risk Pricing: A Literature Survey," *Finanzmarkt und Portfolio Management*, 11(4), 398–412.

COSSIN, D., AND H. PIROTTE (1997): "Swap Credit Risk: An Empirical Investigation on Transaction Data," *Journal of Banking and Finance*, 21, 1351–1373.

COSSIN, D., AND H. PIROTTE (1998): "How Well Do Classical Credit Risk Pricing Models Fit Swap Transaction Data," *European Financial Management*, 4(1), 65–77.

COX, J. C., AND C.-F. HUANG (1989): "Option Pricing and Its Applications," in *Theory of Valuation*, ed. by S. Bhattacharya, and G. M. Constantinides. Rowman & Littlefield, Savage, M.D.

COX, J. C., J. E. INGERSOLL, AND S. A. ROSS (1980): "An Analysis of Variable Rate Loan Contracts," *Journal of Finance*, 35(2), 389–403.

COX, J. C., J. E. INGERSOLL, AND S. A. ROSS (1985): "A Theory of the Term Structure of Interest Rates," *Econometrica*, 36(4), 385–407.

COX, J. C., S. A. ROSS, AND M. RUBINSTEIN (1979): "Option Pricing: A Simplified Approach," *Journal of Financial Economics*, 7(3), 229–263.

DALANG, R. C., A. MORTON, AND W. WILLINGER (1990): "Equivalent Martingale Measures and No-Arbitrage in Stochastic Securities Market Models," *Stochastics and Stochastics Reports*, 29, 185–201.

DAS, S. R. (1995): "Credit Risk Derivatives," *Journal of Derivatives*, 2(3), 7–23.

DAS, S. R., AND R. K. SUNDARAM (1998): "A Direct Approach to Arbitrage-Free Pricing of Credit Derivatives," Discussion paper, National Bureau of Economic Research.

DAS, S. R., AND P. TUFANO (1996): "Pricing Credit-Sensitive Debt when Interest Rates, Credit Ratings and Credit Spreads are Stochastic," *Journal of Financial Engineering*, 5(2), 161–198.

DELBAEN, F. (1992): "Representing Martingale Measures When Asset Prices Are Continuous and Bounded," *Mathematical Finance*, 2(2), 107–130.

DELBAEN, F., AND W. SCHACHERMAYER (1994a): "Arbitrage and Free Lunch with Bounded Risk for Unbounded Continuous Processes," *Mathematical Finance*, 4(4), 343–348.

DELBAEN, F., AND W. SCHACHERMAYER (1994b): "A General Version of the Fundamental Theorem of Asset Pricing," *Mathematische Annalen*, 300(3), 463–520.

DELBAEN, F., AND W. SCHACHERMAYER (1995): "The No-Arbitrage Property under a Change of Numéraire," *Stochastics and Stochastics Reports*, 53, 213–226.

DHARAN, V. G. (1997): "Pricing Path-Dependent Interest Rate Contingent Claims Using a Lattice," *Journal of Fixed Income*, 6(4), 40–49.

DREZNER, Z. (1978): "Computation of the Bivariate Normal Integral," *Mathematics of Computation*, 32, 277–279.

DUFFEE, G. R. (1998): "The Relation Between Treasury Yields and Corporate Bond Yield Spreads," *Journal of Finance*, 53(6), 2225–2241.

DUFFEE, G. R. (1999): "Estimating the Price of Default Risk," *Review of Financial Studies*, 12(1), 197–226.

DUFFEE, G. R., AND C. ZHOU (1996): "Credit Derivatives in Banking: Useful Tools for Loan Risk Management?," Discussion paper, Federal Reserve Board, Washington, D.C. 20551.

DUFFIE, D. (1996): *Dynamic Asset Pricing Theory*. Princeton University Press, Princeton, N.J., 2nd edn.

DUFFIE, D., AND M. HUANG (1996): "Swap Rates and Credit Quality," *Journal of Finance*, 51(3), 921–949.

DUFFIE, D., M. SCHRODER, AND C. SKIADAS (1996): "Recursive Valuation of Defaultable Securities and the Timing of Resolution of Uncertainty," *Annals of Applied Probability*, 6(4), 1075–1090.

DUFFIE, D., AND K. J. SINGLETON (1995): "Modeling Term Structures of Defaultable Bonds," Discussion paper, Graduate School of Business, Stanford University.
DUFFIE, D., AND K. J. SINGLETON (1997): "An Econometric Model of the Term Structure of Interest-Rate Swap Yields," *Journal of Finance*, 52(4), 1287–1321.
DUFFIE, D., AND K. J. SINGLETON (1999): "Modeling Term Structures of Defaultable Bonds," *Review of Financial Studies*, 12(4), 687–720.
DUTT, J. E. (1975): "On Computing the Probability Integral of a General Multivariate t," *Biometrika*, 62, 201–205.
DYBVIG, P. H., AND C.-F. HUANG (1988): "Nonnegative Wealth, Absence of Arbitrage, and Feasible Consumption Plans," *Review of Financial Studies*, 1(4), 377–401.
EBERHART, A., W. MOORE, AND R. ROENFELDT (1990): "Security Pricing and Deviations from the Absolute Priority Rule in Bankruptcy Proceedings," *Journal of Finance*, 45(5), 1457–1469.
FALLOON, W. (1995): "Who's Missing from the Picture?," *RISK*, 8(4), 19–22.
FIGLEWSKI, S. C. (1994): "The Birth of the AAA Derivatives Subsidiary," *Journal of Derivatives*, 1(4), 80–84.
FISCHER, E. O., AND A. GRÜNBICHLER (1991): "Riskoangepasste Prämien für die Einlagensicherung in Deutschland: Eine empirische Studie," *Zeitschrift für betriebswirtschaftliche Forschung*, 43(9), 747–758.
FLESAKER, B., L. HUGHSTON, L. SCHREIBER, AND L. SPRUNG (1994): "Taking all the Credit," *RISK*, 7(9), 104–108.
FONS, J. S. (1994): "Using Default Rates to Model the Term Structure of Credit Risk," *Financial Analysts Journal*, 50(5), 25–32.
FRANKS, J., AND W. TOROUS (1994): "A Comparison of Financial Recontracting in Distressed Exchanges and Chapter 11 Reorganizations," *Journal of Financial Economics*, 35(3), 349–370.
FRANKS, J. R., AND W. N. TOROUS (1989): "An Empirical Investigation of U.S. Firms in Reorganization," *Journal of Finance*, 44(3), 747–769.
FREY, R., AND D. SOMMER (1998): "The Generalization of the Geske-Formula for Compound Options to Stochastic Interest Rates Is Not Trivial — A Note," *Journal of Applied Probability*, 35(2), 501–509.
GEANAKOPLOS, J. (1990): "An Introduction to General Equilibrium with Incomplete Asset Markets," *Journal of Mathematical Economics*, 19(1), 1–38.
GEMAN, H., N. E. KAROUI, AND J.-C. ROCHET (1995): "Changes of Numéraire, Changes of Probability Measure and Option Pricing," *Journal of Applied Probability*, 32, 443–458.
GESKE, R. (1977): "The Valuation of Corporate Liabilities as Compound Options," *Journal of Financial and Quantitative Analysis*, 12(4), 541–552.
GESKE, R., AND H. E. JOHNSON (1984): "The Valuation of Corporate Liabilities as Compound Options: A Correction," *Journal of Financial and Quantitative Analysis*, 19(2), 231–232.
GILSON, S. (1997): "Transactions Costs and Capital Structure Choice: Evidence from Financially Distressed Firms," *Journal of Finance*, 52(1), 161–197.
GRANDELL, J. (1976): *Doubly Stochastic Poisson Processes*, vol. 529 of *Lecture Notes in Mathematics*. Springer Verlag, Berlin, Heidelberg, New York.
GRÜNBICHLER, A. (1990): "Zur Ermittlung risikoangepasster Versicherungsprämien für die betriebliche Altersvorsorge," *Zeitschrift für Betriebswirtschaft*, 60(3), 319–341.
HAND, J., R. HOLTHAUSEN, AND R. LEFTWICH (1992): "The Effect of Bond Rating Announcements on Bond and Stock Prices," *Journal of Finance*, 47(2), 733–750.

HARRISON, J., AND S. R. PLISKA (1981): "Martingales and Stochastic Integrals in the Theory of Continuous Trading," *Stochastic Processes and Their Applications*, 11, 215–260.

HARRISON, J., AND S. R. PLISKA (1983): "A Stochastic Calculus Model of Continuous Trading: Complete Markets," *Stochastic Processes and Their Applications*, 15, 313–316.

HARRISON, J. M., AND D. M. KREPS (1979): "Martingales and Arbitrage in Multiperiod Securities Markets," *Journal of Economic Theory*, 20, 381–408.

HART, D. (1995): "Managing Credit and Market Risk as a Buyer of Credit Derivatives," *Journal of Commercial Lending*, 77(6), 38–43.

HART, O. D. (1975): "On the Optimality of Equilibrium When the Market Structure Is Incomplete," *Journal of Economic Theory*, 11, 418–443.

HEATH, D., AND R. JARROW (1987): "Arbitrage, Continuous Trading, and Margin Requirements," *Journal of Finance*, 42(5), 1129–1142.

HEATH, D., R. JARROW, AND A. MORTON (1990): "Bond Pricing and the Term Structure of Interest Rates: A Discrete Time Approximation," *Journal of Financial and Quantitative Analysis*, 25(4), 419–440.

HEATH, D., R. JARROW, AND A. MORTON (1992): "Bond Pricing and the Term Structure of Interest Rates: A New Methodology for Contingent Claims Valuation," *Econometrica*, 60(1), 77–105.

HELWEGE, J. (1999): "How Long Do Junk Bonds Spend in Default?," *Journal of Finance*, 54(1), 341–357.

HENN, M. (1997): "Valuation of Credit Risky Contingent Claims," Unpublished Dissertation, Universität St.Gallen.

HO, T. S., AND R. F. SINGER (1982): "Bond Indenture Provisions and the Risk of Corporate Debt," *Journal of Financial Economics*, 10(4), 375–406.

HO, T. S., AND R. F. SINGER (1984): "The Value of Corporate Debt with a Sinking-Fund Provision," *Journal of Business*, 57(3), 315–336.

HO, T.-S., R. C. STAPLETON, AND M. G. SUBRAHMANYAM (1993): "Notes on the Valuation of American Options with Stochastic Interest Rates," Discussion paper, Stern School of Business, New York University.

HO, T.-S., R. C. STAPLETON, AND M. G. SUBRAHMANYAM (1995): "Multivariate Binomial Approximations for Asset Prices with Nonstationary Variance and Covariance Characteristics," *Review of Financial Studies*, 8(4), 1125–1152.

HO, T. S. Y., AND S.-B. LEE (1986): "Term Structure Movements and Pricing Interest Rate Contingent Claims," *Journal of Finance*, 41(5), 1011–1029.

HOWARD, K. (1995): "An Introduction to Credit Derivatives," *Derivatives Quarterly*, 2(2), 28–37.

HSUEH, L., AND P. CHANDY (1989): "An Examination of the Yield Spread between Insured and Uninsured Debt," *Journal of Financial Research*, 12, 235–344.

HÜBNER, G. (2001): "The Analytic Pricing of Asymmetric Defaultable Swaps," *Journal of Banking and Finance*, 25, 295–316.

HULL, J. (1997): *Options, Futures, and Other Derivatives*. Prentice-Hall, Upper Saddle River, N.J., 3rd edn.

HULL, J., AND A. WHITE (1990): "Pricing Interest Rate Derivative Securities," *Review of Financial Studies*, 3(4), 573–592.

HULL, J., AND A. WHITE (1992): "The Price of Default," *RISK*, 5(8), 101–103.

HULL, J., AND A. WHITE (1993a): "Efficient Procedures for Valuing European and American Path-Dependent Options," *Journal of Derivatives*, 1(1), 21–31.

HULL, J., AND A. WHITE (1993b): "One-Factor Interest Rate Models and the Valuation of Interest Rate Derivative Securities," *Journal of Financial and Quantitative Analysis*, 28(2), 235–254.

HULL, J., AND A. WHITE (1995): "The Impact of Default Risk on the Prices of Options and Other Derivative Securities," *Journal of Banking and Finance*, 19(2), 299–322.

HURLEY, W. J., AND L. D. JOHNSON (1996): "On the Pricing of Bond Default Risk," *Journal of Portfolio Management*, 22(2), 66–70.

INGERSOLL, J. E. (1987): *Theory of Financial Decision Making.* Rowman & Littlefield, Savage, M.D.

INTERNATIONAL SWAPS AND DERIVATIVES ASSOCIATION (1988-1997): *ISDA Market Survey.*

IRVING, R. (1996): "Credit Derivatives Come Good," *RISK*, 9(7), 22–26.

JAMSHIDIAN, F. (1989): "An Exact Option Formula," *Journal of Finance*, 44(1), 205–209.

JAMSHIDIAN, F. (1991a): "Bond and Options Evaluation in the Gaussian Interest Rate Model," *Research in Finance*, 9, 131–170.

JAMSHIDIAN, F. (1991b): "Forward Induction and Construction of Yield Curve Diffusion Models," *Journal of Fixed Income*, 1(1), 62–74.

JAMSHIDIAN, F. (1993): "Options and Futures Evaluation with Deterministic Volatilities," *Mathematical Finance*, 3(2), 149–159.

JARROW, R. A., D. LANDO, AND S. M. TURNBULL (1997): "A Markov Model for the Term Structure of Credit Risk Spreads," *Review of Financial Studies*, 10(2), 481–523.

JARROW, R. A., AND D. B. MADAN (1991): "A Characterization of Complete Security Markets on a Brownian Filtration," *Mathematical Finance*, 1(3), 31–43.

JARROW, R. A., AND D. B. MADAN (1995): "Option Pricing Using the Term Structure of Interest Rates to Hedge Systematic Discontinuitites in Asset Returns," *Mathematical Finance*, 5(4), 311–336.

JARROW, R. A., AND S. M. TURNBULL (1992a): "Drawing the Analogy," *RISK*, 5(9), 63–70.

JARROW, R. A., AND S. M. TURNBULL (1992b): "A Unified Approach for Pricing Contingent Claims on Multiple Term Structures," Discussion paper, Johnson Graduate School of Management, Cornell University.

JARROW, R. A., AND S. M. TURNBULL (1995): "Pricing Derivatives on Financial Securities Subject to Credit Risk," *Journal of Finance*, 50(1), 53–85.

JARROW, R. A., AND S. M. TURNBULL (1996a): "The Impact of Default Risk on Swap Rates and Swap Values," Discussion paper, Queen's University.

JARROW, R. A., AND S. M. TURNBULL (1996b): "An Integrated Approach to the Hedging and Pricing of Eurodollar Derivatives," Discussion paper, Johnson Graduate School of Management, Cornell University.

JEFFREY, A. (1995): "Single Factor Heath-Jarrow-Morton Term Structure Models Based on Markov Spot Interest Rate Dynamics," *Journal of Financial and Quantitative Analysis*, 30(4), 619–642.

JENSEN, M. (1991): "Corporate Control and the Politics of Finance," *Journal of Applied Corporate Finance*, 4(2), 13–33.

JENSEN, M., AND W. MECKLING (1976): "Theory of the Firm: Managerial Behavior, Agency Costs, and Ownership Structure," *Journal of Financial Economics*, 3(4), 305–360.

JOHNSON, H., AND R. STULZ (1987): "The Pricing of Options with Default Risk," *Journal of Finance*, 42(2), 267–280.

JOHNSON, R. (1967): "Term Structures of Corporate Bond Yields as a Function of Risk of Default," *Journal of Finance*, 22, 313–345.

JONES, E. P., S. P. MASON, AND E. ROSENFELD (1984): "Contingent Claim Analysis of Corporate Capital Structures: An Empirical Investigation," *Journal of Finance*, 39(3).

KABANOV, Y., AND D. KRAMKOV (1994): "No-Arbitrage and Equivalent Martingale Measures: An Elementary Proof of the Harrison-Pliska Theorem," *Theory of Probability and Its Applications*, 39(3), 523–527.

KABANOV, Y. M., AND D. O. KRAMKOV (1998): "Asymptotic Arbitrage in Large Financial Markets," *Finance and Stochastics*, 2(2), 143–172.

KARATZAS, I., AND S. E. SHREVE (1991): *Brownian Motion and Stochastic Calculus*. Springer Verlag, Berlin, Heidelberg, New York, 2nd edn.

KARATZAS, I., AND S. E. SHREVE (1998): *Methods of Mathematical Finance*. Springer Verlag, Berlin, Heidelberg, New York.

KAU, J., AND D. KEENAN (1995): "An Overview of the Option-Theoretic Pricing of Mortgages," *Journal of Housing Research*, 6, 217–244.

KIJIMA, M., AND K. KOMORIBAYASHI (1998): "A Markov Chain Model for Valuing Credit Risk Derivatives," *Journal of Derivatives*, 6(1), 97–108.

KIM, J., K. RAMASWAMY, AND S. SUNDARESAN (1993): "Does Default Risk in Coupons Affect the Valuation of Corporate Bonds?: A Contingent Claim Model," *Financial Management*, 22(3), 117–131.

KLEIN, P. (1996): "Pricing Black-Scholes Options with Correlated Credit Risk," *Journal of Banking and Finance*, 20(7), 1211–1129.

KLEIN, P., AND M. INGLIS (1999): "Valuation of European Options Subject to Financial Distress and Interest Rate Risk," *Journal of Derivatives*, 6(3), 44–56.

KREPS, D. M. (1981): "Arbitrage and Equilibrium in Economies with Infinitely Many Commodities," *Journal of Mathematical Economics*, 8(1), 15–35.

LANDO, D. (1997): "Modelling Bonds and Derivatives with Default Risk," in *Mathematics of Financial Derivatives*, ed. by M. Dempster, and S. Pliska. Cambridge University Press, Cambridge, U.K.

LANDO, D. (1998): "On Cox Processes and Credit Risky Securities," *Review of Derivatives Research*, 2(2/3), 99–120.

LELAND, H. E. (1994a): "Bond Prices, Yield Spreads, and Optimal Capital Structure with Default Risk," Discussion paper, University of California at Berkeley.

LELAND, H. E. (1994b): "Corporate Debt Value, Bond Covenants, and Optimal Capital Structure," *Journal of Finance*, 49(4), 1213–1252.

LELAND, H. E., AND K. B. TOFT (1996): "Optimal Capital Structure, Endogenous Bankruptcy, and the Term Structure of Credit Spreads," *Journal of Finance*, 51(3), 987–1019.

LI, A., P. RITCHKEN, AND L. SANKARASUBRAMANIAN (1995): "Lattice Models for Pricing American Interest Rate Claims," *Journal of Finance*, 50(2), 719–737.

LI, H. (1998): "Pricing of Swaps with Default Risk," *Review of Derivatives Research*, 2(2/3), 231–250.

LITTERMAN, R., AND T. IBEN (1991): "Corporate Bond Valuation and the Term Structure of Credit Spreads," *Journal of Portfolio Management*, 17(3), 52–64.

LONGSTAFF, F. A., AND E. S. SCHWARTZ (1994): "A Simple Approach to Valuing Risky Fixed and Floating Rate Debt and Determining Swap Spreads," Discussion paper, Anderson Graduate School of Management, University of California at Los Angeles.

LONGSTAFF, F. A., AND E. S. SCHWARTZ (1995a): "A Simple Approach to Valuing Risky Fixed and Floating Rate Debt," *Journal of Finance*, 50(3), 789–819.

LONGSTAFF, F. A., AND E. S. SCHWARTZ (1995b): "Valuing Credit Derivatives," *Journal of Fixed Income*, 5(1), 6–12.

MADAN, D. B., AND H. UNAL (1998): "Pricing the Risks of Default," *Review of Derivatives Research*, 2(2/3), 121–160.

MARGRABE, W. (1978): "The Value of an Option to Exchange One Asset for Another," *Journal of Finance*, 33(1), 177–186.

MASON, S. P., AND S. BHATTACHARYA (1981): "Risky Debt, Jump Processes, and Safety Covenants," *Journal of Financial Economics*, 9(3), 281–307.

MELLA-BARRAL, P., AND W. PERRAUDIN (1997): "Strategic Debt Service," *Journal of Finance*, 52(2), 531–556.

MERTON, R. C. (1973): "Theory of Rational Option Pricing," *Bell Journal of Economics an Management Science*, 4, 141–183.

MERTON, R. C. (1974): "On the Pricing of Corporate Debt: The Risk Structure of Interest Rates," *Journal of Finance*, 2(2), 449–470.

MERTON, R. C. (1977): "An Analytic Derivation of the Cost of Deposit Insurance and Loan Guarantees: An Application of Modern Option Pricing Theory," *Journal of Banking and Finance*, 1(1), 3–11.

MERTON, R. C. (1978): "On the Cost of Deposit Insurance When There are Surveillance Costs," *Journal of Business*, 51, 439–452.

MUSIELA, M., AND M. RUTKOWSKI (1997): *Arbitrage Pricing of Derivative Securities: Theory and Applications.* Springer Verlag, Berlin, Heidelberg, New York.

MYERS, S. C. (1977): "Determinants of Corporate Borrowing," *Journal of Financial Economics*, 5(2), 147–175.

NELSON, D. B., AND K. RAMASWAMY (1990): "Simple Binomial Processes as Diffusion Approximations in Financial Models," *Review of Financial Studies*, 3(3), 393–430.

NIELSEN, L. T., J. SAÀ-REQUEJO, AND P. SANTA-CLARA (1993): "Default Risk and Interest-Rate Risk: The Term Structure of Default Spreads," Discussion paper, INSEAD.

NIELSEN, S. S., AND E. I. RONN (1997): "The Valuation of Default Risk in Corporate Bonds and Interest Rate Swaps," *Advances in Futures and Options Research*, 9, 175–196.

OFFICE OF THE COMPTROLLER OF THE CURRENCY (1997-2000): *Quarterly Derivatives Fact Sheets.*

PEARSON, N., AND T.-S. SUN (1994): "Exploiting the Conditional Density in Estimating the Term Structure: An Application to the Cox, Ingersoll, Ross Model," *Journal of Finance*, 49, 1279–1304.

PIERIDES, Y. A. (1997): "The Pricing of Credit Risk Derivatives," *Journal of Economic Dynamics and Control*, 21(10), 1479–1611.

PITTS, C., AND M. SELBY (1983): "The Pricing of Corporate Debt: A Further Note," *Journal of Finance*, 38(4), 1311–1313.

PLISKA, S. R. (1997): *Introduction to Mathematical Finance: Discrete Time Models.* Blackwell Publishers, Malden, M.A.

PRESS, W. H., W. T. VETTERLING, S. A. TEUKOLSKY, AND B. P. FLANNERY (1992): *Numerical Recipes in C: The Art of Scientific Computing.* Cambridge University Press, Cambridge, U.K., 2nd edn.

PROTTER, P. (1990): *Stochastic Integration and Differential Equations.* Springer Verlag, Berlin, Heidelberg, New York.

RAMASWAMY, K., AND S. SUNDARESAN (1986): "The Valuation of Floating-Rate Instruments: Theory and Evidence," *Journal of Financial Economics*, 17(2), 251–272.

REVUZ, D., AND M. YOR (1994): *Continuous Martingales and Brownian Motion.* Springer Verlag, Berlin, Heidelberg, New York, 2nd edn.

RITCHKEN, P., AND L. SANKARASUBRAMANIAN (1995): "Volatility Structures of Forward Rates and the Dynamics of the Term Structure," *Mathematical Finance*, 5(1), 55–72.

RODRIGUEZ, R. J. (1988): "Default Risk, Yield Spreads, and Time to Maturity," *Journal of Financial and Quantitative Analysis*, 23(1), 111–117.

ROGERS, L. (1994): "Equivalent Martingale Measures and No-Arbitrage," *Stochastics and Stochastics Reports*, 51, 41–49.

RONN, E. I., AND A. K. VERMA (1986): "Pricing Risk-Adjusted Deposit Insurance: An Option-Based Model," *Journal of Finance*, 41(4), 871–895.

RUBINSTEIN, M. (1991): "Somewhere over the Rainbow," *RISK*, 4(10), 63–66.

RUBINSTEIN, M. (1994): "Return to Oz," *RISK*, 7(11), 67–71.

SARIG, O., AND A. WARGA (1989): "Some Empirical Estimates of the Risk Structure of Interest Rates," *Journal of Finance*, 44(5), 1351–1360.

SCHACHERMAYER, W. (1994): "Martingale Measures for Discrete-Time Processes with Infinite Horizon," *Mathematical Finance*, 4(1), 25–55.

SCHICH, S. T. (1997): "An Option-Pricing Approach to the Costs of Export Credit Insurance," *Geneva Papers on Risk and Insurance Theory*, 22(1), 43–58.

SCHÖNBUCHER, P. J. (1998): "Term Structure Modelling of Defaultable Bonds," *Review of Derivatives Research*, 2(2/3), 161–192.

SELBY, M., AND S. HODGES (1987): "On the Evaluation of Compound Options," *Management Science*, 33(3), 347–355.

SHIMKO, D. C., N. TEJIMA, AND D. R. V. DEVENTER (1993): "The Pricing of Risky Debt When Interest Rates Are Stochastic," *Journal of Fixed Income*, 3(2), 58–65.

SHLEIFER, A., AND R. VISHNY (1992): "Liquidation Values and Debt Capacity: A Market Equilibrium Approach," *Journal of Finance*, 47(4), 1343–1366.

SMITHSON, C. (1995): "Credit Derivatives," *RISK*, 8(12), 38–39.

SOLNIK, B. (1990): "Swap Pricing and Default Risk: A Note," *Journal of International Financial Management and Accounting*, 2(1), 79–91.

SORENSEN, E. H., AND T. F. BOLLIER (1994): "Pricing Swap Default Risk," *Financial Analysts Journal*, 50(3), 23–33.

SOSIN, H. B. (1980): "On the Valuation of Federal Loan Guarantees to Corporations," *Journal of Finance*, 35(5), 1209–1221.

SUN, T. S., S. SURESH, AND W. CHING (1993): "Interest Rate Swaps: An Empirical Investigation," *Journal of Financial Economics*, 34(1), 77–99.

SUNDARESAN, S. (1991): "Valuation of Swaps," in *Recent Developments in International Banking and Finance*, ed. by S. J. Khoury, chap. 12. Elsevier (North-Holland).

TAQQU, M. S., AND W. WILLINGER (1987): "The Analysis of Finite Security Markets Using Martingales," *Advances in Applied Probability*, 19, 1–25.

TIAN, Y. (1992): "A Simplified Binomial Approach to the Pricing of Interest Rate Contingent Claims," *Journal of Financial Engineering*, 1(1), 14–37.

TITMAN, S., AND W. TOROUS (1989): "Valuing Commercial Mortgages: An Empirical Investigation of the Contingent Claims Approach to Pricing Risky Debt," *Journal of Finance*, 44(2), 345–373.

VASICEK, O. (1977): "An Equilibrium Characterization of the Term Structure," *Journal of Financial Economics*, 5(2), 177–188.

WEINSTEIN, M. (1983): "Bond Systematic Risk and the Option Pricing Model," *Journal of Finance*, 38(5), 1415–1429.

WEISS, L. (1990): "Bankruptcy Resolution: Direct Costs and Violation of Priority of Claims," *Journal of Financial Economics*, 27(2), 285–314.

WILLIAMS, D. (1991): *Probability with Martingales*. Cambridge University Press, Cambridge, U.K.

WILMOTT, P., S. HOWISON, AND J. DEWYNNE (1995): *The Mathematics of Financial Derivatives*. Cambridge University Press, Cambridge, U.K.

ZHENG, C. (2000): "Understanding the Default-Implied Volatility for Credit Spreads," *Journal of Derivatives*, 7(4), 67–77.

ZHOU, C. (1997): "A Jump-Diffusion Approach to Modeling Credit Risk and Valuing Defaultable Securities," Discussion paper, Federal Reserve Board, Washington, D.C.

ZIMMERMANN, H. (1998): *State-Preference Theorie und Asset Pricing: Eine Einführung*. Physica-Springer Verlag, Heidelberg.

List of Figures

List of Tables

Index